Microorganisms and Autoimmune Diseases

INFECTIOUS AGENTS AND PATHOGENESIS

Series Editors: Mauro Bendinelli, *University of Pisa*
Herman Friedman, *University of South Florida*

COXSACKIEVIRUSES
A General Update
Edited by Mauro Bendinelli and Herman Friedman

DNA TUMOR VIRUSES
Oncogenic Mechanisms
Edited by Giuseppe Barbanti-Brodano, Mauro Bendinelli, and
Herman Friedman

ENTERIC INFECTIONS AND IMMUNITY
Edited by Lois J. Paradise, Mauro Bendinelli, and Herman Friedman

FUNGAL INFECTIONS AND IMMUNE RESPONSES
Edited by Juneann W. Murphy, Herman Friedman, and
Mauro Bendinelli

MICROORGANISMS AND AUTOIMMUNE DISEASES
Edited by Herman Friedman, Noel R. Rose, and Mauro Bendinelli

MYCOBACTERIUM TUBERCULOSIS
Interactions with the Immune System
Edited by Mauro Bendinelli and Herman Friedman

NEUROPATHOGENIC VIRUSES AND IMMUNITY
Edited by Steven Specter, Mauro Bendinelli, and Herman Friedman

PSEUDOMONAS AERUGINOSA AS AN OPPORTUNISTIC
PATHOGEN
Edited by Mario Campa, Mauro Bendinelli, and Herman Friedman

PULMONARY INFECTIONS AND IMMUNITY
Edited by Herman Chmel, Mauro Bendinelli, and Herman Friedman

VIRUS-INDUCED IMMUNOSUPPRESSION
Edited by Steven Specter, Mauro Bendinelli, and Herman Friedman

---

A Continuation Order Plan is available for this series. A continuation order will bring
delivery of each new volume immediately upon publication. Volumes are billed only
upon actual shipment. For further information please contact the publisher.

Microorganisms and Autoimmune Diseases

Edited by

Herman Friedman

University of South Florida
College of Medicine
Tampa, Florida

Noel R. Rose

The Johns Hopkins University
School of Hygiene and Public Health
Baltimore, Maryland

and

Mauro Bendinelli

University of Pisa
Pisa, Italy

Springer Science+Business Media, LLC

Library of Congress Cataloging-in-Publication Data

On file

© 1996 Springer Science+Business Media New York
Originally published by Plenum Press, New York in 1996
Softcover reprint of the hardcover 1st edition 1996

ISBN 978-1-4613-8009-2   ISBN 978-1-4613-0347-3 (eBook)
DOI 10.1007/978-1-4613-0347-3
10 9 8 7 6 5 4 3 2 1

Contributors

S. M. ANDERTON • Department of Immunology, Institute of Infectious Diseases and Immunology, Faculty of Veterinary Medicine, University of Utrecht, 3508 TD Utrecht, The Netherlands

ROBERT E. BAUGHN • V. A. Medical Center, Houston, Texas 77030-4298

VAGN BONNEVIE • Department of Medical Microbiology, University of Odense, 5000 Odense C, Denmark

MADELEINE W. CUNNINGHAM • Department of Microbiology and Immunology, University of Oklahoma Health Sciences Center, Oklahoma City, Oklahoma 73190

THOMAS DYRBERG • Diabetes Immunology Department, Novo Nordisk A/S, 2880 Bagsvaerd, Denmark

LISA M. ESOLEN • Department of Medicine, Division of Infectious Diseases, The Johns Hopkins University School of Medicine, Baltimore, Maryland 21205

LAURA FERNANDES • Jefferson Medical College, Thomas Jefferson University, Philadelphia, Pennsylvania 19107

ROBERT S. FUJINAMI • Department of Neurology, University of Utah, Salt Lake City, Utah 84132

CARLO GARZELLI • Department of Biomedicine, University of Pisa, I-56127 Pisa, Italy

ANDREW F. GECZY • NSW Red Cross Blood Transfusion Service, Sydney 2000 NSW, Australia

DIANE E. GRIFFIN • Department of Molecular Microbiology and Immunology, The Johns Hopkins School of Hygiene and Public Health, Baltimore, Maryland 21205

ALLAN KARLSEN • Steno Diabetes Center, 2820 Gentofte, Denmark

HELEN KOMINEK • Laboratory of Viral and Immunopathogenesis of Diabetes, Julia McFarlane Diabetes Research Centre, Department of Microbiology and Infectious Diseases, Faculty of Medicine, University of Calgary, Calgary, Alberta T2N 4N1, Canada

PETER MACKAY • Diabetes Immunology Department, Novo Nordisk A/S, 2880 Bagsvaerd, Denmark

BIRGITTE MICHELSEN • Hagedorn Research Institute, 2820 Gentofte, Denmark

JACOB PETERSEN • Hagedorn Research Institute, 2820 Gentofte, Denmark

A. B. J. PRAKKEN • Department of Immunology, Institute of Infectious Diseases and Immunology, Faculty of Veterinary Medicine, University of Utrecht, 3508 TD Utrecht, The Netherlands

CARLOS M. RIPOLL • Department of Regional Pathology, Laboratory of Public Health, San Salvador de Jujuy, Jujuy, 4600 Argentina

NOEL R. ROSE • Departments of Pathology and of Molecular Microbiology and Immunology, The Johns Hopkins Medical Institutions, Baltimore, Maryland 21205

CHARLES A. SANTOS-BUCH • Department of Pathology, Cornell University Medical College, New York, New York 10021

H. RALPH SCHUMACHER • Department of Medicine, University of Pennsylvania School of Medicine, Philadelphia, Pennsylvania 19104

JOHN S. SULLIVAN • NSW Red Cross Blood Transfusion Service, Sydney 2000 NSW, Australia

ANTONIO R. L. TEIXEIRA • Laboratory of Multidisciplinary Research on Chagas Disease, Department of Pathology, Faculty of Health Sciences, University of Brasilia, Brasilia, 70-919-970 Brazil

KENNETH E. UGEN • Department of Medical Microbiology and Immunology, University of South Florida College of Medicine, Tampa, Florida 33612

R. VAN DER ZEE • Department of Immunology, Institute of Infectious Diseases and Immunology, Faculty of Veterinary Medicine, University of Utrecht, 3508 TD Utrecht, The Netherlands

W. VAN EDEN • Department of Immunology, Institute of Infectious Diseases and Immunology, Faculty of Veterinary Medicine, University of Utrecht, 3508 TD Utrecht, The Netherlands

BIN WANG • Department of Pathology and Laboratory Medicine, University of Pennsylvania School of Medicine, Philadelphia, Pennsylvania 19104

ANTHONY P. WEETMAN • Department of Medicine, Clinical Sciences Centre, Northern General Hospital, Sheffield S5 7AU, England

DAVID B. WEINER • Department of Pathology and Laboratory Medicine, University of Pennsylvania School of Medicine, Philadelphia, Pennsylvania 19104

CYNTHIA T. WELSH • Department of Pathology, University of Utah, Salt Lake City, Utah 84132

WILLIAM V. WILLIAMS • Department of Medicine, University of Pennsylvania School of Medicine, Philadelphia, Pennsylvania 19104

JI-WON YOON • Laboratory of Viral and Immunopathogenesis of Diabetes, Julia McFarlane Diabetes Research Centre, Department of Microbiology and Infectious Diseases, Faculty of Medicine, University of Calgary, Calgary, Alberta T2N 4N1, Canada

Preface to the Series

The mechanisms of disease production by infectious agents are presently the focus of an unprecedented flowering of studies. The field has undoubtedly received impetus from the considerable advances recently made in the understanding of the structure, biochemistry, and biology of viruses, bacteria, fungi, and other parasites. Another contributing factor is our improved knowledge of immune responses and other adaptive or constitutive mechanisms by which hosts react to infection. Furthermore, recombinant DNA technology, monoclonal antibodies, and other newer methodologies have provided the technical tools for examining questions previously considered too complex to be successfully tackled. The most important incentive of all is probably the regenerated idea that infection might be the initiating event in many clinical entities presently classified as idiopathic or of uncertain origin.

Infectious pathogenesis research holds great promise. As more information is uncovered, it is becoming increasingly apparent that our present knowledge of the pathogenic potential of infectious agents is often limited to the most noticeable effects, which sometimes represent only the tip of the iceberg. For example, it is now well appreciated that pathological processes caused by infectious agents may emerge clinically after an incubation of decades and may result from genetic, immunologic, and other indirect routes more than from the infecting agent itself. Thus, there is a general expectation that continued investigation will lead to the isolation of new agents of infection, the identification of hitherto unsuspected etiologic correlations, and eventually, more effective approaches to prevention and therapy.

Studies on the mechanisms of disease caused by infectious agents demand a breadth of understanding across many specialized areas, as well as much cooperation between clinicians and experimentalists. The series *Infectious Agents and Pathogenesis* is intended not only to document the state of the art in this fascinating and challenging field but also to help lay bridges among diverse areas and people.

M. Bendinelli
H. Friedman

Preface

Over the last few decades there has been an explosion of information concerning the immune response. We know not only that different cell types have specialized functions in immunity (i.e., T and B cells as well as macrophages and other supporting cells) but also that there are many subsets of these different cell types with unique characteristics and functions important in normal functioning of the immune system. It is also well known that different B cells produce a wide variety of immunoglobulins categorized in several major classes with many subgroups. Traditional and newer molecular biology techniques have identified specific chemical mediators involved in the interaction of immune cells among themselves and with cells in other systems. Many of these mediators (cytokines) are produced by the immune system but affect cells in the central nervous system as well as in different organs of the body. Production of at least 15 distinct molecular categories of cytokines may occur following stimulation by various activators, including many infectious agents and their products. The targets of these cytokines include other cells of the immune system, cells from other systems, and target cells that are either normal or malignant cells, as well as foreign invading organisms or host cells infected by such organisms.

There is much interest concerning the nature and mechanisms of immunity, including molecular signaling events and the identity of receptors of immune cells that recognize foreign versus self antigens. Furthermore, it is now widely accepted that evolutionary development of the immune system in higher organisms occurred under the pressure of infectious agents, which concurrently developed mechanisms to avoid factors produced by the host. Thus, infectious agents have developed many strategies to avoid attack by the immune system of the host, including alterations or decrease in antigenicity so that the agents would be "invisible" to the host as well as the ability to depress or deregulate the host immune response. Induction of an unresponsive state by a variety of means, including suppressive interaction with immune cells of the host, neonatally or postnatally, also provides survival advantage for the microorganism. In turn, the host has developed varied immune defense mechanisms to overcome the virulence of microbial pathogens.

Nevertheless, microorganisms in general have continued to exert pressure on the immune system of individuals. One intriguing strategy adopted by some infectious agents is to mimic the antigens of the host, thus taking advantage of self-tolerance. It is to the infecting agent's "advantage" to share antigenic determinants with a host so it is less likely the host will respond immunologically to the invader, since the host should be tolerant to its own tissue. However, similarities in antigenic make-up of host cells and a microbe may lead to immune recognition not only of the microorganism, but also of the host, and result in autoimmune reactions based on antibodies, T cells, or both.

Over the last few decades, many investigators have suggested that autoimmune responses initiated by immune stimulation by an infectious agent or its products may be due to shared immunogenic determinants between the agent and the host. Microbial infections, even subclinical ones, may compromise the delicate balance that exists between immune responsiveness and the control of such responses. Thus, an immune response to microorganisms or their antigens may cause a loss of self-tolerance. However, there is only minimal information as to the possible mechanisms involved and very little evidence that conclusively shows that an infection causes a chronic or acute disease due to autoimmunity or autoallergy. The chapters in this volume, written by experts in the field of microbiology and autoimmunity, address the question of whether and how microorganisms may be associated with development of specific categories of autoimmune disease.

The first chapters deal with possible bacterial induction of diseases considered autoimmune in nature. Dr. W. van Eden and colleagues discuss general phenomena of bacterial infection and autoimmune-like diseases, especially chronic diseases such as rheumatoid arthritis and even lupus erythematosus, which are thought by some to be associated or induced by microorganisms as diverse as mycobacteria, borellia, or other agents that cause chronic infections. Autoimmune-like etiologic reactions have been postulated for these connective tissue diseases. Dr. Madeleine Cunningham describes the role of streptococcal infection in rheumatic fever. It is now almost axiomatic that streptococcal infection is a prerequisite for rheumatic fever. Similarly, Drs. John Sullivan and Andrew Geczy describe the possible role of *Klebsiella* in development and progression of ankylosing spondylitis.

Dr. Robert Baughn then discusses the autoimmune phenomena associated with syphilis and describes how diseases due to spirochetes have a chronic component associated with immune responses to self induced by spirochetal infection.

The next group of contributions describes the role of viruses in possible induction of a wide variety of diseases considered either to be autoimmune in nature or to have an autoimmune component. Dr. Thomas Dyrberg and associates describe the possible role of viruses associated with induction and maintenance of diabetes mellitus, while Drs. Ji-Won Yoon and Helen Kominek focus on the role of Coxsackie B viruses in the pathogenesis of this important health affliction. Drs. Cynthia Welsh and Robert Fujinami describe viruses involved in encephalitis-associated diseases, including autoimmune-like diseases such as multiple sclerosis. Drs. Lisa Esolen and Diane Griffin discuss measles and encephalomyelitis. Dr. Carlo Garzelli describes the effects of Epstein-Barr virus in the induction of a variety of autoimmune-type diseases. Dr. Kenneth Ugen and associates discuss the autoimmune component of retrovirus infection, including autoimmunity associated with acquired immunodeficiency induced by HIV.

Dr. Charles Santos-Buch and associates describe the involvement of an autoimmune phenomenon in parasitic diseases, namely Chagas disease, and Dr. Anthony Weetman discusses endocrine autoimmune responses following infections. In the final chapter, Dr. Noel Rose revisits the possible mechanisms by which infection can serve as the precursor to autoimmunity and describes one mouse model, Coxsackie virus–induced myocarditis, in which the major self antigen has been identified.

It is hoped and anticipated by the editors and the authors of the individual chapters that this volume will stimulate interest in the role of infectious agents, be they bacteria, parasites, or viruses, in inducing immunologic responses that recognize or react with self antigens and thus induce an autoimmune-like disease. The editors thank Ms. Ilona Friedman for excellent editorial assistance in coordinating and assisting in the preparation of the manuscripts for this volume.

Herman Friedman
Noel R. Rose
Mauro Bendinelli

Contents

Bacterial Heat-Shock Proteins and Autoimmune Disease

W. VAN EDEN, S. M. ANDERTON, A. B. J. PRAKKEN, and R. VAN DER ZEE

1. INTRODUCTION

Of all the microorganisms, bacteria, especially, seem to exert continuous pressure on the immune system. The epithelial surfaces of the skin and the intestinal tract are colonized by a wide variety of bacterial species, forcing the immune system into a situation where it has to scan continuously an immense repertoire of foreign antigenic determinants. It may be self-evident that as much as the healthy immune system does not develop an aggressive response directed against self antigens, or autoantigens, the same immune system is not tuned to mount continuous and aggressive responses to antigens present in the resident bacterial flora. Nonetheless, the capacity to respond must be there. Within every healthy immune system, cells specific for autoantigens are present, and their potential for causing disease has been demonstrated in a variety of experimental autoimmune disease models. The mechanisms by which the immune

W. VAN EDEN, S. M. ANDERTON, A. B. J. PRAKKEN, and R. VAN DER ZEE • Department of Immunology, Institute of Infectious Diseases and Immunology, Faculty of Veterinary Medicine, University of Utrecht, 3508 TD Utrecht, The Netherlands.

Microorganisms and Autoimmune Diseases, edited by Herman Friedman *et al.* Plenum Press, New York, 1996.

system controls the presence of such cells in what can be called the "autoimmune carrier state" are under intense investigation.[1] It is probable that similar mechanisms control the eagerness of the immune system to respond to relatively safe foreign antigens present in the resident flora. From this it can be anticipated that bacterial infections may severely compromise the delicate balance between responsiveness and control of responsiveness. And indeed, in various infectious diseases of bacterial origin signs of loss of self-tolerance, such as production of rheumatoid factors, are evident. However, there are few proven cases of induction of lasting autoimmune diseases by bacterial infections. Probably the best candidates are arthritic diseases such as reactive arthritis and Lyme disease, both of which are consequences of infection with identified bacterial species. However, in both diseases the possibility that continued presence of bacterial antigen at the sites of inflammation is essentially driving the disease process has not been formally ruled out.[2] In other words, triggering of an autonomous autoimmune process by mimicry or otherwise, the so-called hit-and-run scenario, has not been proven in these cases.

In the case of ankylosing spondylitis no single bacterial species has been identified with certainty as the cause. However, since both reactive arthritis and ankylosing spondylitis are strongly associated with the same major histocompatibility complex (MHC) determinant HLA-B27, it is tempting to assume a similar involvement of bacterial antigens in ankylosing spondylitis. Therefore, it is possible that ankylosing spondylitis is a true case of the hit-and-run scenario, where bacterial antigens cause a permanent break of tolerance, leading to a chronic and progressive autoimmune disease. The best support for the hit-and-run scenario, however, seems to be from experimental animal models. Immunization with mycobacterial antigen, streptococcal cell walls, or bacterial peptidoglycans in suitable adjuvants can elicit chronic and destructive forms of arthritis in rodent species such as rats. The significance of the bacterial antigens, however, is not exclusive, since in some strains of inbred animals, similar forms of arthritis are induced by the mere administration of oily adjuvants such as pristane, avridine, and mineral oil. Despite this, the successful passive transfer of arthritis by T cells specific for bacterial antigens[3-5] has clearly demonstrated that bacterial antigens have the potential to trigger autoimmune arthritic T cells.

From the same experimental models good evidence has now been obtained that exposure to bacterial antigens may cause resistance to the induction of disease. In this manner, disease and resistance to disease seem to be two sides of the same coin, that is, the delicate balance between responsiveness and control of responsiveness. In the case of mycobacterial-induced arthritis, the mycobacterial 65-kDa protein, a member of the heat-shock protein 60 (hsp60) family of proteins, has now been identified as an antigen that seems to unite the two sides of the coin. Distinct epitopes within the molecule have been shown to have the capacity to either induce arthritis or induce resistance to arthritis.

These findings, combined with the findings of the immunodominant nature of bacterial hsp's suggest that immunity to bacterial antigens such as hsp's is a principal element in the control of autoimmunity through the maintenance of *peripheral tolerance.*

2. PERIPHERAL TOLERANCE, AUTOIMMUNITY, AND ENVIRONMENTAL BACTERIA

Findings from experimental models of autoimmunity and transgenic animals expressing "neo-self" antigens have caused a rather drastic change in our perception of immunological tolerance. While in the past the role of the thymus was considered to be absolute in creating a repertoire tolerant for self by eliminating potentially self-reactive T cells from the system (central tolerance), it has recently become clear that potentially self-reactive T cells might be central to a properly functioning immune system.[6]

Despite the presence of self-reactive T cells and B cells as an integral part of the immunological repertoire, the healthy individual appears to be tolerant of his or her self antigens. Apparently, the immune system has a built-in capacity to contain such self-reactive cells by functionally regulating their activity and not by eliminating them. This condition is called peripheral tolerance.[7] The mechanism of peripheral tolerance is currently being explored and may turn out to be a composite of regulatory activities such as suppression due to the suppressive action of certain cytokines, anergy at the level of single T cells, and idiotypic cell interactions leading to feedback downregulatory events. The existence of such built-in regulatory mechanisms, ensuring an active and flexible state of functional regulation, may reinforce tolerance in situations where a failure of self-tolerance leads to autoimmune disease.

The possibility that the presence of bacterial flora contributes to the establishment or maintenance of peripheral tolerance has been suggested by experimental findings in germ-free animals. In the case of arthritis, experiments performed in inbred Fisher rats have been most illustrative.[8] Fisher rats are relatively resistant to the induction of arthritis following immunizations with mycobacteria or streptococcal cell walls. However, germ-free bred Fisher rats are susceptible to a degree similar to

Lewis rats. Colonization of the gut with *Escherichia coli* or with the bacterial flora as present in conventionally bred animals was shown to lead to arthritis resistance, despite the fact that the animals had a history of germ-free development.

That environmental microflora may be of influence in arthritis susceptible Lewis rats was indicated by findings that microbial environmental pressure contributed to the known resistance against the reinduction of arthritis that normally develops after disease remission (Cohen, personal communication). It appeared that when a colony of Lewis rats were transferred from a conventional "dirty" facility into a "clean," specific pathogen–free (SPF) animal housing facility, the animals remained susceptible to mycobacteria-induced arthritis, even after having experienced and having recovered from a first bout of the disease.

It is possible, however, that the impact of bacterial environment on arthritis susceptibility varies with the nature of the arthritis induced. In a model of mineral oil–induced arthritis in dark Agouty (DA) rats the influence of the presence of bacterial flora was found to be minimal or absent (Klareskog, personal communication), and in a model of pristane-induced arthritis in mice, the presence of bacterial flora was found to enhance disease susceptibility.[9]

In diabetes models, a positive contribution of environmental flora to resistance was observed. In non-obese diabetic (NOD) mice an increase in disease incidence (up to 100%) was observed in germ-free animals and in animals kept under SPF conditions.[10] Similarly, it was found that complete Freund's adjuvant immunizations prevented diabetes in Bio-Breeding (BB) rats[11] and that bacillus Calmette-Guerin (BCG) may be effective in the treatment of diabetes in humans.[12]

Altogether, it seems that mechanisms of peripheral tolerance as a hedge against autoimmunity are dependent, at least in part, on interactions with the environmental microflora. When such interactions can be determined by antigen-specific recognition events, vaccination and similar specific immunostimulatory triggering may well be exploited in the control of autoimmunity.

3. EXPERIMENTAL MODELS SUPPORT A CRITICAL ROLE OF SPECIFIC IMMUNITY IN AUTOIMMUNE DISEASES

One of the best studied models of arthritis is autoimmune arthritis (AA). The disease is induced by immunization with mycobacterial antigens and is transferable by T cells alone. Holoshitz *et al.*[4] have shown that passive transfer of a single T-cell clone that is reactive to mycobacteria can also lead to induction of AA. Subsequent analyses have shown that this arthritogenic T-cell clone, called A2b, recognizes the 180–188 sequence in mycobacterial (hsp60)[13] and responds, to a lesser extent, to cartilage proteoglycan.[14] This was the first demonstration that autoimmune T cells with a single antigenic specificity, in this case "self" cross-reactive specificity, can induce the full pathology of clinically overt arthritis. Further experiments have shown that the same activated T cells and their specific antigens (or epitopes in the form of synthetic peptides) can be used to induce a specific form of protection (i.e., vaccination) against AA. In this way AA has shown the principle of regulating autoimmunity at the level of single T-cell specificities. Arthritis induced by collagen type II, streptococcal cell walls, and pristane, as well as other models, have yielded similar experimental evidence for the involvement of antigen-specific mechanisms. The obvious question is whether the models are representative of human arthritic diseases. In general one can easily admit that every model is artificial by its nature and only capable of shedding light upon certain mechanistic aspects of pathological processes. So, one might rephrase the question by asking whether the antigen-specific regulation of autoimmunity in AA is representative of the antigen-specific regulatory events in human diseases.

One aspect is the antigen-induced nature of models, while human arthritis seems "spontaneous." Except for the lupus-prone MRL lpr/lpr mouse in which arthritis develops along with lymphoproliferative disorders, arthritis does not seem to occur spontaneously in rodents, at least not in a way amenable to experimentation. A second aspect of models is that they are induced in animals selected for genetic susceptibility.

One may argue, however, that the human "model" may be similar in both of these aspects to artificial models. In arthritis especially (i.e., reactive arthritis, Lyme disease, acute rheumatic fever), the possibility of induction of disease by unconventional exposure to, for instance, microbial antigens, does exist. With respect to genetic predisposition (i.e., HLA, gender, etc.), humans and rodents may have similar pathogenic factors. Thus, it may seem fair to conclude, given the relative ease with which autoimmune arthritis can be induced in selected animals and given the relative multitude of suitable triggers, that disease induction in humans may depend on similar, specific stimuli at least at its initiation. It is probable that there are many more ways to trigger disease than we know of now. Likewise, there should be many ways to interrupt the development of disease. Non-obese diabetic mice, an inbred strain of animals featuring spontaneous type I diabetes, are illustrative in this respect.

Diabetes in NOD mice was found by chance in a genetic breeding program that was carried out by selecting for features unrelated to diabetes. Apparently, spontaneous diabetes occurred in one fortuitous constellation of genes. Extensive studies now have shown that almost every immunological intervention, from immune stimulation to immune suppression, both antigen-specific and nonspecific, affects the development of the disease, usually in a disease-suppressive manner.[10] This shows, at least in the case of NOD diabetes, that disease may depend on a delicate interplay of many distinct, genetically determined factors. Any intervention that critically affects one of the essential factors involved will lead to inhibition of disease development. It is tempting to assume that the same is true for autoimmune arthritis. It suggests that more than one approach could lead to successful clinical intervention, irrespective of the fact that each individual may have a personal set of predisposing genes and immunological anamneses.

The initial trigger leading to disease may vary from individual to individual, whereas in the progression of pathology many factors are common. Obviously, such common features are of particular interest for therapeutic intervention. On the basis of the attractive successes obtained in the experimental models, one may take the optimistic view that it will be possible to define specific antigens that take part in processes of inflammatory arthritis, irrespective of the initial triggering antigen. Antigens such as collagen type II and hsp60 have shown their activity in animals of various genetic backgrounds, including mice and rats. They may not be primary autoantigens, but they may become involved in an ancillary way as the disease progresses.

4. HSP60 AND THE INHIBITION OF EXPERIMENTAL ARTHRITIS

Hsp60 was defined as an arthritis-associated antigen by the fact that the arthritogenic T-cell clone A2b, raised against whole *Mycobacterium tuberculosis* in the AA model, responded to recombinant mycobacterial hsp60 as cloned in *E. coli*.[13] Having identified mycobacterial hsp60 as a critical antigen in the induction of AA in Lewis rats, experiments were initiated to see whether arthritis was inducible by immunization with recombinant hsp60. However, no arthritis was induced after hsp60 immunization, and immunized animals were found to have become resistant to subsequent induction of AA with whole mycobacteria. As mentioned above, mycobacterial hsp60 was also found to protect against arthritis induced by streptococcal cell walls and pristane (oil) and to a variable degree against arthritis induced by a lipoidal amine also called CP20961, and collagen type II.[14,15]

This seemingly general activity of mycobacterial hsp60 in preventing experimentally induced arthritis also in models not induced with mycobacteria was compatible with the possibility that "self" hsp60 was an autoantigen critically involved in every from of autoimmune arthritis.

As reviewed elsewhere,[16] heat-shock proteins are intracellular proteins with critical functions in the protein housekeeping machinery of every living cell. Probably due to such essential cellular functions, their evolutionary variation has remained remarkably limited. As a consequence, extensive amino acid sequence identities exist between microbial and mammalian hsp's. Despite their similarities to self (host) hsp's, microbial hsp's have been found to be strong immunogens. Therefore, the possibility existed that immunization with mycobacterial hsp60 (also called the 65-kDa molecule of mycobacteria) led to induction of responses directed at self hsp60 as an autoantigen.

An important feature of hsp's is their increased synthesis when cells are subjected to stress (they are also called stress proteins). Inflamed synovium, apparently a condition characterized by local cellular stress, was shown to feature a more intense staining in immunohistology using hsp60-specific antibodies than healthy synovium.[17] In the experimental arthritis models, an increase in expression, as well as T-cell reactivity against hsp60, was observed in diseased animals, irrespective of how the arthritis was induced. Thus, together, the observations made in various models suggested that immunization with mycobacteria led to regulatory responses involving the endogenous self hsp60 as expressed in the inflamed synovium. Recent observations in the AA model are supportive of this hypothesis.[18,19]

Lewis rats were immunized with either whole M. tuberculosis (Mt) or recombinant mycobacterial hsp60 in a suitable adjuvant. Primed lymph node cells taken from immunized rats were analyzed for their proliferative responses in the presence of sets of overlapping peptides spanning the whole mycobacterial hsp60 molecule. Immunization with Mt led to weak responses to various peptides indicating several epitopes located on the hsp60 molecule. However, responses to a peptide that included amino acids 180–188, formerly demonstrated to be the epitope recognized by arthritogenic T cell A2b, dominated the response pattern obtained. Immunization with mycobacterial hsp60 led to responses to these and some additional epitopes, and interestingly no exclusive dominance of 180–188 was apparent. All epitopes involved were mapped in detail using epitope-specific T-cell lines. Upon testing these T-cell lines for responses to peptides based on homologous rat hsp60 sequences, only one T-cell line also responded to the homologous rat hsp60 peptide. As expected from the cross-recognition between the mycobacterial and rat sequence, this particular epitope turned out to be strongly conserved.[19]

The various peptides that included our newly defined T-cell epitopes were tested for their capacity to protect Lewis rats against AA. Remarkably, only peptides containing the conserved rat hsp60 cross-reactive epitope were protective. These findings suggested that mycobacterial hsp60 was dependent on activity of T-cell epitopes in the molecule, which induced responses that cross-reacted with rat self hsp60.

More recent experiments have shown that non-mycobacterial-induced arthritis is also effectively prevented by preimmunization with the conserved peptide. Teleologically arguing, it seems attractive to suppose that the cross-recognition of rat self hsp60 is the basis of the protection observed. As discussed above, recognition of self hsp60 may well be subjected to the regulatory pathways of peripheral tolerance. Temporarily compromising such tolerance by forcing the immune system to respond to self hsp60 could lead to a counteractive regulatory (suppressive) response targeted at the expressed self hsp60 molecule. Such suppressive regulation may well contribute to the control of local inflammatory processes. Some of the findings made in children suffering from juvenile chronic arthritis (JCA) seem to be compatible with these possibilities.

5. RESPONSES TO HUMAN HSP60 IN JUVENILE CHRONIC ARTHRITIS

Mechanisms of peripheral tolerance are likely to be involved in recovery from arthritis and the subsequent resistance to newly induced arthritis as seen in the animal models. In humans, in the case of rheumatoid arthritis (RA), disease is usually progressive, and in children disease may remit in many cases. Probably with the exception of very early cases of disease, proliferative responses to hsp60 are not seen in the majority of patients with RA,[20] although nonproliferative responses, such as production of certain cytokines after stimulation,[21] have been demonstrated in the presence of hsp60. In children with JCA, however, proliferative responses have been obtained in lymphocytes taken from both the peripheral blood and the synovial compartment. Most significant and reliable responses were seen by stimulating the cells with self antigen human hsp60.[22]

Since most of the patients, however, did respond to mycobacterial hsp60 as well, it seems that also in this case, conserved epitopes, equally present in both the human and mycobacterial hsp60, are recognized by patient T cells. From comparison of patients from distinct clinical subgroups, it became evident that responders had oligoarticular (OA) forms of JCA, whereas nonresponders had polyarticular or systemic JCA. In other words, those with a remitting form of disease responded, whereas those with a nonremitting form did not.[23] Longitudinal studies indicated that (temporary) remission of OA-JCA coincided with disappearance of responses and that a clinical relapse coincided with reappearance of responses. Furthermore, it was found that *in vitro* priming of nonresponder cells led to positive secondary responses only in the case of OA-JCA (remission phase) and not in the case of other clinical subgroups of JCA.

Altogether, the data obtained in JCA patients have shown that responses to human hsp60 as a self antigen do occur and that they are associated with relatively benign forms of arthritis. The presence of responses during the active phase of disease preceding remission suggests, in line with the observations made in the AA model, that responses to self hsp60 may contribute positively to mechanisms leading to disease remission. If so, possibilities for immunological intervention in JCA, and possibly also RA, may be found in strategies aimed at manipulating peripheral tolerance through vaccination with hsp60 or peptides containing defined epitopes. For these purposes hsp60 may well serve as a useful ancillary autoantigen.

6. LESSONS FOR THE DEVELOPMENT OF SPECIFIC IMMUNOTHERAPY IN AUTOIMMUNITY

As argued above, immunity to bacterial antigens such as hsp's may contribute to maintenance of self-tolerance as a hedge against autoimmunity. To achieve a lasting restoration of such tolerance in the case of disease, it seems most appropriate to target immunotherapy at the reinforcement of natural mechanisms that contribute to such maintenance of self-tolerance. In other words, exposure of the immune system to bacterial antigens such as hsp's may well stimulate the immune system to resume control over unwanted self-reactive clones. In line with the known contribution of bacterial gut flora to tolerance, it seems best to effect such exposure through oral administration of bacterial antigens. Although the effectivity of such an approach is not supported by work in experimental disease models, our experience in human medicine thus far does lend such support. OM-Laboratories (Geneva) has been producing *E. coli* bacterial lysates, which are used among other agents for the treatment of RA. They are administered orally and have shown in multiple trials in RA patients an effectivity comparable to that of gold.[24] Recent analyses have revealed that the *E. coli* hsp60 molecule (Gro-EL) is one of the more prominent immunogens present in this material.[25] This would suggest that *E. coli* hsp60, when administered orally, may trigger a T-cell regulatory event that contributes to the control of RA, similar to the effect of mycobacterial hsp60 in models of experimental arthritis. For obvious reasons it would be of great interest to analyze such mechanisms at the level of T-cell responses in patients treated in this way.

Positive findings of such an analysis could then lead to the development of better defined pharmaceutical compounds such as synthetic peptides. Such peptides could be composed of conserved sequences of bacterial hsp's and be used to stimulate the frequency or activity of T cells with the potential to recognize self hsp molecules expressed at sites of inflammation.

Alternatively, so-called altered peptide ligands[26] that are aimed at modifying or controlling responses to "dangerous" epitopes associated with autoimmunity can be designed. Recently, we have introduced so-called competitor-modulator peptides, with the capacity on one hand to block MHC presentation of these dangerous epitopes by means of superior MHC binding qualities and on the other hand to modulate responses with specificity for these dangerous epitopes by means of their immunogenicity and close antigenic relationship to these epitopes.[27] In the model of adjuvant arthritis, a peptide based on the 180–188 sequence of mycobacterial hsp60, with a subtle change of the leucine at position 183 into an alanine residue, was found to have a unique capacity to prevent development of AA when administered at the time of Mt immunization for arthritis induction.[28] It can be anticipated that in the not too distant future specific peptides with carefully designed immunogenic qualities will be evaluated for their therapeutic potential in the clinical situation. The identification of antigens with critical involvement in autoimmunity has to precede the development of such peptides. For reasons discussed above, bacterial antigens are attractive candidates in the search for antigens with critical involvement in autoimmunity.

ACKNOWLEDGMENTS. S. M. Anderton is supported by a Wellcome Trust traveling postdoctoral research fellowship to Western Europe. We thank Mrs. Jona Gianotten for expert editorial assistance.

REFERENCES

1. Cohen, I. R., 1986, Regulation of autoimmune disease: Physiological and therapeutic, *Immunol. Rev.* 94:5–21.
2. Toivanen, A., and Toivanen, P., 1991, Reactive vs. rheumatoid arthritis: What is to be learned?, in: *Trends in Rheumatoid Arthritis Research* (P. Hedquist, J. R. Kalden, R. Müiller-Peddinghaus and D. R. Robinson, eds.), Eular Publishers, Basel, Switzerland, pp. 29–35.
3. Whitehouse, D. J., Whitehouse, M. W., and Pearson, C. M., 1969, Passive transfer of adjuvant induced arthritis and allergic encephalomyelitis in rats using thoracic duct lymphocytes, *Nature* 224:1322–1324.
4. Holoshitz, J., Naparstek, Y., Ben-Nun, A., and Cohen, I. R., 1983, Lines of T lymphocytes induce or vaccinate against autoimmune arthritis, *Science* 219:56–58.
5. Klasen, I. S., Kool, J., Melief, M. J., and Hazenberg, M., 1992, Arthritis autoreactive T cell lines obtained from rats after injection of intestinal bacterial cell wall fragments, *Cell Immunol.* 139:455–467.
6. Cohen, I. R., 1992, The cognitive paradigm and the immunological homunculus, *Immunol. Today* 13:490–494.
7. Hammerling, G. J., Schonrich, G., Ferber, I., and Arnold, B., 1993, Peripheral tolerance as a multi-step mechanism, *Immunol. Rev.* 133:93–103.
8. Kohashi, O., Kohashi, Y., Ozawa, A., and Shigematsu, N., 1986, Suppressive effect of *E. coli* on adjuvant-induced arthritis in germ-free rats, *Arthritis Rheum.* 29:547–552.
9. Thompson, S. J., and Elson, C. J., 1993, Susceptibility to pristane-induced arthritis is altered with changes in bowel flora, *Immunol. Lett.* 36:227–230.
10. Bowman, M. A., Leiter, E. H., and Atkinson, M. A., 1994, Prevention of diabetes in the NOD mouse: Implications for therapeutic intervention in human disease, *Immunol. Today* 15:115–120.
11. Sadelain, M. W. J., Qin, H. Y., Sumoski, W., Parfrey, N., Singh, B., and Rabinovitch, A., 1990, Prevention of diabetes in the BB rat by early immunotherapy using Freund's adjuvant, *J. Autoimmun.* 3:671–680.
12. Shehadh, N., Calcinaro, F., Bradley, B. J., Brucklim, I., Vardi, P., and Lafferty, K. J., 1994, Effect of adjuvant therapy on development of diabetes in mouse and man, *Lancet* 343:706–707.
13. Van Eden, W., Thole, J. E. R., van der Zee, R., Noordzij, A., van Embden, J. D. A., Hensen, E. J., and Cohen, I. R., 1988, Cloning of the mycobacterial epitope recognized by T lymphocytes in adjuvant arthritis, *Nature* 331:171–173.
14. Van Eden, W., Hogervorst, E. J., Hensen, E. J., van der Zee, R., van Embden, J. D. A., and Cohen, I. R., 1989, A cartilage-mimicking T cell epitope on a 65K mycobacterial heat shock protein: Adjuvant arthritis as a model for human rheumatoid arthritis, *Curr. Top. Microbiol. Immunol.* 145:27–43.
15. Van Eden, W., 1991, Heat-shock proteins and the immune system, *Immunol. Rev.* 121: 5–28.
16. Welch, W. J., 1993, How cells respond to stress, *Sci. Am.* 268:56–62.
17. Boog, C. J. P., de Graeff-Meeder, E. R., Lucassen, M. A., van der Zee, R., Voorhorst-Ogink, M. M., van Kooten, P. J. S., Geuze, H. J., and van Eden, W., 1992, Two monoclonal antibodies generated against human hsp60 show reactivity with synovial membranes of patients with juvenile chronic arthritis, *J. Exp. Med.* 175:1805–1810.
18. Anderton, S. M., van der Zee, R., Noordzij, A., and van Eden, W., 1994, Differential

mycobacterial 65-kDa hsp T cell epitope recognition after AA inducing or protective immunization protocols, *J. Immunol.* **152**:3656.

19. Anderton, S. M., and W. van Eden, 1996, T lymphocyte recognition of hsp60 in experimental arthritis, in: *Stress Proteins in Medicine* (W. can Eden and D. Young, eds.), Marcel Dekker, New York, pp. 73–92.

20. Res, P. C. M., Telgt, D., Laar, J. M., Oudkerk Pool, M., Breedveld, F. C., and de Vries, R. R. P., 1990, Increased antigen reactivity of mononuclear cells from sites of chronic inflammation, *Lancet* **336**:1406–1408.

21. Wilbrink, B., Holewijn, M., Bijlsma, J. W. J., van Roy, J. L. A. M., den Otter, W., and van Eden, W., 1993, Suppression of human cartilage proteoglycan synthesis by rheumatoid synovial fluid mononuclear cells activated with mycobacterial 60 kD heat-shock protein, *Arthritis Rheum.* **36**:514–518.

22. de Graeff-Meeder, E. R., van der Zee, R., Rijkers, G. T., Schuurman, H. J., Kuis, W., Bijlsma, J. W. J., Zegers, B. J. M., and van Eden, W., 1991, Recognition of human 60 kD heat shock protein by mononuclear cells from patients with juvenile chronic arthritis, *Lancet* **337**:1368–1372.

23. de Graeff-Meeder, E. R., van Eden, W., Rijkers, G. T., Kuis, W., Voorhorst-Ogink, M. M., van der Zee, R., Schuurman, H.-J., Helders, P. J. M., and Zegers, B. J. M., Juvenile chronic arthritis: T-cell reactivity to human hsp60 in patients with a favorable course of arthritis, *J. Clin. Invest.* **95**:934–940.

24. Rosenthal, M., Bahous, I., and Ambrosini, G., 1991, Long term treatment of rheumatoid arthritis with OM-8980. A retrospective study, *J. Rheum.* **18**:1790–1793.

25. Vischer, T. L., 1988, A double blind multi center study of OM-8980 and auranofin in rheumatoid arthritis, *Ann. Rheum. Dis.* **47**:582–587.

26. Evavold, B. D., Sloan-Lancaster, J., and Allen, P. H., 1993, Tickling the TcR selective T-cell functions stimulated by altered peptide ligands, *Immunol. Today* **14**:602–609.

27. Van Eden, W., Wauben, M. H. M., Boog, C. J. P., and van der Zee, R., 1993, Inhibition of autoimmune T cells by "competitor modulator" peptides, in: *Progress in Immunology* (J. Gergely, ed.), Springer Verlag, Budapest, pp. 587–594.

28. Wauben, M. H. M., Boog, C. J. P., van der Zee, R., Joosten, I., Schlief, A., and van Eden, W., 1992, Disease inhibition by MHC binding peptide analogues of disease associated epitopes: More than blocking alone, *J. Exp. Med.* **176**:667–677.

2

Streptococci and Rheumatic Fever

MADELEINE W. CUNNINGHAM

1. INTRODUCTION

Rheumatic fever occurs as a delayed sequel to group A streptococcal infection. The disease is usually manifested as an inflammation of the joints or heart, but symptoms of chorea, subcutaneous nodules, or erythema marginatum may be present. The importance of rheumatic fever is due to the involvement of the heart in the disease, which can be fatal in the acute stage or can lead to chronic rheumatic heart disease with scarring and deformity of the heart valves. Although rheumatic fever has declined in decades past, the current resurgence of the disease has kept us aware of its presence. The pathogenesis of the disease is related to the host immune response made simultaneously against the streptococci and the heart and host tissues. During the past decade, progress has been made in our understanding of rheumatic fever as an autoimmune disease. Figure 1 highlights the components of the disease. The target antigens in the heart, including cardiac myosin, have been identified, and the streptococcal M protein epitopes recognized by antimyosin antibodies in acute rheumatic fever have been determined. The class I M protein epitope was found on group A streptococcal M protein serotypes that were associated with acute rheumatic fever outbreaks. Antibodies

MADELEINE W. CUNNINGHAM • Department of Microbiology and Immunology, University of Oklahoma Health Sciences Center, Oklahoma City, Oklahoma 73190.

Microorganisms and Autoimmune Diseases, edited by Herman Friedman *et al.* Plenum Press, New York, 1996.

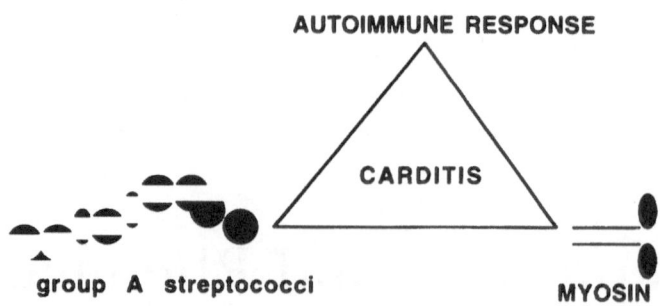

FIGURE 1. Relationship between the group A streptococcus and heart in acute rheumatic fever. The autoimmune responses in the host are produced against streptococci and myosin and may lead to carditis, which is the most serious manifestation of rheumatic fever.

against the streptococcal group A carbohydrate antigen were shown to persist in valvulitis, and antistreptococcal antibodies have been shown to recognize epitopes on the heart-cell surface or valve endothelium. The cross-reactive antibodies may direct inflammation to specific regions of the heart. Sequences of the V region genes encoding the human cross-reactive antistreptococcal/antimyosin antibodies are being determined and the molecular basis of the cross-reactivity of the antibody molecules defined. An antimyosin antibody idiotype, MyI, has been associated with antimyosin antibody responses in acute rheumatic fever, systemic lupus erythematosus and Sjögren's syndrome. In the susceptible host, a B-cell alloantigen was identified as a genetic marker for rheumatic fever, and acute rheumatic fever was associated with certain class II HLA-DR antigens. Despite the advances that have been made in our understanding of rheumatic fever, there remain gaps in our knowledge. Since the resurgence of acute rheumatic fever in the United States in the 1980s, genetic studies to identify streptococcal virulence determinants or clones associated with the disease are under way. Pep M proteins are under investigation as superantigens that may play a role in altering the immune response in rheumatic fever. Although the antibody response against streptococcal M protein and myosin has been well characterized, the cellular immunology of the disease is not well defined. Regions of the M protein, the major virulence determinant of the group A streptococci, are under study to determine cross-reactive and pathogenic regions of the M protein molecule which may participate in the development of rheumatic carditis. With the current knowledge of the pathogenesis and immunology of rheumatic fever, it could be proposed that the antibody response in acute rheumatic fever is responsible for the arthritic and other transient manifestations of the disease but that the permanent heart damage is due to the cellular immune response. The object of recent studies has been to determine the roles of antibodies as well as T cells in the disease. Now that T cells have been classified into subsets based on cytokine patterns and specific cell-surface markers, the elucidation of the types of T cells producing the heart damage in rheumatic fever will be possible. If acute rheumatic fever is an autoimmune disease as proposed, then it is an important model to study the role of infectious agents in the development of autoimmunity in man.

2. HISTORICAL PERSPECTIVE

Since the early 1800s, scientists and physicians have observed and investigated acute rheumatic fever. A most comprehensive and detailed history of the discovery of rheumatic fever and rheumatic heart disease is outlined by Stollerman.[1] The first accounts of rheumatic fever describe the "rheumatism" as early as the 1600s,[2] and Syndenham described chorea in 1686.[3] However, it was not until the late 1700s that the first descriptions of the symptoms of rheumatic fever were correlated with cardiac lesions and involvement.[4] Stollerman describes the findings of Bouillaud,[5] Sokolsky,[6] and Wells[7] as being the most outstanding early accounts of acute rheumatic fever.[1] By the late 1800s Cheadle had assimilated the group of divergent manifestations into a rheumatic syndrome that was called "Cheadle's cycle." The Jones criteria,[8] which are still used in diagnosis of acute rheumatic fever, were patterned after the manifestations described by Cheadle.[9] In 1904 Aschoff[10] discovered the myocardial lesion that is considered to be a hallmark of rheumatic fever, the Aschoff body or the Aschoff-Talalaev granuloma.[11] The acute cardiac failure observed in patients with rheumatic fever was believed to be caused by these changes described in the myocardium rather than a result of the valvulitis. Around the same time, physicians made the connection between a sore throat or pharyngitis and the rheumatic episode.[12,13]

3. CLINICAL FEATURES

The characteristics of acute rheumatic fever can vary greatly depending on the severity of the disease and the body systems that are involved. There is a latent period between the onset of streptococcal pharyngitis and the appearance of rheumatic fever. According to Stollerman,[1] the average latent period is 18.6 days, but it can range from 1 to 5 weeks following streptococcal pharyngitis. Rheumatic fever is now rare in the United States, affecting approximately 1 or 2 per 100,000 children and teenagers.[14] A recent resurgence in particular areas of the United States has heightened awareness of the disease and will be discussed later in this chapter. The rate of occurrence in developing regions such as India, Africa, Southeast Asia, and South and Central America is 100–150 per 100,000, significantly higher than the incidence in the United States.[15] The clinical features of acute rheumatic fever include the criteria developed by Jones,[8] which were updated in 1992 by the American Heart Association's special Writing Group of the Committee on Rheumatic Fever, Endocarditis, and Kawasaki's Disease of the Council on Cardiovascular Disease in the Young. In the 1992 update, the guidelines for the diagnosis of rheumatic fever were described in a special report by the committee.[16] The first guidelines were published in 1944 by T. Duckett Jones[8] and have been revised over the years by the American Heart Association.

The Jones criteria include major and minor manifestations (Table I). Two major manifestations or two minor manifestations and one major manifestation must be observed in order to make the diagnosis of acute rheumatic fever. The major manifestations are suggestive of an immune disorder and include carditis, the most serious manifestation, and arthritis, the most common manifestation. Rheumatic fever classically pre-

TABLE I
Jones Criteria:
Major and Minor Manifestations of Rheumatic Fever

Major manifestations
  Carditis
  Polyarthritis
  Chorea
  Erythema marginatum
  Subcutaneous nodules
Minor manifestations
  Fever
  Arthralgia
  Previous rheumatic fever or rheumatic heart disease
  Elevated erythrocyte sedimentation rate or positive C-reactive protein
  Prolonged PR interval

sents as a migrating polyarthritis affecting one or more of the joints.[16] The arthritis never causes permanent damage and is therefore not like rheumatoid arthritis. Its appearance is coincident with the peak titer of antistreptococcal antibodies.[1]

The most serious manifestation of rheumatic fever is carditis. Studies have shown that patients with severe polyarthritis have lower incidence and severity of cardiac involvement and that there is an inverse relationship between development of the most severe forms of arthritis and development of rheumatic heart disease.[1] In addition, Stollerman points out that "cardiac involvement may occur in the mildest form of the disease," leaving the patient with undiagnosed carditis and development of unnoticed rheumatic heart disease.[1] The incidence of carditis in acute rheumatic fever ranges from 30–50%. In general, carditis presents as a heart murmur due to mitral and/or aortic regurgitation. In severe cases, pericarditis with enlargement of the heart and congestive heart failure may also appear. Heart failure is seen far less frequently today than in decades past and has been found to be more common during a recurrence of rheumatic fever.[1] Permanent heart damage is related to the extent of the valvular, pericardial, and myocardial involvement.

Other major manifestations of rheumatic fever are chorea, subcutaneous nodules, and erythema marginatum.[16] Syndenham's chorea is a neurological disorder which may occur with symptoms of involuntary movements and weakness, and it can be the only major manifestation present.[16] The choreiform movements affect all muscles but primarily affect the hands, feet, and face. In postpubescent subjects, chorea is seen only in females. The appearance of chorea may occur after a lengthy latency period (1–6 months) when antistreptococcal antibody titers are not increased.[1] Chorea has declined over the years as a manifestation of acute rheumatic fever in comparison to carditis and arthritis. Erythema marginatum is a rarely observed, transient, red, macular, circinate skin rash found on the trunk or proximal extremities. It may be observed any time during the disease course but usually occurs in the early stages of the disease.[1] The blanching rash is quite distinct and has been observed in a few other conditions such as sepsis, drug reactions, and glomerulonephritis. Subcutaneous nodules are painless, freely movable nodules that are 0.5–2.0 cm in diameter found over the extensor surfaces of the knee, elbow, wrist, and other joints, and over the spinous processes of the vertebrae.[1,16] Although subcutaneous nodules are rare, they are most often seen following the onset of carditis. Subcutaneous nodules are also seen associated with rheumatoid arthritis and systemic lupus erythematosus and are therefore not pathognomonic of acute rheumatic fever.[1]

Minor manifestations (Table I) include fever, elevated erythrocyte sedimentation rate, abnormal C-reactive protein, leukocytosis, cardiac abnormalities, arthralgia, and evidence of a streptococcal infection, including positive antistreptococcal antibody serology such as an elevated antistreptolysin O (ASO) antibody titer.[1,16]

4. ETIOLOGY: THE GROUP A STREPTOCOCCUS

4.1. Identification and Association of Group A Streptococci with Rheumatic Fever

The group A streptococcus (*Streptococcus pyogenes*) has long been recognized as the primary initiating factor in the development of acute rheumatic fever in the susceptible host.[14] Rheumatic fever is associated only with streptococcal throat infections and not with infection of the skin.[1,17] The strongest evidence supporting the hypothesis that rheumatic fever is a sequel to group A streptococcal infection is that the incidence of rheumatic fever parallels the occurrence of streptococcal infections.[1] Wannamaker considered the group A streptococcus as "the chain that links the heart to the throat."[17] Clinicians who studied the rheumatic fever outbreaks concluded on the basis of microbiologic and epidemiologic data that the group A streptococcus was the etiologic agent in acute rheumatic fever.[18,19] In addition, antibodies against streptococcal components and enzymes were shown to appear approximately 10 days following pharyngeal infection. In acute rheumatic fever elevated titers of antistreptolysin O antibodies were described by Todd.[20] The titer of ASO antibodies was elevated in both streptococcal infection and acute rheumatic fever. Antistreptolysin O titer is now used to document antecedent streptococcal infection in suspected cases of acute rheumatic fever. The elevation of these antibodies is associated with a rise in titer of other antistreptococcal antibodies such as anti-DNAse B or antihyaluronidase. In acute rheumatic fever an overall hyperresponsiveness is observed against streptococcal antigens,[17] and the attack rate is related to the overall magnitude of the hyperresponsiveness.[21,22]

The group A streptococcus is covered with an outer hyaluronic acid capsule,[23] while the group A carbohydrate antigen and the type-specific M protein are attached to the bacterial cell wall and membrane as shown in Fig. 2.[24] Both the M protein and the capsule are considered to be virulence factors conferring antiphagocytic properties upon the streptococcal cell.[24,25] Antibodies made by the host against the cell surface

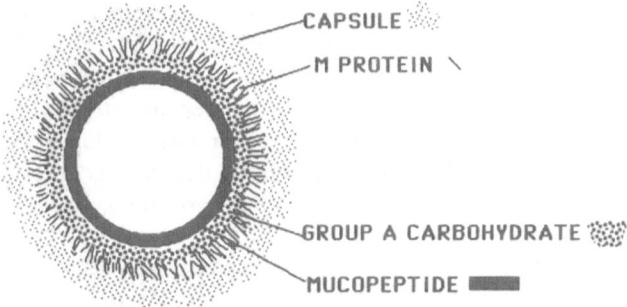

FIGURE 2. Group A streptococcus displayed with surface hyaluronic acid capsule, serotype-specific M protein, group A carbohydrate, and mucopeptide.

antigens such as the M protein or the hyaluronate capsule opsonize the streptococci and enhance phagocytosis and clearance of the bacteria by the host. The identification and classification of the group A streptococcus is attributed to Lancefield[26] and Griffith,[27] who developed serological grouping scheme of streptococci based on their group carbohydrate antigen. The group A carbohydrate antigen of S. *pyogenes* is composed of repeating units of rhamnose linked to N-acetyl-β-D-glucosamine.[28] Further work by Lancefield divided the group A streptococci into M serotypes based on the immunological specificity of the streptococcal M protein antigen.[29] Lancefield determined that the M protein was an important virulence factor of the group A streptococcus and that it induced protective immunity in the host against homologous type organisms.[30,31]

4.2. Group A Streptococcal M Protein

The group A streptococcal M protein has been studied extensively since its discovery. A recent comprehensive review by Fischetti describes the characteristics of the M protein molecule in detail.[24] Antigen for serological typing of the group A streptococcus into M types is obtained by boiling hydrochloric acid extraction of the streptococcus, releasing the M protein molecule.[29] The extract is used as the M antigen in a reaction with different rabbit antisera, each against a different M type. A positive capillary precipitin reaction indicates the M protein serotype. Over 80 different M serotypes have been identified, with many strains that are untypable.[24]

Over the past 60 years new procedures have been developed to obtain purified M protein. Beachey and Cunningham and their co-workers investigated the treatment of M protein with pepsin at subopti-mal pH,[32] which led to the development of a pepsin-extracted M protein, Pep M protein.[33] Methods of M protein extraction developed by Fischetti and colleagues included a type of non-ionic detergent extraction[34] and the group C phage-associated lysin which acts on the cell wall to release the M protein molecule.[35,36]

Structural studies of the streptococcal M protein were begun in the late 1970s and early 1980s by Beachey[37-39] and Fischetti[40-42] when the technology of protein chemistry was used to obtain the amino acid sequence of peptide fragments of the M proteins. The M protein types 24, 5, and 6 were investigated first and their sequences elucidated. It was from amino acid sequence data that Fischetti and Manjula defined the alpha-helical coiled-coil structure of the M proteins and their heptad repeating structure, which was quite similar to the alpha-helical coiled-coil struc-ture in host tissue proteins such as tropomyosin and the keratin-desmin-vimentin family of molecules.[43-45] The age of molecular biology brought cloning technology to the study of streptococcal antigens, and the *emm* genes from M types 24, 5, 6, 12, and group G were cloned and the nucleotide sequence deduced.[46-51] The cloning and sequencing of the *emm* genes[46-51] revealed repeating sequence motifs within 1) the N-termi-nal region, 2) the mid-molecule, pepsin-sensitive region, and 3) the conserved carboxy-terminal region. The N-terminal region, called the A repeat region, confers serotype specificity to the group A streptococcus and was found to be highly variable between M protein serotypes. The mid-region was also variable and was called the B repeat region.[24] The carboxy-terminal region also contained amino acid sequence repeats which extend throughout the carboxy-terminal third of the molecule. Figure 3 illustrates the repeating regions and pepsin cleavage site of the M

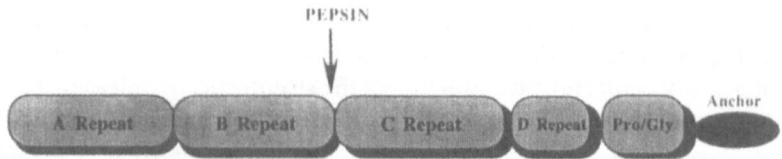

FIGURE 3. Group A streptococcal M protein containing A, B, C, and D repeat regions with a proline-glycine-rich region and an anchor for insertion into the cell membrane. The diagram is adapted from work by V. A. Fischetti.[24]

protein molecule. Using molecular probes, Hollingshead, Fischetti, and Scott[52] determined that the highly conserved carboxy-terminal region of the M protein contained sequence homology shared among most of the M protein serotypes. Vaccine efforts have targeted both the N-terminal region and the more highly conserved carboxy-terminal region of the M protein molecule.[24] The M protein has been and will continue to be the subject of intensive investigation due to its role as a major virulence factor and its potential as a vaccine against streptococcal infections.

The M proteins have been divided into class I and class II molecules.[53] The division of the M proteins into two classes is based on their reaction with antibodies (such as anti–M protein mAb 10B6) against the C repeat region of M protein. Class I M proteins are reported to contain a surface-exposed epitope on whole group A streptococci reactive with the antibodies against the C repeat region. Streptococcal strains containing the class II M proteins do not react with these antibodies and do not contain the class I epitope.[53] In addition, the class I M protein serotypes are opacity factor–negative, while the class II serotypes are opacity factor–positive. In studies of 130 streptococcal isolates, there was a strong correlation between serotypes known to produce rheumatic fever and the presence of the class I epitope.[53] These data correlate with the M-associated protein (MAP) I and II antigen profiles and may be the basis of the MAP reactivity previously studied by Widdowson and coworkers.[54,55] Antibodies against heart in rheumatic fever were associated with the MAP I antigen. Bisno has reported that there is a strong correlation between streptococcal serotype and the occurrence of rheumatic fever.[56] The epidemiologic data has led to the proposal that group A streptococci associated with acute rheumatic fever outbreaks harbor a unique antigen or epitope that is associated with the development of acute rheumatic fever and carditis in the susceptible host.

5. EPIDEMIOLOGY AND HOST SUSCEPTIBILITY

5.1. Rheumatic Fever: Epidemiology during 1900–1980

The incidence of acute rheumatic fever in the early 1900s was 100–200 cases per 100,000 in the United States.[57,58] However, in the 1940s rheumatic fever began to decline and was reported at 50 cases per 100,000.[57] Since this time the incidence of rheumatic fever in the United States and Western Europe fell steadily until it reached 0.5 cases per 100,000 in the early 1980s.[57] In contrast, the number of cases of childhood rheumatic fever in India and other developing countries was estimated at 6 million, with 250,000 new cases in school-age children each year.[59]

5.2. Resurgence of Rheumatic Fever during 1980–1990

The 1980s and 1990s have seen a resurgence of rheumatic fever in eight outbreaks documented throughout the United States.[60–62] The first and one of the largest outbreaks was reported from Salt Lake City, Utah, in 1985.[61] Ayoub succinctly summarizes the outbreaks in a recent review article and states, "the resurgence in the United States has rekindled interest in the disease and has served notice to the medical community." Other outbreaks have been reported in Pennsylvania, Ohio, Tennessee, West Virginia, and at the Naval Training Center in San Diego, California.[62] A feature of the recent outbreaks has included a change in the population at risk where the disease appeared in children of high- to middle-income families with good access to medical care. However, the families were larger than average, and as before crowding was found to be a factor. Previous epidemiologic data from the rheumatic fever outbreaks in the past had shown that the disease was associated with children from the inner city and low-income families with little access to medical care.[1] The reasons for the increase in rheumatic fever as well as an increase in more serious group A streptococcal infections over the past ten years remains unknown.[61]

5.3. Features of Group A Streptococci Associated with Rheumatic Fever Outbreaks

Kaplan et al. report that the streptococcal M protein serotypes involved with the new outbreaks of rheumatic fever include M types 1, 3, 5, 6, and 18.[63] These M serotypes were reported to be associated with previous outbreaks.[64] The association of certain M protein serotypes with rheumatic fever over the last 50 years[56] suggests that the M protein may be a decisive risk factor in the development of the disease in the susceptible host. Many strains recovered from acute rheumatic fever patients were found to have a mucoid appearance. It has not been determined if the hyaluronic acid capsule, the M protein, or some other virulence factor of the group A streptococcus has played a role in the recent resurgence of acute rheumatic fever.

The group A streptococci produce a number of toxins and extracellular products that may play a role in the virulence and pathogenesis of streptococcal infection or sequelae.[65] Figure 4 illustrates streptococcal extracellular enzymes and toxins. The erythrogenic toxins, particularly

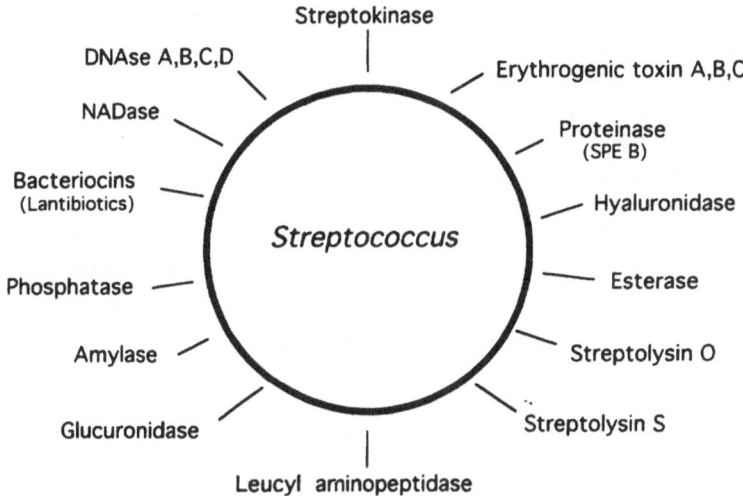

FIGURE 4. Extracellular products and toxins produced by group A streptococci. The illustration is a gift from J. J. Ferretti and adapted from previous work.[66]

erythrogenic toxin A, have been found to be of considerable importance due to their association with alteration of the immune system, enhancement of endotoxic shock, and cardiotoxicity.[6] Ferretti and colleagues have studied the genes for these toxins.[65-68] Using toxin gene probes, streptococcal isolates associated with rheumatic fever outbreaks were tested for the presence of the toxin genes. The data suggest that the erythrogenic toxin A gene (*SpeA*) is present in strains associated with scarlet fever and rheumatic fever at a significantly higher frequency than that found for group A streptococcal isolates associated with other types of disease.[67,68] Streptolysin O has also been shown to be cardiotoxic and proposed to play a role in streptococcal diseases and sequelae.[69] However, the roles of the toxins in the pathogenesis of rheumatic fever remain to be defined.

It is possible that global regulators of virulence genes expressed only in the host are turned on and are important in the persistent colonization of susceptible individuals who develop acute rheumatic fever. Only one positive regulator gene, virR, or Mry, which is known to control expression of M protein,[70-72] C5a peptidase,[73] and the Fc receptor protein[74] of the group A streptococcus, has been identified. The regulator enhances expression of M protein manyfold. Whether environmental factors play a role in controlled expression of M protein and other virulence factors in the group A streptococcus is not known. The M protein is thought to be one of the most important streptococcal virulence factors in the development of acute rheumatic fever.

Recently a streptococcal gene was cloned for a 60-kDa protein associated with hyperresponsiveness and reactivity with antimyosin antibodies in acute rheumatic fever patients.[75,76] The gene was present only in pathogenic groups A, C, and G streptococci.[76] Upon sequence the gene was found to have sequence homology with class II HLA antigens such as DR, DP, and DQ in humans (Fig. 5).[76] The role of the 60-kDa protein in the pathogenesis of acute rheumatic fever remains to be determined.

5.4. Genetic Susceptibility to Rheumatic Fever

The role that the group A streptococcal antigens play in the development of acute rheumatic fever is most likely associated with an aberrant

```
        Similarity: 48%              Identity: 19%
                      .                         .
DQ     1 DFVYQFKGMCYFT...NGTERVRLVTRYIYNREEYARFDSDVGVYRAVTP 47
         : :. |.::.. .   :|.|:.:|.     .||  .:|:: ::||.:..
67KD  50 ERIHLFEELPLAGGSLDGIEKPHLGFVTRGGREMENHFECMWDMYRSIPS 99

              .                     .               .
DQ    48 LGRPDA....EYWNSQKEVLERTRAELDTVCRHNYEVELRTTLQRRVEPT 93
         |: |:|    |::  |::  :....|    :. : : . ||.:. ..
67KD 100 LEIPGASYLDEFYWLDKDDPNSSNCRLIHKRGNRVDDDGQYTLGKQSKEL 149

             .        .                    .         .
DQ    94 VTISPSRTEALNHHNLLVCSVTDFYPAQIKV.................. 124
         | :  . .|.|..:.:      .||:...:.:
67KD 150 VHLIMKTEESLGDQTIEEFFSEDFFKSNFWIYWATMFAFEKWQFCCRNAG 199

                .                          .        .
DQ   125 ..RWFRNDQEETAGVVSTPLIRNGDW............... 148
         .:| :. :: ::..| .: : .:|
67KD 200 YAMRFIHHIDGLPDFTSLKFNKYNQYDSMVKPIIAYLESHDVDIQFDTKV 249

                              .               .
DQ   149 ..............TFQILV.....MLEMTPQ..........RGDVYT 167
                       |::: |        :|:||:          :..|.
67KD 250 TDIQVEQTAGKKVAKTIHMTVSGEAKAIELTPDDLVFVTNGSITESSTYG 299

                            .            .               .
DQ   168 CHVEHPSLESPITVEWR......AQSE............SAQSKMLSGIG 199
         :| | :. ..:...|.       |||:           .|:|.::|:.:
67KD 300 SHHEVAKPTKALGGSWNLWENLAAQSDDFGHPKVFYQDLPAESWFVSATA 349

             .                .
DQ   200 GFVLGLIFLGLGLIIHHRSQKGLL 223
         .:  . |   :: :.|: :.| :
67KD 350 TIKHPAIEPYIERLTHRDLHDGKV 373
```

FIGURE 5. Homology between the amino acid sequences of human class II major histocompatibility antigen DQ and deduced amino acid sequence of a 67-kDa streptococcal gene found only in streptococcal groups A, C, and G.[76] (With permission from *Infection and Immunity*): **I**, identity; **:**, conserved substitution; **.**, functional substitution.

immune response in the susceptible host. The response of the susceptible host to group A streptococcal antigens has been shown to be associated with a hyperresponsiveness to these antigens.[1] Antigen processing is thought to differ among individuals with different HLA phenotypes, and in individuals with acute rheumatic fever, immune responses to streptococcal antigens may differ from those of individuals with uncomplicated infections. Studies of responses against particular M protein epitopes have revealed overall differences in the antibody responses of rheumatic fever patients in comparison with normals and uncomplicated disease.[77,78] Furthermore, an idiotype (My1) expressed on antimyosin antibodies produced in rheumatic fever is present only in streptococcal sequelae and in the autoimmune diseases systemic lupus erythematosus and Sjögren's syndrome.[79] The idiotype is not elevated above normal levels in uncomplicated streptococcal disease, myocarditis, IgA nephropathy, Chagas disease, or rheumatoid arthritis (Fig. 6). The identification of the My1 idiotype associated with rheumatic fever and other autoimmune diseases suggests that these diseases are linked in expression of particular antibody molecules with the My1 idiotype. Acute rheumatic fever has many features of an autoimmune disease.

Despite the prevalence of streptococcal pharyngitis in populations, only a very small percentage develop acute rheumatic fever.[80] The susceptible host that contracts rheumatic fever produces a greater immune response to streptococcal antigens than the nonrheumatic host. Hyperresponsiveness to streptococcal antigens is characteristic of the patient with rheumatic fever. Studies in the past[81-84] have suggested that the rheumatic host was genetically predisposed to disease, and familial studies[84] have shown that the genetic factor has limited penetrance. Other autoimmune and rheumatic diseases have often been associated with host expression of major histocompatibility (MHC) genes of a particular human leukocyte antigen (HLA) phenotype. Ayoub has recently described higher frequency of HLA-DR2 expression in black patients with rheumatic fever and HLA-DR4 in Caucasian patients with rheumatic fever.[85] In South African black populations with rheumatic fever, HLA-DR1 and Drw6 were observed with increased frequency.[86] The association of different HLA phenotypes with rheumatic fever in different ethnic groups suggests that the immunogenetics of the susceptibility is complex. Twins do not usually both develop rheumatic fever, indicating that environmental and host factors are important in susceptibility to the disease. Repeated exposure to streptococcal infections is believed to be necessary for the immune responses to reach the level of disease production.

Environmental factors such as exposure of the immune system to

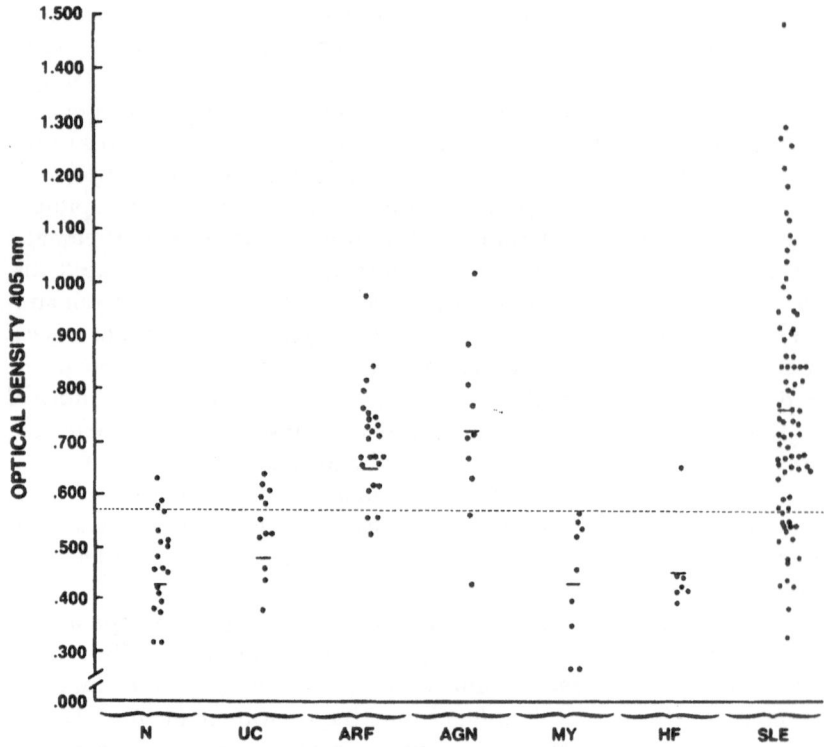

FIGURE 6. The Myl idiotype detected in sera from normal subjects (N) and from patients with uncomplicated group A streptococcal disease (UC), acute rheumatic fever (ARF), acute glomerulonephritis (AGN), myocarditis (MY), heart failure (HF), or systemic lupus erythematosus (SLE).[79] The Myl idiotype was significantly elevated in ARF, AGN, SLE, and Sjögren's syndrome (not shown). (Reproduced with permission from *Journal of Infectious Diseases*).

specific antigens during neonatal life may play a role in the immune responses of the genetically predisposed host at a later stage in life such as early childhood through adolescence when rheumatic fever peaks at ages 5–15. Work by Vakil and Kearney and their coworkers shows that neonatal mice exposed to pneumococcal antigen do not develop a normal antibody response against pneumococci.[87,88] Animals exposed neonatally to pneumococcal polysaccharide were unable to produce the idiotype associated with normal immune responses to pneumococci when challenged later in life. Differences in exposure to common pathogens and normal flora during development of the immune system in early infancy may change the responsiveness of the host.[87,88] Environmentally driven changes in the immune repertoire linked with the expression of an influential HLA phenotype which may present potentially pathogenic epitopes to the immune system may elicit hyperresponsiveness, delayed type hypersensitivity, and inflammatory responses in target tissues.

Idiotypes which may image the host/bacterial determinant could act in concert with predisposing factors in the host to initiate or exacerbate disease (Fig. 7). The presence of an antimyosin idiotype (Myl) was found in the serum of patients with streptococcal sequelae, rheumatic fever, and acute glomerulonephritis, but elevated levels were not found in normal or uncomplicated streptococcal infections (Fig. 6).[79] Studies of anti-myosin antibodies in rheumatic fever show that the idiotype (Myl) is present on antibodies in streptococcal sequelae. and two other auto-immune diseases, systemic lupus erythematosus (Fig. 6) and Sjögren's syndrome (not shown). [79] These surprising data may link predisposing and environmental factors with the autoimmune diseases expressing the Myl idiotype. The Myl idiotype was not elevated in rheumatoid arthritis, Chagas disease, IgA nephropathy, myocarditis, and heart failure (Fig. 6). The data show that the Myl idiotype was not present in heart failure (Fig. 6) and other states[79] and support the findings of Zabriskie,[101] who showed that the antiheart antibodies in postcardiotomy or myocardial infarction

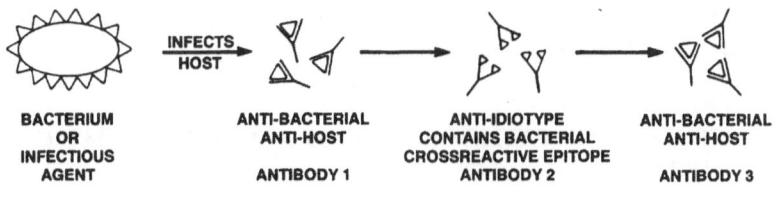

BACTERIUM OR INFECTIOUS AGENT   ANTI-BACTERIAL ANTI-HOST ANTIBODY 1   ANTI-IDIOTYPE CONTAINS BACTERIAL CROSSREACTIVE EPITOPE ANTIBODY 2   ANTI-BACTERIAL ANTI-HOST ANTIBODY 3

△ BACTERIA/HOST EPITOPE

Y ANTIBODY MOLECULES

FIGURE 7. Idiotype–antiidiotype interactions. The bacterium may express epitopes capable of inducing and reacting with antibacterial and antihost antibodies (antibody 1). The idiotype of the cross-reactive antibody would hypothetically induce an antiidiotype response (antibody 2) against antibody 1. An anti-antiidiotypic response (antibody 3) is then produced against antibody 2. Antibody 2 could perpetuate the production of cross-reactive, antibacterial, antihost antibodies. The idiotypic site on antibody 2 hypothetically would mimic the epitope originally presented to the host on the bacterium or infectious particle. (From Ref. 134.)

sera were not absorbed by group A streptococci and were different from the antiheart/antistreptococcal antibodies seen in acute rheumatic fever.

Host susceptibility has been defined further in recent work by Zabriskie and colleagues.[89,90] They identified a strong association between the presence of a non-HLA B cell marker, known as 883 or D8/17, and the occurrence of rheumatic fever. Using antibodies against the B-cell antigen, B cells from patients with a history of rheumatic fever were positive with approximately 33% of their B cells staining with the D8/17 antibody. A low level of staining (5–7%) was observed in the control groups including a group of patients with acute poststreptococcal glomerulonephritis. In familial studies using the antibody to follow the B-cell marker, it was found that the percentage of positive parents and siblings in rheumatic fever families was consistent with an autosomal recessive mode of inheritance.[89] Unaffected siblings or twins had approximately 15% of their B cells expressing the D8/17 marker. The D8/17 B-cell antigen was present in unaffected individuals and controls at a lower level of expression (5–7% of B cells) than in rheumatic fever patients, and the marker was elevated in patients with rheumatic fever from all ethnic groups. These studies suggest that there is a genetic factor(s) involved in susceptibility to rheumatic fever. The nature of the D8/17 antigen is unknown. Studies are currently in progress to further elucidate the D8/17 antigen as well as HLA antigens associated with the disease. Like many of the autoimmune diseases, acute rheumatic fever is a multifactorial disease due to a combination of susceptible host, group A streptococcal infection, and other environmental and host factors affecting the immune response.

6. AUTOIMMUNE PATHOGENESIS OF ACUTE RHEUMATIC FEVER

6.1. Antibody-Mediated Mechanisms

6.1.1. Antiheart Antibody

In 1945 Cavelti reported[91] one of the first descriptions of antibodies against the heart in rheumatic fever, and in 1969 with the use of fluorescein-labeled antibodies Kaplan and Freugley demonstrated labeling of the heart with antibodies from acute rheumatic fever sera.[92] The studies continued and heart-reactive antibodies were reported to be elevated in the sera of patients with acute rheumatic fever.[93] Antibody and complement were found deposited in the hearts of rheumatic fever patients at autopsy.[94] These data supported the hypothesis that acute rheumatic fever has an autoimmune origin. Further studies[95-97] demonstrated that rabbit antisera against group A streptococcal cell walls reacted with human heart and that antibodies prepared against human heart reacted with streptococcal antigens. Antibodies against group A streptococci in human serum could be absorbed with human heart preparations, and conversely, antiheart antibodies could be absorbed from the sera with group A streptococci or their membranes.[98,99] Rheumatic fever sera or anti–group A streptococcal antisera labeled rheumatic or nonrheumatic human heart or skeletal muscle tissue with equal intensity.[100]

Titers of antiheart antibodies appeared to persist in patients with rheumatic recurrences. The studies suggested that there was a relationship between the elevated antiheart antibody titers and susceptibility to a second attack. Usually heart-reactive antibodies declined within 5 years of the initial rheumatic episode.[100] Zabriskie suggests that repeated episodes of streptococcal infections are required to induce elevated levels of heart-reactive antibodies and expression of the symptoms of rheumatic fever. However, not all agree[80] that autoantibodies play a role in the development of rheumatic heart disease, and therefore, the role of antibodies in acute rheumatic fever remains speculative and unclear. Due to the conflicting clinical evidence, the approaches to study the autoantibodies have been molecular and have dissected the cross-reactive antibody response. The autoantibodies are likely to deposit in tissues and create inflammation, but the permanent heart damage seen in rheumatic fever is related to immune cells that infiltrate the heart.

6.1.2. Streptococcal Cross-Reactive Antigens

Early work by Kaplan and colleagues[94-98] implicated the cell wall containing M protein as the cross-reactive antigen recognized by the antiheart antibodies, while studies by Zabriskie and Freimer[99,100] implicated the streptococcal membrane. Streptococcal antigens such as the M-associated non-type-specific antigen and the MAP antigen were studied by Beachey and Stollerman[101] and by Widdowson and Maxted,[102] respectively. Finally, highly purified peptide components from the group A streptococcal membrane were shown to react with antiheart antibody in rheumatic fever sera.[103] However, the consensus was that the streptococcal cell wall and membrane contained antigens that were reactive with antiheart antibodies.

Studies by Goldstein and colleagues[104] defining the streptococcal heart cross-reactive antigens suggested that the group A polysaccharide and the glycoproteins of heart valves containing N-acetyl-glucosamine were responsible for the antistreptococcal antibody cross-reactivity with heart. Others such as Lyampert[105,106] published studies suggesting that the group A streptococcal polysaccharide induced responses against host tissues, and Sandson[107] reported that the hyaluronic acid in the streptococcal capsule might induce immune responses against the joint tissues. The studies in the 1960s and 1970s left little doubt that the antistreptococcal immune response could induce autoantibodies against the heart and other tissues and that several group A streptococcal antigens were involved in the process.

6.1.3. Myosin: Autoantigen Target in Human Heart

In the early 1980s with the advent of monoclonal antibodies (mAbs) and the Western immunoblot, Cunningham and colleagues[108,109] developed mouse antistreptococcal mAbs which reacted with heart tissue sections (Fig. 8) or with heart tissue extracts in the enzyme-linked immuno-

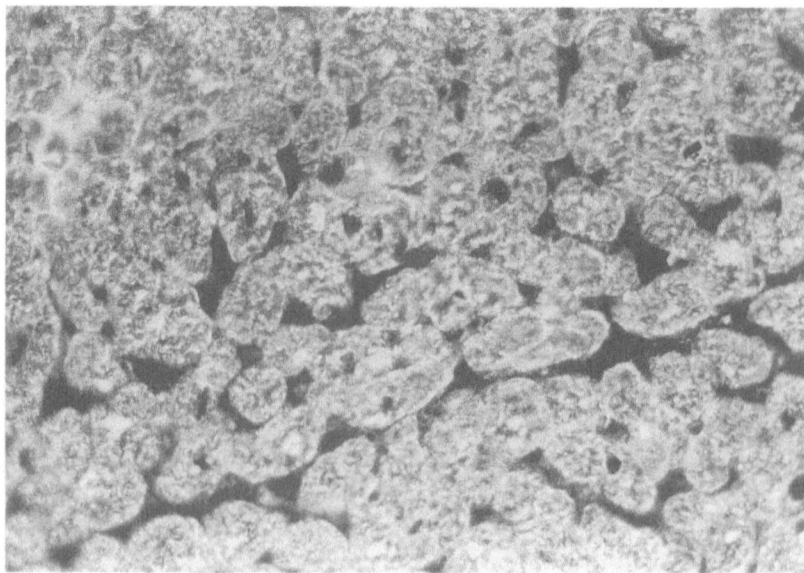

FIGURE 8. Indirect immunofluorescence staining of a section of human heart tissue by antistreptococcal monoclonal antibody 112.2.2. (With permission from the *Journal of Immunology.*)

sorbent assay. In 1985 Krisher and Cunningham[110] reported that the heart antigen recognized by the antistreptococcal mAbs was cardiac myosin. Subsequently, the streptococcal M protein,[111-113] as well as streptococcal membranes,[75,114] were shown to contain heart and myosin cross-reactive epitopes that were recognized by antimyosin antibodies in acute rheumatic fever and by murine and human antistreptococcal mAbs. The studies using mAbs revealed that different streptococcal antigens, including the M protein present in the cell wall and a 60-kDa antigen present in the streptococcal membrane, were important in antibody cross-reactivity with streptococci and myosin.[75,111-114] The data clarified the observations of Zabriskie and Kaplan[94-99] and showed that both the streptococcal cell wall and membrane contained heart cross-reactive antigens that reacted with the antimyosin antibodies in acute rheumatic fever and with the antiheart/antimyosin/antistreptococcal mAbs.[75,109-116] The cross-reactive specificities of the mouse monoclonal antibodies are shown in Table II.

The production of the human monoclonal antibodies from patients with rheumatic fever has been instrumental in identifying the cross-reactive antigens and proving that antiheart antibodies can recognize both streptococcal and heart antigens. Table III summarizes the cross-reactivities of the human antistreptococcal mAbs. Studies using affinity-purified human antimyosin antibodies from acute rheumatic fever sera have been an important correlate to the mAbs studies, because they have confirmed that the reactivities of the mAbs are similar or the same as

TABLE II
Antigen Binding of Cross-Reactive
Murine Monoclonal Antibody Probes

mAb	Antigen specificity
6.5.1	Actin/M proteins 5,6
8.5.1	Myosin/streptococci
24.1.2	Myosin/tropomyosin/keratin/M protein 5,6
36.2.2[a]	Myosin/tropomyosin/keratin/actin/lamanin/M proteins 1,5,6
40.4.1	Myosin/M protein 5
49.8.9	Vimentin/M protein 5,6
54.2.8	Myosin/tropomyosin/vimentin/DNA
101.4.1	Myosin/streptococci
112.2.2	Myosin/tropomyosin/M proteins 5,6
113.2.1	Actin/M proteins 5,6
654.1.1	Myosin/tropomyosin/DNA/M proteins 5,6

[a]Cytotoxic for heart cells in ^{51}Cr release assay.

TABLE III
Cross-Reactivity of Human
Antistreptococcal Monoclonal Antibodies

Clone	Cross-reactive proteins
10.2.3	Human cardiac myosin/M protein/GlcNac[b]
10.2.5	Human cardiac myosin/M protein/GlcNac
1.C8	Myosin/vimentin/keratin/M protein/ GlcNac
1.H9	Human cardiac myosin/keratin/M protein/ GlcNac
4.F2	Rabbit skeletal myosin/keratin/M protein/ GlcNac
5.G7[a]	Rabbit skeletal myosin/keratin/M protein/ GlcNac
5.G3	Rabbit skeletal myosin/keratin/M protein/ GlcNac
1.C3	Laminin/keratin/GlcNac
9.B12	Keratin/GlcNac
2.H11	Keratin/GlcNac

[a]Rheumatoid factor.
[b]N-acetyl-β-D-glucosamine.

those found in acute rheumatic fever sera.[111,116] Antimyosin antibodies purified from acute rheumatic fever sera were shown to react with M protein.[111,116] One of the myosin cross-reactive epitopes of M5, a protein recognized by the antimyosin antibodies in acute rheumatic fever sera, is located near the pepsin cleavage site in both M5 and M6 proteins. The epitope contains the amino acid sequence Gln-Lys-Ser-Lys-Gln.[77]

The human cardiac myosin molecule is composed of one heavy chain and at least two light chains which form a dimeric alpha-helical coiled coil structure (Fig. 9). The reactivity of the antistreptococcal mAbs and the affinity-purified antimyosin antibodies from acute rheumatic fever sera react with the heavy chain of human cardiac myosin[117] or skeletal myosin[115] and do not react with the light chains of myosin. Figure 9 shows the myosin molecule and its subfragments. Proteolytic fragments of human cardiac myosin were used to identify sites of cross-reactivity of the antistreptococcal/antiheart mAbs. Antistreptococcal mAb reactivity was observed with either the heavy or light meromyosin (HMM and LMM) subfragments of the myosin heavy chain.[117] The proteolytic fragments were produced by cleavage with chymotrypsin. The LMM and S2 fragments contain the alpha-helical rod region of the heavy chain,

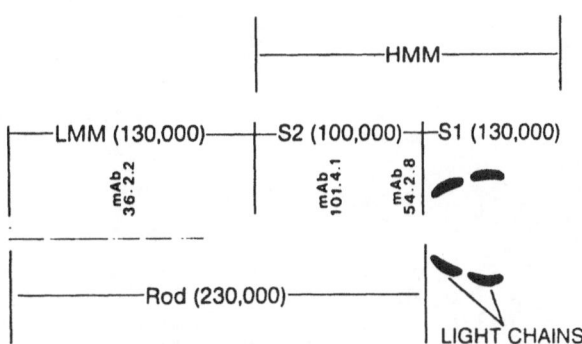

FIGURE 9. Myosin heavy chains and light chains. The diagram illustrates the proteolytic fragments of myosin, including the globular head containing the ATPase activity, the heavy meromyosin (HMM) containing the S2 hinge fragment, and the light meromyosin (LMM) tail of the α-helical rod region. The epitopes recognized by antistreptococcal monoclonal antibodies 36.2.2, 101.4.1, and 54.2.8 are positioned in the rod region.

whereas the S1 contains the globular headpiece and the ATPase activity of myosin. The globular headpiece or S1 can be proteolytically separated from the rod by papain digestion.[117] Sites of localized mouse antistreptococcal/antimyosin mAb reactivity with the myosin rod is shown in Fig. 9. A chymotrypsin fragment was used to identify the 54.2.8 site.[117] Our subsequent studies have localized the sites of human antistreptococcal/antimyosin antibody reactivity in the LMM rod region using 49 overlapping synthetic peptides of human cardiac LMM (unpublished data). Although the antistreptococcal antibody target in the heart is myosin, the data do not preclude the reaction of antimyosin antibodies with heart cell-surface molecules which mimic epitopes of myosin. Heart cell-surface molecules may contribute to cytotoxicity and overall damage to the heart. Studies of antistreptococcal mAb 36.2.2 show that this highly cytotoxic mAb reacts with the heart cell-surface protein laminin.

6.1.4. Antistreptococcal Antibodies React with DNA: Implications in Autoimmunity

Rheumatic fever characteristically does not present with antinuclear antibody, which is a distinguishing feature of systemic lupus erythematosus. Furthermore, our human antistreptococcal/antimyosin mAbs did not show antinuclear reactivity, but some did cross-react with DNA. However, studies of murine mAbs indicated that antistreptococcal/antiheart

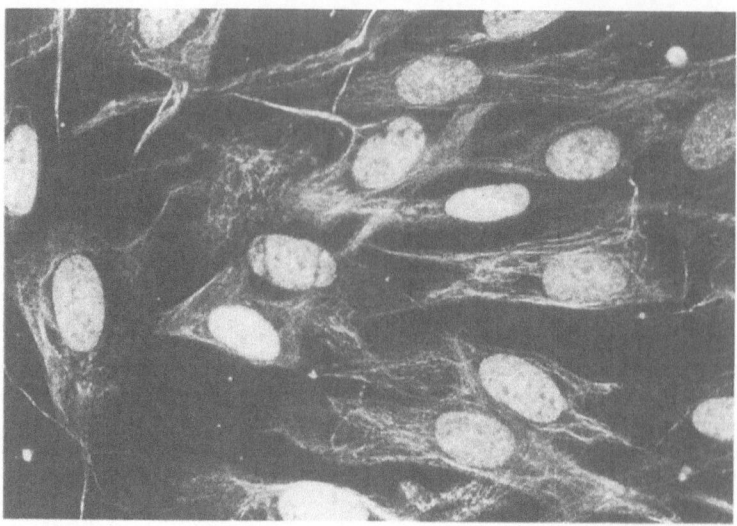

FIGURE 10. Indirect immunofluorescence staining of acetone-fixed fibroblasts after reaction with antistreptococcal monoclonal antibody 54.2.8. The strong reaction of monoclonal antibody 54.2.8 with fibroblasts shows the intense antinuclear reactivity and the staining of the vimentin cytoskeleton. (Taken from Ref. 112 with permission from the *Journal of Experimental Medicine.*)

antibodies in the mouse contained a population which not only reacted with DNA but were strongly antinuclear (Fig. 10).[112] These mAbs reacted with M protein and myosin as well as with sections of human heart (Fig. 8). In support of our findings, antistreptococcal/antinuclear antibodies were identified among anti-DNA mAbs from mice with systemic lupus erythematosus.[118,119] Furthermore, our studies demonstrated that antistreptococcal/antinuclear antibodies had profiles of cross-reactivity with synthetic polynucleotides that were virtually identical to the antinuclear/anti-DNA antibodies from humans and mice developing systemic lupus erythematosus.[112] These data suggested that the antistreptococcal cross-reactive antibodies had important implications in the development of autoimmunity.[112]

By the mid 1980s the polyspecificity of antibodies and the idea that an antibody molecule could specifically react with more than one antigen was recognized.[120] In addition, cross-reactivity of antistreptococcal antibodies was resolved using mouse and human mAbs (Tables II and III). Antibodies with cross-reactive specificities were found in normal sera, and investigators[121,122] demonstrated that low levels of cross-reactive antibodies or autoantibodies were present in normal individuals. Since these discoveries, the antibody variable (V) region genes encoding cross-reactivities and autoantibody specificities have been intensely studied for their nucleotide sequences.[123] There is no consensus hypothesis on the molecular basis of antibody cross-reactivity, but it is most likely associated with the antigen-binding, complementarity-determining regions (CDRs) of the immunoglobulin molecules which are encoded by the V region, diversity (D), or joining (J) segment genes.[124] The nucleotide sequence of the V_H region gene from a human antistreptococcal/antimyosin antibody has recently been determined (Fig. 11).[125] The sequence reveals that the V_H gene was nearly identical to the V_H26 germ-line gene sequence. The V_H26 germ-line gene is known to encode anti-DNA antibodies and rheumatoid factors.[126] However, the molecular basis of the cross-reactivity of the antistreptococcal antibodies is still under investigation. As more sequences of the human and mouse antistreptococcal cross-reactive mAb V-D-J genes become known, it will be possible to identify specific regions responsible for the cross-reactivity.

We have postulated that the purpose of antibody cross-reactivity in the host is to protect the host against a plethora of infectious agents.[127] An antibody molecule that can neutralize several different infectious agents is an important first line of defense. This is shown in studies by Cunningham and colleagues where monoclonal antistreptococcal antibodies were demonstrated to neutralize specific enteroviruses.[127] In the study it was shown, however, that one of the antistreptococcal antibodies (mAb 36.2.2) with broad cross-reactivity was highly cytotoxic in the presence of complement against heart cells and fibroblasts. The data suggested that the cross-reactivity or polyspecificity could lead to detrimental cytotoxic side effects on the host although initially well intended to neutralize infectious microorganisms. Table IV lists the bacterial, viral, and host proteins with which the streptococcal M protein shares immunological characteristics. Included in this group is the mycobacterial heat-shock protein hsp65, which is thought to play a role in adjuvant arthritis. Shared epitopes of strong microbial antigens may be important in triggering autoimmune diseases.

6.2. Molecular Mimicry of Host Antigens

Molecular mimicry may be due to an identical sequence shared between host and bacterial or viral proteins,[128,129] or to structural similarities shared between proteins with homologous but nonidentical sites.[127] In addition, molecular mimicry has been shown to occur between dissimilar molecules such as proteins and carbohydrates,[130–132] proteins

```
                A   I   L   K   G   V   Q   C   E   V   Q   Q   L   V   E   S   G   G   G   L   V   Q   P
                                                1                              10
mAb 10:2.3     GCT ATT TTA AAA GGT GTC CAG TGT GAG GTG CAG CAG CTG GTG GAG TCT GGG GGA GGC TTG GTA CAG CCT
germline VH26   -   -   -   -   -   -   -   -   -   -   -   -  T--  -   -   -   -   -   -   -   -   -   -

                G   G   S   L   R   L   S   C   A   A   S   G   F   T   F   S   S   Y   A   M   S   W
                                L                              30      |---CDR-1---------------|
                               20
mAb 10.2.3     GGG GGG TCC CTG AGA CTC TCC TGT GCA GCC TCT GGA TTC ACC TTT AGC AGC TAT GCC ATG AGC TGG
germline VH26   -   -   -   -   -   -   -   -   -   -   -   -   -   -   -   -   -   -   -   -   -   -

                V   R   Q   A   P   G   K   G   L   E   W   V   S   A   I   S   G   G   G   S   T
                            A                                      |---CDR-2-----------------------
                           40                                      50
mAb 10.2.3     GTC CGC CAG GCT CCA GGG AAG GGG CTG GAG TGG GTC TCA GCT ATT AGT GGT GGT GGT AGC ACA
germline VH26   -   -   -   -   -   -   -   -   -   -   -   -   -   -   -   -   -   -   -   -   -

                Y   Y   A   D   S   V   K   G   R   F   T   I   S   R   D   N   S   K   N   T   L   Y
                -------|           60                      70
mAb 10.2.3     TAC TAC GCA GAC TCC GTG AAG GGC CGG TTC ACC ATC TCC AGA GAC AAT TCC AAG AAC ACG CTG TAT
germline VH26   -   -   -   -   -   -   -   -   -   -   -   -   -   -   -   -   -   -   -   -   -   -

                L   Q   M   N   S   L   R   A   E   D   T   A   V   Y   Y   C   A   K   Q   K   R   T
               80                                                              90
mAb 10.2.3     CTG CAA ATG AAC AGC CTG AGA GCC GAG GAC ACG GCC GTA TAT TAC TGT GCG AAA CGT ACT GTA TTA
germline VH26   -   -   -   -   -   -   -   -   -   -   -   -   -   -   -   -   -   -   -   -   -   -

                R   P   W   S   G   R   S   S   P   D   A   F   D
               |-------------CDR-3----------------------------|
mAb10.2.3      CGA TTT TTG GAG CCG AAG TCC TGA TAG TGC TTT TGA TA     primer
germline DXP1   -   -A-  --T  --C  --- TTA TTA  --A  C
germline JH3                                                -G       germline JH3
```

FIGURE 11. Nucleotide sequence of V_H gene and CDR3 region of human monoclonal antibody 10.2.3. The germ-line gene V_H26 is shown to be identical to the V_H sequence of 10.2.3.[125]

TABLE IV
Proteins with Immunological Similarity
to Streptococcal M Protein

Host antigens
  Cardiac myosin
  Skeletal myosin
  Tropomyosin
  Keratin
  Vimentin
  Laminin
  Retinal S antigen
  DNA
Microbial antigens
  Coxsackievirus capsid proteins-VP1,2,3
  Mycobacterial heat-shock protein-hsp65
  Streptococcal 60-kDa actin-like protein
  Streptococcal group A carbohydrate (N-Ac-Gln)

and DNA,[112] and carbohydrate structures and DNA.[133] Molecular mimicry between bacterial and host antigens has been recently reviewed.[134] The streptococcal antigens implicated in cross-reactivity with host tissues have been investigated on a molecular level over the past 10 years. The findings reveal the importance of the streptococcal M protein and its similarities with a number of host antigens present in the myocardium, valve, joint, and skin as summarized in Table IV. The data suggest that the host antigen targets myosin, tropomyosin, laminin, vimentin, and keratin, are at least in part, the basis of the clinical manifestations observed in rheumatic fever. Antistreptococcal antibodies that bind to these host antigens may target inflammation to the specific tissues. Cell adhesion or extracellular matrix molecules such as laminin may act like a sieve to trap cross-reactive antibodies in tissues and induce inflammatory responses. Inflammation caused by the antibodies alone might be expected to be self-limiting and transient such as that seen in arthritis and chorea in rheumatic fever. However, more permanent damage such as that in the valve in rheumatic carditis may result from cell-mediated immune responses against the streptococci and host antigens.

6.2.1. Alpha-Helical Coiled-Coil Molecules

The structure of alpha-helical coiled-coil molecules such as myosin and streptococcal M protein is composed of a heptad repeat pattern which contains a periodicity of seven amino acid residues where specific amino acid residues are positioned to produce an alpha-helix.[135] In the helical coil conformation, hydrophobic nonpolar residues, such as leucine, seek the buried environment away from the solvent, while the polar residues, such as lysine, are solvent-oriented. The continuous heptad repeat pattern generates an alpha-helical coiled-coil conformation with a rod region that forms dimers. Although the amino acid sequence of the heptad repeat may differ, helix-promoting residues are present in the correct positions to achieve the alpha-helical structure. Dimers form as the hydrophobic side of the molecule joins with another like molecule along its hydrophobic face. The alpha-helical coiled-coil molecules involved in mimicry between streptococcus and host include myosin, tropomyosin, keratin, vimentin, and laminin.

  6.2.1a.   *Streptococcal M Protein: Heart Cross-Reactive Epitopes.*   The primary structure of the streptococcal M protein has been determined over the past 15 years.[37–53] The definition of the true primary structure of M protein has been instrumental in identification of immunologically and structurally similar proteins in the host. Manjula and Fischetti[40–45] investigated the primary sequence of the streptococcal M protein for sequence identity with host proteins before the M protein genes were cloned and their nucleotide sequence determined.[46–52] Determination of the M protein sequence confirmed its alpha-helical coiled-coil structure, resembling host proteins such as tropomyosin and the desmin-keratin family of host alpha-helical coiled-coil molecules.[43–45] At the same time, antistreptococcal mAb studies by Cunningham and colleagues revealed that the target of antistreptococcal antibodies in human heart was myosin.[110–117] Dale and Beachey discovered that PepM5 induced antibodies to myosin in rabbits.[111] They also found that a myosin cross-reactive epitope of M protein was present in M6 and M19 serotypes.[111]

  The proposed model of the M protein[24] shows it as a coiled coil rod extending from the streptococcal cell wall. The M protein is anchored in the membrane, and its fimbriae extend through the peptidoglycan and group carbohydrate at the cell wall (Fig. 2). Once the complete nucleotide sequence of the M proteins was known, it was determined that the M protein molecule was divided into regions containing amino acid sequence repeats which were designated the A, B, C, and D repeat regions (Fig. 2).[24]

  The first epitope mapping studies of the myosin cross-reactive epitopes of M protein were performed using peptides of the PepM5 protein. These studies identified two regions in the M protein which were reactive with antimyosin antibodies. The Gln-Lys-Ser-Lys-Gln site was found to react with antimyosin antibodies affinity-purified from acute rheumatic fever patients.[77] A synthetic M5 peptide containing this sequence (residues 164–197) induced antibodies against the sarcolemmal membranes of heart.[136] The other site was located at residues 84–116 of the PepM5 molecule.[77,136] Furthermore, a site at the N-terminus of M19 was shown to cross-react with heart tissue.[137] The determination of these cross-reactive sites began the studies to identify sites in the M protein that might be responsible for manifestations observed in acute rheumatic fever.[77,136–138] Once the genes for the M proteins were cloned and the nucleotide sequence determined, overlapping synthetic peptides were produced spanning the A, B, and C repeat regions of the M5 molecule (Table V and Fig. 12). Recent studies[139] using these overlapping M5 peptides have revealed B-cell and T-cell epitopes of the M protein which are cross-reactive with host tissues. Table VI summarizes the. M protein sequences that are cross-reactive with myosin. Sites of myosin cross-reactivity have been associated with a region in the A repeat (NT3-7), B repeat (B2B3B), and C repeat (see Tables V and VI) (unpublished data). The B2B3B peptide contains the QKSKQ sequence previously reported to react with antimyosin antibodies affinity-purified from acute rheumatic fever sera.[77] The B2B3B peptide is a site in the B repeat region which induces antimyosin antibody responses in mice.

The synthetic overlapping M5 peptides have been utilized in the ELISA, immunodot blots, and ELISA inhibition to map the sites recognized by the cross-reactive antistreptococcal/antimyosin mAbs. The mAb mapping data suggest that homologies are shared among regions within the M protein (unpublished data). Furthermore, the repeated regions may contribute to the cross-reactivity of the antibody molecules and enhance induction of cross-reactive antibodies. The M protein serotypes that contain amino acid sequence repeats identical or similar to those of host antigens may enhance cross-reactive and autoantibody production in the host. Antistreptococcal mAb reactivity with multiple sites in particular regions of the M molecule (unpublished data) may be attributed to the repeated amino acid sequences found in the overlapping peptides and to the heptad repeat pattern necessary to maintain the alpha-helical coiled coil. Therefore, repeated amino acid sequences found in the A, B, or C repeat regions of the M protein may be important in the induction of immune cross-reactivity in the host and in antigenic reactions with cross-reactive antibodies. Specific repeats have been observed in the M5 and M6 proteins, and the M24 serotype contains a hexapeptide from tropomyosin repeated five times.[24] It is possible that certain repeated primary amino acid sequences influence the level and specificity of cross-reactive immune responses and antigen processing in the susceptible host.

We have found at least two major sites within the M5 protein to

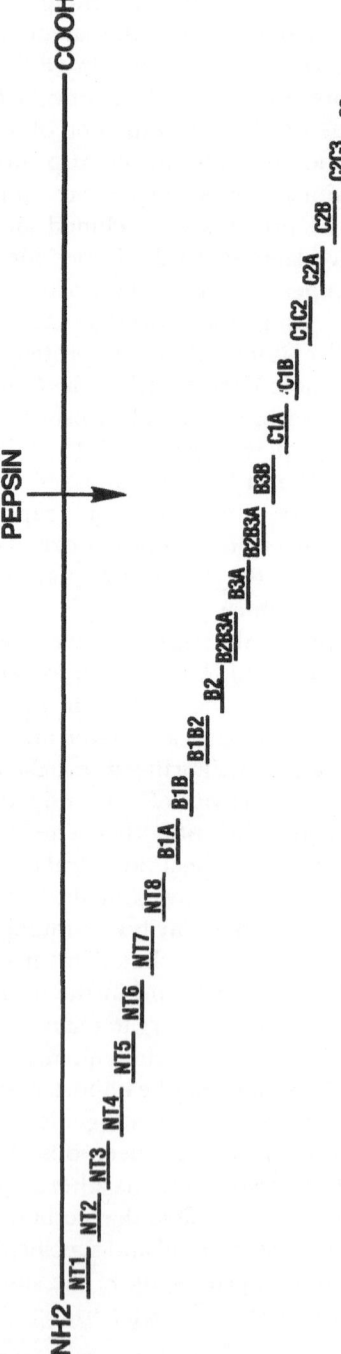

FIGURE 12. Overlapping synthetic peptides of the M5 protein representing the A, B, and C repeat regions. The exact amino acid sequences of each peptide and overlap are shown in Table V. The pepsin cleavage site of M5 protein is shown in the figure.

TABLE V
Overlapping Synthetic Peptides
of Streptococcal M5 Protein[a]

Peptide	Sequence	Amino acid residue
NT1	AVTRGTINDPQRAKEALD	1–18
NT2	KEALDKYELENHDLKTKN	14–31
NT3	LKTKNEGLKTENEGLKTE	27–44
NT4	GLKTENEGLKTENEGLKTE	40–58
NT5	KKEHEAENDKLKQQRDTL	59–76
NT6	QRDTLSTQKETLEREVQN	72–89
NT7	REVQNTQYNNETLKIKNG	85–102
NT8	KIKNGDLTKELNKTRQEL	98–115
B1A	TRQELANKQQESKENEKAL	111–129
B1B	ENEKALNELLEKTVKDKI	124–141
B1B2	VKDKIAKEQENKETIGTL	137–154
B2	TIGTLKKILDETVKDKIA	151–167
B2B3A	KDKIAKEQENKETIGTLK	163–180
B3A	IGTLKKILDETVKDKLAK	176–193
B2B3B	DKLAKEQKSKQNIGALKQ	189–206
B3B	GALKQELAKKDEANKISD	202–219
C1A	NKISDASRKGLRRDLDAS	215–232
C1B	DLDASREAKKQLEAEHQK	228–245
C1C2	AEHQKLEEQNKISEASRK	241–258
C2A	EASRKGLRRDLDASREAK	254–271
C2B	SREAKKQLEAEQQKLEEQ	267–284
C2C3	KLEEQNKISEASRKGLRR	280–297
C3	KGLRRDLDASREAKKQ	293–308

[a]From Ref. 215.

induce antibody reactivity with myosin when BALB/c mice or Lewis rats were immunized with each of the 23 overlapping M5 peptides (Tables V and VI; Fig. 12) (unpublished data). The NT4 peptide (M5[40–58]) contains a sequence repeated in the N-terminal or A repeat region of M5 protein which shares 80% identity with a site in the LMM region of human cardiac myosin (Fig. 13).[140] Immunization of BALB/c mice with the NT4 peptide alone or linked to keyhole limpet hemocyanin (KLH) resulted in the production of antimyosin antibody (unpublished data). Immunization with NT4 produced inflammatory infiltrates in the hearts of BALB/c mice (Table VI) and MRL mice.[140] In addition, the M protein peptide NT4 was found to contain a myosin cross-reactive T-cell epitope.[139]

A second site, M5 peptide C3 (M5[293–308]), is important because it contains the class I M protein epitope believed to be associated with

TABLE VI
M5 Peptide Sequences Responsible
for Induction of Antimyosin Antibody
in Balb/c Mice,[a] Lewis Rats, or Humans

A Repeat region	B Repeat region	C Repeat region
NT3	B2B3B	C1A
NT4[b]		C1B
NT5[b]		C2A
NT6[b]		C2B
NT7[b]		C2C3
		C3

[a]Immunization of mice with 500 μg peptide in CFA with three boosters in IFA at 2-week intervals. Mice were euthanized at 8 weeks following immunization. Sera were tested in Western blot and ELISA and compared with prebleed and CFA adjuvant controls which were negative.
[b]Produced inflammatory lesions in the myocardium of Balb/c mice.

rheumatogenic serotypes.[53,78] The evidence strongly implicates C3 as a myosin cross-reactive site in M5 protein. First, affinity-purified antimyosin antibodies from acute rheumatic fever serum recognize M5 peptide C3. Second, anti-C3 antibodies affinity-purified from acute rheumatic fever serum recognize human cardiac myosin.[78] Antibodies against peptide C3 (M5[293–308]) are not present in normal sera and are not significantly elevated in uncomplicated streptococcal infections.[78] The peptide C3 (M5[293–308]) site contains a sequence repeated throughout the C repeat region of M5 protein. The C3 site has identity with a site in human cardiac myosin as shown in Fig. 14. If this site is important in the abnormal immune responses in rheumatic fever, as the data would suggest, the susceptible host response to and presentation of this site in M protein to the immune system might be a key factor in development of disease. Although the C3 peptide does not appear to induce cardiac inflammation

```
NT4    GLKTENEG--LKTENEG--LKTE
       |  |||     |  |||     |  ||
MYO    KLQTENGE   LQTENGE   LQTE
```

FIGURE 13. Homology between the NT4 peptide sequence and a sequence in myosin. The sequence repeat in NT4 repeats four times in the M5 A repeat region, but it does not repeat in myosin. The NT4 sequence is a B-cell and T-cell epitope of M5 protein that induces antimyosin responses in animals as well as cardiac inflammation. The repeated sequences within M proteins, which are similar to sequences in myosin, may be important in inducing an autoimmune response against myosin.

QKMRRDLE   HUMAN CARDIAC MYOSIN [1168-1175]
 . :::: .
KGLRRDLDASREAK   M5 PEPTIDE C3 [282-299]

FIGURE 14. Homology between human cardiac myosin and M5 peptide C3. The M5 sequence RRDL is within the class I epitope in the M5 protein. Class I M protein serotypes such as type 5 are associated with rheumatic fever outbreaks. Peptide C3 contains an epitope of M protein which not only reacts with antimyosin antibodies, but induces immune responses against myosin in animals.

in BALB/c mice or Lewis rats (unpublished data), it may be responsible in part for the antimyosin antibody responses in animals and man.[141,142] Although antistreptococcal antibody deposition may play an important role in development of cardiac murmurs and rheumatic heart disease, cell-mediated immunity and heart-reactive T cells are implicated in the development of the permanent heart damage that is discussed in Section 6.4. Transient manifestations such as arthritis, chorea, and erythema marginatum may result from the deposition of cross-reactive antibodies in tissues.

6.2.1b. Host Antigens. Studies using human and mouse antistreptococcal mAbs[110–112,115–117,125,127,130–132,136–139,142,143] have shown that the cross-reactive proteins in the host include myosin, tropomyosin, laminin, vimentin, and keratin. Cunningham and colleagues have defined epitopes of myosin recognized by human and mouse antistreptococcal monoclonal antibodies. Dell[117] first described the fragments of human cardiac myosin which reacted with antistreptococcal mAbs, and Quinn[125] utilized a panel of 49 overlapping synthetic peptides of human cardiac LMM to map human mAb 10.2.3, which reacted with M protein and myosin. The sites of cross-reactivity between M protein and myosin may share as much as 80% identity. However, some sites have very little amino acid sequence homology, suggesting that conformational constraints on the peptides as well as the primary structure may affect antibody cross-reactivity with M protein peptides.

Although the role of molecular mimicry in the pathogenesis of rheumatic fever is not entirely clear, our studies suggest that certain cross-reactive antistreptococcal antibodies are cytotoxic.[127] Most of the cross-reactive alpha-helical host antigens are cytoskeletal proteins located within host cells and are unlikely to be cell-surface targets that would contribute to the cytotoxicity of antistreptococcal antibodies. However, a highly cytotoxic antistreptococcal/antimyosin antibody reacted strongly with laminin, a heart cell-surface molecule.[139,142] Laminin is an alpha-helical molecule located on the surface of muscle cells, including those in the myocardium. Laminin is composed of three chains that form a large 800-kDa structure in the extracellular matrix. Domains I and II of each of the three laminin chains, A, B1, and B2, are alpha-helical (Fig. 15).[144] The trimer is maintained through an alpha-helical coiled coil structure within domains I and II. The M protein shares homology with these alpha-helical regions. The importance of laminin or other similar extracellular matrix proteins in rheumatic fever may be their recognition by antistreptococcal antibodies on the heart cell surface. This was demonstrated using mouse antistreptococcal mAb 36.2.2 which was shown to be highly cytotoxic in the presence of complement for heart cells and fibroblasts but not cytotoxic for liver cells that did not express laminin.[139,142] Only the cytotoxic antistreptococcal mAb recognized laminin, while the other antistreptococcal mAbs did not react with laminin and were not cytotoxic. Cytotoxic antistreptococcal antibodies recognizing host cell-surface epitopes may be, in part, an explanation for inflammation and tissue damage in acute rheumatic fever.

6.2.2. Group A Streptococcal Carbohydrate Antigen

Recent studies have shown that a subset of cross-reactive antistreptococcal mAbs reacted with the N-acetyl-β-D-glucosamine epitope of group A streptococcal carbohydrate.[130–132] Antibodies against group A carbohydrate are important due to their persistence in serum of acute rheumatic fever patients with valvulitis.[145] The group A carbohydrate consists of repeating polysaccharide units of rhamnose O-linked to N-acetyl-β-D-glucosamine.[146] McCarty suggested that the terminal O-linked sugar might cross-react with antibodies against the group A carbohydrate and host tissues.[147] N-acetyl-glucosamine is found in host tissue glycoproteins and mucopolysaccharides.[148] Previous work by Goldstein and colleagues[104] demonstrated cross-reactivity of anti–cardiac valve antibodies with N-acetyl-glucosamine and the streptococcal group A carbohydrate. Lyampert et al.[105,106] demonstrated that antibodies produced in rabbits to group A streptococcal carbohydrate reacted with thymus and skin. Other studies have implicated the anti–group A carbohydrate antibodies in cross-reactions with epithelium of various tissues.[148,149] Specific cross-reactions of anti-N-acetyl-glucosamine antibodies was recently investigated by Shikhman et al.[130–132] First, a subset of antistreptococcal mouse mAbs was found to recognize N-acetyl-β-D-glucosamine and the cytoskele-

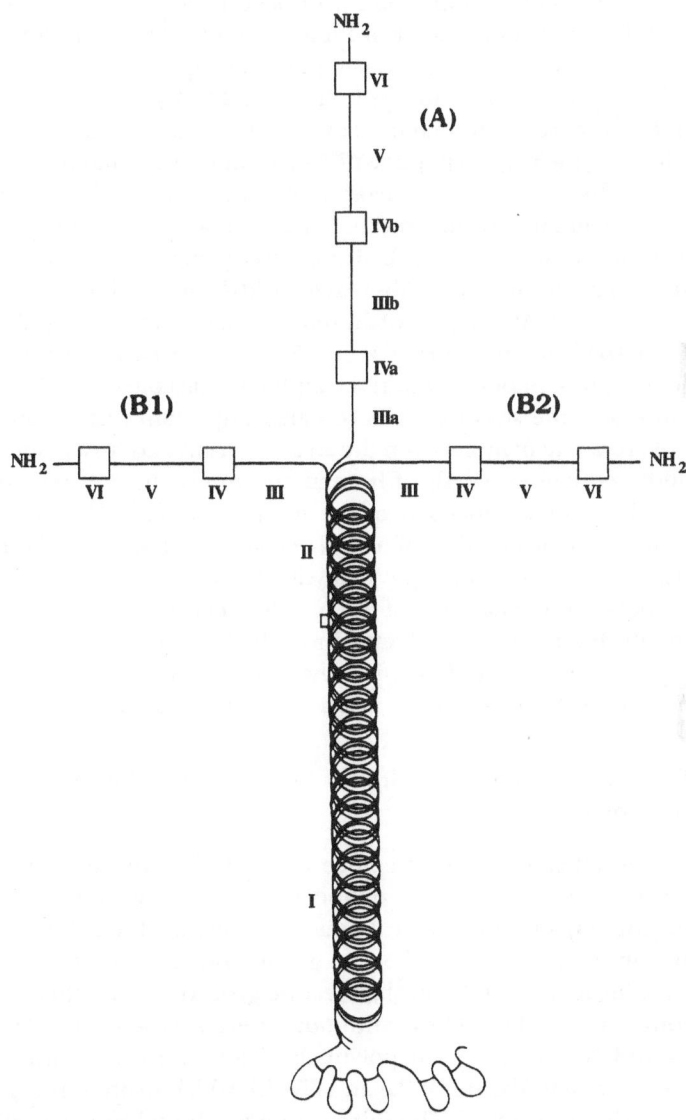

FIGURE 15. Diagram of laminin. The domains I and II are shown in their coiled coil conformation. (Adapted from Ref. 144.)

tal proteins myosin and keratin.[130] Once human mAbs against the N-acetyl-glucosamine epitope were produced from rheumatic fever patients,[131] it was shown that many of them recognized keratin, which may be the basis of the previously reported anti–group A carbohydrate antibody reactivity with skin and epithelium (Table II). Epitopes of cytokeratin 14 were recently mapped to specific sites within the keratin molecule using the Geysen peptide pin technique.[131] The cumulative data proved that the anti-N-acetyl-glucosamine antibodies cross-reacted with peptide determinants of the keratin molecule as well as with peptides from M protein and other microbial and host antigens.[130,131] Interestingly, all of the human antimyosin mAbs reacted with N-acetyl-glucosamine and with sites in the LMM region of human cardiac myosin (unpublished data, Table III). The discovery that anti-N-acetyl-glucosamine antibodies can also recognize peptides within the alpha-helical coiled coil molecules of the host and the streptococcus revealed important insights into the nature of cross-reactivity between the streptococcus and host tissues. The nucleotide sequence analysis of human and mouse mAb V regions will provide a base for comparison of the immunoglobulin V region gene families used for production of anti-N-acetyl-glucosamine antibodies in humans. Furthermore, idiotypic analysis of the human antibodies will lead to a better understanding of the profiles of cross-reactive antibodies in rheumatic fever and in unaffected individuals. Considering the data of Dudding and Ayoub,[145] pathogenic cross-reactive antibodies in rheumatic fever may result from responses against the group A carbohydrate.

6.3. Summary: Three Major Groups of Heart Cross-Reactive Antibodies

In the investigation of antistreptococcal antibodies that recognize heart tissue, three major groups were delineated. The first group recognized myosin, tropomyosin, or α-helical coiled coil molecules in host tissues. Cytotoxic mAb 36.2.2 was in the first group and recognized laminin, a distinguishing feature of this mAb. A second group reacted with DNA and was strongly antinuclear. These antibodies were lupus-like in their reactivities with DNA and synthetic polynucleotides. Prototype antinuclear/antistreptococcal mAbs are 54.2.8 and 654.1.1, which strongly recognized epitopes within M protein. The third group was found to react with the group A carbohydrate epitope N-acetyl-glucosamine. Human antistreptococcal/antiheart mAbs do not demonstrate antinuclear activity, are not in the second group, and demonstrate characteristics of group 3.

6.4. Cell-Mediated Immunity and the Rheumatic Lesion

6.4.1. Cellular Immune Responses in Rheumatic Fever

Cell-mediated immune responses against the group A streptococcus in rheumatic fever are not as well defined as antibody responses and their cross-reactivity with host antigens. However, it is well known that lymphocytes from acute rheumatic fever patients are hyperresponsive to group A streptococcal antigens. Skin tests for delayed-type hypersensitivity demonstrated that acute rheumatic fever patients were more reactive than controls when streptococcal components were used for skin testing.[150] Beachey *et al.* used better defined streptococcal M protein to demonstrate delayed-type hypersensitivity in guinea pigs and man.[151] Abnormal responses were observed by Read and Zabriskie and colleagues[152-154] and by Gray *et al.*[155] in *in vitro* tests of lymphocyte responsiveness to streptococcal membranes or extracellular products in acute rheumatic fever. Gibofsky *et al.*[156] have recently reviewed the importance of cell-mediated immunity in rheumatic fever.

6.4.2. Cellular Immune Responses against Myosin and Streptococcal M Protein

A few studies of T-cell responses in animals to streptococcal and cardiac antigens[157-159] have suggested that lymphocytes from humans with acute rheumatic heart disease or from animals sensitized with group A streptococcal antigens were cytotoxic for cardiac myofibers *in vitro*. Further work has shown that streptococcal PepM5 protein stimulated T lymphocytes that were cytotoxic for several human cell lines.[160] Recent work to identify T-cell epitopes of M protein crossreactive with cardiac myosin that one T-cell epitope of M protein, type 5, resides in the N-terminal region in the M5 peptide NT4 (M5[27-44]).[141] BALB/c mice were immunized with human cardiac myosin and the myosin-sensitized lymphocytes were found to be stimulated by M protein peptides NT4 (M5[27-44]) and B1B2 (M5[138-155]). The NT4 peptide also induced production of antibodies against myosin. Studies of T-cell clones responsive to M5 protein indicate that the clones respond to regions in the N-terminus[161] (M5 residues 1-35) or the C-terminus[162,163] (M5 residues 300-319). T-cell lines/clones cross-reactive with M5 peptides and heart proteins have been produced from rheumatic heart valves of humans.[179]

6.4.3.   Streptococcal Superantigens

Superantigens are bacterial or viral proteins that differ from typical antigens in several aspects.[164,165] The superantigen recognizes the beta chain of the T-cell receptor, while other parts of the receptor play little or no role in recognition and are not processed as conventional antigens. The unprocessed superantigen is recognized and presented by the class II MHC molecule, and the entire complex of superantigen and MHC is recognized by certain beta chains (Vβ) of the T-cell receptor. The super-antigen binds outside the conventional site or groove of the MHC molecule, where recognition of processed peptide antigens usually occurs. When this complex is recognized by the Vβ of the T-cell receptor, the T cells are stimulated nonspecifically, and large numbers of T cells can become activated unlike conventional antigens that specifically stimulate small numbers of T cells in a population. Although the superantigens are not true mitogens, they behave very much like mitogens to activate large populations of T cells.[164-166]

The most potent of the streptococcal superantigens are streptococcal erythrogenic toxins A, B, and C, which stimulate the Vβ8+ subset of T cells.[66,166] The erythrogenic toxin A is most well known for its role in the symptoms of scarlet fever and toxic shock–like syndrome during streptococcal infections.[65,66] In addition to the streptococcal toxins, PepM protein has been shown by Kotb and colleagues[167,168] to act as a superantigen only in humans by stimulating T-cell subsets Vβ 2, 4, and 8. It has been proposed that superantigen exposure could lead to activation of self-reactive T cells and to autoimmunity and tissue damage.[165] It is possible that cross-reactive T cells in rheumatic fever are activated by streptococcal superantigens. The hyperresponsiveness to all streptococcal antigens could be due to the exposure of the host to streptococcal superantigens.

6.4.4.   Nature of the Rheumatic Cardiac Lesion

Characterization of the rheumatic lesion has been important in our current understanding of the permanent heart damage in rheumatic fever. Although the cardiac lesions that develop in rheumatic carditis can involve all parts of the heart, the valves are affected most often.[1] The characteristic lesion described in the rheumatic heart was the Aschoff body or a focal lesion which developed in myocardial tissue.[10] Although pathologists may still disagree as to the nature of the lesion, recent studies of rheumatic hearts by McManus and colleagues[169] suggest that the cells in the lesion stain with macrophage markers and have the characteristics of cells of the macrophage lineage. The components of the Aschoff body have been in dispute since Murphy[170] suggested that the cells in Aschoff nodules were degenerating muscle cells and Wagner[171] suggested that they were more like connective tissue cells than muscle fibers. In one study[172] biopsy specimens from a patient with rheumatic carditis revealed the predominance of T lymphocytes of both CD4+ and CD8+ subsets, although macrophages, B lymphocytes, and mast cells were present. Stollerman has pointed out that interstitial myocarditis seen in rheumatic fever was more likely to be the cause of heart failure than the focal Aschoff bodies.[1,173] The myocardial lesions contained predominantly infiltrates of lymphocytes. In the heart muscle, myocardial fibers are damaged, with the greatest damage observed near Aschoff bodies and around blood vessels.[1,173] This is not surprising since the interaction of circulating lymphocytes with the endothelial lining of blood vessels precedes cellular infiltration into tissues.[174] Figure 16 illustrates the hypothetical steps leading to heart damage in acute rheumatic fever.

Cardiac failure in acute rheumatic fever is most likely due to damage of the heart muscle by an immune response against myosin and other heart autoantigens. Muscular portions of arteries may also be involved in the inflammatory changes. Valvulitis with subsequent fibrosis is also an important part of the development of the scarred valves, which can no longer function properly, and can lead to heart failure and valvular insufficiency.[1] Studies by Gulizia et al.[175] suggest that cross-reactive anti-streptococcal antibodies react with specific sites on the valve, including the endothelial lining where inflammation signals upregulation of integrins, MHC class II molecules, and subsequent infiltration of tissue.[174] Increased numbers of valvular interstitial cells in the rheumatic valve may be important in the scarring process.[175] However, Stollerman states that in rheumatic fever there is a "remarkable tendency for the disease to heal rather than to scar tissues it affects, with the exception of some cardiac lesions."[173] Although antibodies may target the immune and inflammatory response to the heart, joints, brain, and skin, the infiltration of the target organ by T-cell subsets may be the pivotal point leading to tissue damage or healing. One hypothesis for the pathogenesis of acute rheumatic fever is that antibodies mediate many of the disease manifestations but T-cell subsets that mediate delayed-type hypersensitivity and/or cytotoxicity become activated and produce the chronic lesions in the valve (Fig. 16). The T-cell response to particular epitopes by the susceptible host may lead to production of influential cytokines by specifically activated T-helper subsets which can lead to inflammation and tissue damage.[176,177] Cytokines may play a pivotal role in the regulation of tissue-

STREPTOCOCCAL INFECTION

|

**AUTOANTIBODY RESPONSE AGAINST THE HEART
(ANTI-MYOSIN ANTIBODIES)**

|

**ANTIBODY DEPOSITION ON HEART VALVE ENDOTHELIUM AND IN THE
MYOCARDIAL VESSELS AND TISSUES**

|

**CYTOKINE PRODUCTION
(γIFN AND IL-2)**

|

**UPREGULATION
OF MHC CLASS II
EXPRESSION ON ENDOTHELIUM**

|

**EXPRESSION OF
CELL ADHESION
MOLECULES ON
VASCULAR ENDOTHELIUM**

|

**INFILTRATION OF MYOCARDIUM AND
VALVULAR REGION WITH MACROPHAGES
AND AUTOREACTIVE T LYMPHOCYTES**

|

**CD4+ TH1 SUBSET
(γI N F)**

|

CD8+ CYTOTOXIC SUBSET

|

**DESTRUCTION OF MYOFIBERS
(CARDIOMYOCYTES IN MYOCARDIUM)**

|

**VALVULITIS, SCARRING,
COLLAGEN DEPOSITION
(HEART MURMUR)**

specific injury by inducing vascular cell adhesion molecules (VCAMs) which leads to infiltration of tissue by lymphocytes.[174,178,182-185] Local production of cytokines may be very important in the course of permanent heart damage in rheumatic heart disease. Much has been learned about cytokine patterns of T-cell helper subsets Th1 and Th2 and their functions in delayed-type hypersensitivity reactions and in antibody production, respectively.[180,181] A new understanding of the T-cell subsets and their cytokine patterns should help in the investigation of the T cells that are activated in rheumatic fever. T-cell clones, cross-reactive with M protein and myosin from M protein–immunized rats developing myocarditis, produce interleukin-2 (IL-2) and interferon-γ (Quinn and Cunningham, unpublished data). These data suggest that the cross-reactive T cells may be the Th1 phenotype which produce the proinflammatory cytokines IFN-γ and IL-2. The continued investigation of T-cell clones and the pathogenetic mechanisms in rheumatic fever will provide insights into its pathogenesis as well as other related rheumatic and autoimmune diseases.

7. ANIMAL MODELS OF RHEUMATIC HEART DISEASE

Although many would say that there has not been a good model for rheumatic fever, models of the disease have been investigated in rabbits, mice, rats, and monkeys.[186-195] A review on rheumatic carditis describes many studies using streptococcal components, different types of immunization protocols, and different animal species.[196] Studies in decades past used whole streptococci and crude streptococcal preparations, lacked

---

FIGURE 16. Hypothetical scheme of the pathogenesis of rheumatic carditis. *In vitro* data support the hypothesis that antistreptococcal antibodies bind to valvular endothelium,[175] and may deposit in myocardium to extracellular matrix proteins such as laminin or to extruded myosin. Cytokine production and upregulation of MHC class II molecules in localized tissue may lead to the expression of cell adhesion molecules on vascular endothelium[174] and promote infiltration of the valve and myocardium by trafficking autoreactive lymphocytes. Both CD4+ and CD8+ lymphocytes have been demonstrated in rheumatic valves.[169] Observation of the Lewis rat following immunization with recombinant M6 protein shows foci of myocarditis with destruction of myofibers at the site of cellular infiltration. Destruction of myofibers was observed in myocardial lesions in ARF. The production of interferon-γ (γIFN) by CD4+ T lymphocytes of the T helper-1 subset may activate adjacent CD8+ T lymphocytes and lead to their cytotoxic destruction of myocardium. Valvulitis ultimately will lead to scarring of tissue due to collagen deposition within the inflamed valve.

appropriate controls, and had equivocal results. Murphy and Swift[181,182] produced focal cardiac lesions in rabbits which were reproducible by one group but not by another. However, the investigations of Schwab and colleagues[186–190] are highly reproducible and have been studied for many years. In these studies large amounts of streptococcal cell walls or peptidoglycan-polysaccharide complexes were given to mice or rats that developed carditis, arthritis, or uveitis.[186–190,192–197] None of the models have been used to study adequately the immune mechanisms of rheumatic carditis, but antibody and cell-mediated immunity have been studied in the arthritis-uveitis model, and cross-reactivity between streptococcal mAbs and uveitogenic proteins has been noted.[192–198] The hypothesis of Schwab and associates is that the peptidoglycan-polysaccharide and/or cell-wall antigens persist in tissues and act as a chronic stimulus for continued tissue injury and production and deposition of immune complexes.

Recently, Quinn and Cunningham found that recombinant M6 protein produced focal lesions of myocarditis in 50% of Lewis rats tested, while the adjuvant controls did not develop heart lesions (unpublished data). The Lewis rat is known to develop a granulomatous myocarditis when injected with cardiac myosin.[199] Lewis rats given the recombinant M6 protein developed the same signs and symptoms as rats with myosin-induced myocarditis. In a small percentage of the rats, valvulitis was observed (unpublished data). The cellular infiltrates appeared to contain both lymphocytes and macrophages. The model is being used to investigate T-cell clones cross-reactive with M protein and myosin. Further study of the model will determine its usefulness in elucidating the potential mechanisms in rheumatic fever.

8. TREATMENT AND PREVENTION

8.1. Antibiotic Therapy and Prophylaxis

In the 1940s penicillin first became available and was used successfully to treat group A streptococcal infections.[200] The treatment for group A streptococcal pharyngitis or tonsillitis for the past 40 years has been penicillin. In addition, penicillin prophylaxis is used as a measure to prevent recurrences of rheumatic fever.[201] Early treatment of streptococcal infection with penicillin prevents an immune response against streptococcal antigens. Treatment failures have been reported[202,203] but are thought to be due to the coexistence of lactamase-producing bacteria in the tonsillopharynx. Bacteria degrade penicillin in the infected area and allow survival of the group A streptococci. Streptococci can be eradicated in these situations by administering amoxicillin and clavulanate together.[202] Group A streptococci have not developed resistance to penicillin, and they remain exquisitely sensitive. However, some strains have developed a penicillin tolerance by inhibiting the bactericidal effects of the antibiotic.[203] In penicillin-tolerant strains, the minimal bactericidal concentration will be at least 32-fold higher than the minimal inhibitory concentration. Other choices for treatment of streptococcal pharyngitis include cephalosporins and erythromycin, but penicillin remains the drug of choice.

8.2. M Protein Vaccine Strategies

Investigators have worked over the past two decades to develop a safe, efficacious vaccine to be used for immunization against rheumatogenic serotypes. With the resurgence of rheumatic fever and serious streptococcal infections, it is prudent to continue to pursue a group A streptococcal vaccine. The development of a vaccine has always been met with enthusiasm, but it has certain problems. First, one does not want to exacerbate the rheumatic disease that the vaccine would be designed to prevent. Thus, the M protein sites associated with allergy, tissue cross-reactivity, or tissue infiltrates should be avoided and selected sites should be thoroughly tested in animals. Second, the immune response should be long-lived and provide lasting protection. Third, there are over 80 different M serotypes which cause infections, but only a limited number of M protein serotypes is practical for a vaccine. Sites common to all or some of the M proteins might be an effective vaccine against a number of serotypes; however, the serotype specificity is shown to reside at the highly variable N-terminal region of the molecule.

In a recent review, Fischetti[24] discussed the development of M protein vaccines. Many outstanding scientists have made contributions to our understanding of the M protein antigen since its discovery by Lancefield.[30] In the 1970s, Fox and Wittner studied M protein vaccines and their immune response in mice.[204,205] By 1979, Beachey et al. had taken advantage of the pepsin extraction method and produced highly purified PepM24.[206] PepM24 contained the N-terminal half of the M protein molecule and was mixed with alum as an adjuvant and used to immunize a small group of 12 human volunteers. The studies by Beachey et al. were the first studies in humans with a defined synthetic M protein vaccine antigen.[206,207] The volunteers developed opsonic antibody against the type 24 streptococcus but had no delayed-type hypersensitivity reaction, and no heart cross-reactive antibodies were observed by immunofluorescence tests of heart sections.

One of the major problems of immunization with the streptococcal M protein as an effective prevention of group A streptococcal infection is the number (>80) of M protein serotypes. For effective protection against rheumatic fever, a combination of M protein serotypes from rheumatogenic strains would be required. Studies in 1986 by Beachey *et al.*[208] described opsonic antibodies against a hybrid peptide containing copies of the type 5 and type 24 M proteins synthesized in tandem. The peptide induced antibodies against both serotypes of M protein, suggesting that effective immunization with multivalent synthetic vaccines was possible. More recently, Dale *et al.* have examined a multivalent, hybrid, recombinant tetravalent M protein vaccine.[209] Four M protein serotypes were represented in the peptide, which contained the N-terminus of M types 24, 5, 6, and 19. The immune sera from rabbits immunized with the multivalent peptide proved to opsonize all four serotypes of group A streptococci. However, in some instances one of the M serotypes included in the multivalent vaccine would not elicit an immune response. The reasons for this are unclear, but recombinant vaccines containing as many as eight M type sequences have been shown to be effective at inducing opsonic, serotype-specific responses in rabbits (Dale *et al.*, personal communication).

The enhancement of mucosal immunity has also been investigated using synthetic peptides corresponding to conserved epitopes found in the C-terminal region of M protein.[210] The M protein sequences were conjugated with cholera toxin B subunit to enhance immunogenicity. Other studies by Bronze *et al.* also suggest that mucosal immunity will be enhanced by local administration of vaccines.[211] Recently, the C-terminal region of the M protein was expressed in vaccinia virus[212] and in *Streptococcus gordonii*.[213–214] The administration of these mucosal vaccines induced both IgA and IgG antibody in the recipient animals. Animals immunized intranasally were resistant to mucosal infection by *Streptococcus pyogenes*.

The most serious concern is that the M protein vaccines be safe and not contain deleterious epitopes that might generate high levels of heart-reactive antibody, cardiac inflammation, or rheumatic fever–like symptoms in vaccine recipients. Further testing of M protein vaccines will be required to assure their safety for human use.

ACKNOWLEDGEMENTS. Appreciation is expressed to all of my colleagues who have influenced my scientific development and to the students and fellows who made our research possible. Our research has been supported by grants #HL35280 and #HL01913 from the National Heart, Lung and Blood Institute, grants from the American Heart Association (AHA) and the AHA Oklahoma Affiliate, grants from the Presbyterian Health Foundation, and grants from the Oklahoma Center for the Advancement of Science and Technology. I also thank Dr. Ken Jackson, director of the Molecular Biology Resource Facility at the University of Oklahoma Health Sciences Center, for synthesis of many peptides utilized in our research and Michelle Smart for assistance in preparation of this manuscript.

REFERENCES

1. Stollerman, G. H., 1975, *Rheumatic Fever and Streptococcal Infection*, Grune and Stratton, New York, pp. 1–19.
2. Ballonii, G., 1642, De rheumatisma, et pleuritide dorsali, *Opuscula Medica*, Quesnel, Paris, pp. 164–300.
3. Sydenham, T., 1848, *The Works of Thomas Sydenham* (trans. from Latin edition of Dr. Greenhill by R. G. Latham), Vol. 1, Sydenham Society, London, p. 254.
4. Pulteney, R., 1761, Case of a man whose heart was found enlarged to a very uncommon size, *Philos. Trans. R. Soc. London* 52:344–353.
5. Bouillaud, J. P., 1835, *Traite Clinique des Maladies du Coeur*, Bailliere, Paris.
6. Copland, J., 1821, Case of chorea, etc. with an account of postmortem appearances, *London Med. Repos.* 15:23–26.
7. Wells, W. C., 1941, On rheumatism of the heart, Trans. Soc. Improv. Med. Chir., Knowl. London, 1812, 3:373–424, in: *Cardiac Classics* (F. S. Willius and T. E. Keys, eds.), Mosby, St. Louis, MO, pp. 294–312.
8. Jones, T. D., 1944, The diagnosis of rheumatic fever, *J. Am. Med. Assoc.* 126:481–484.
9. Cheadle, W. B., 1889, *Various Manifestations of the Rheumatic State as Exemplified in Childhood and Early Life*, Smith, Elder, London.
10. Aschoff, L., 1904, Zur myocarditisfrage, *Verh. Dtsch. Ges. Pathol.* 8:46–51.
11. Talalaev, V. T., 1929, *Acute Rheumatism* (Pathogenesis, Pathological Anatomy and Clinical Anatomical Classification) (Trans. from Russian), Moscow-Leningrad.
12. Whipham, T., 1888, Reports of the collective investigation committee of the British Medical Association: Report on inquiry No. III. Acute rheumatism, *Br. Med. J.* 1:387–404.
13. Schick, B., 1907, Uber die nachkrankheiten des scharlach, *Jahrb. Kinderheilkd.* 65: 132–175.
14. Pope, R. M., 1989, Rheumatic fever in the 1980s, *Bull. Rheum. Dis.* 38:1–8.
15. Markowitz, M., 1981, Observations on the epidemiology and prevention of rheumatic fever in developing countries, *Clin. Ther.* 4:240–251.
16. Dajani, A. S., 1992, Guidelines for the diagnosis of rheumatic fever (Jones Criteria, 1992 Update), *J. Am. Med. Assoc.* 268:2069–2073.
17. Wannamaker, L. W., 1973, The chain that links the throat to the heart, *Circulation* 48: 9–18.
18. Coburn, A. F., 1931, *The Factor of Infection in the Rheumatic State*, Williams and Wilkins, Baltimore.

19. Rammelkamp, C. H., Denny, F. W., and Wannamaker, L. W., 1952, Studies on the epidemiology of rheumatic fever in the armed services, in: *Rheumatic Fever* (L. Thomas, ed.), University of Minnesota Press, Minneapolis, pp. 72–89.

20. Todd, E. W., 1932, Antihaemolysin titres in haemolytic streptococcal infections and their significance in rheumatic fever, *Br. J. Exp. Pathol.* 13:248–259.

21. Stetson, C. A., 1954, The relation of antibody response to rheumatic fever, in: *Streptococcal Infection* (M. McCarty, ed.), Columbia University Press, New York, pp. 208–218.

22. Rammelkamp Jr., C. H., 1958, The Lewis A. Conner Memorial Lecture. Rheumatic heart disease—A challenge, *Circulation* 17:842–851.

23. Kendall, F. E., Heidelberger, M., and Dawson, M. N., 1941, A serologically inactive polysaccharide elaborated by mucoid strains of group A hemolytic streptococcus, *J. Biol. Chem.* 118:61–69.

24. Fischetti, V. A., 1989, Streptococcal M protein: Molecular design and biological behavior, *Clin. Microbiol. Rev.* 2:286–314.

25. Foley, M. J., and Wood, W. B., 1959, Studies of the pathogenicity of group A streptococci. II. The anti-phagocytic effects of the M protein and its capsular gel, *J. Exp. Med.* 110:617–628.

26. Lancefield, R. C., 1941, Specific relationship of cell composition to biological activity of hemolytic streptococci, *Harvey Lect.* 36:251.

27. Griffith, F., 1926, Types of haemolytic streptococci in relation to scarlet fever, *J. Hyg.* 25:385–397.

28. McCarty, M., and Lancefield, R. C., 1955, Variation in the group specific carbohydrate of group A streptococci I. Immunochemical studies on the carbohydrate of variant strains, *J. Exp. Med.* 102:11–28.

29. Lancefield, R. C., 1928, The antigenic complex of *Streptococcus hemolyticus*. I. Demonstration of a type-specific substance in extracts of *Streptococcus hemolyticus*, *J. Exp. Med.* 47:91–103.

30. Lancefield, R. C., 1959, Persistence of type specific antibodies in man following infection with group A streptococci, *J. Exp. Med.* 110:271–292.

31. Lancefield, R. C., 1962, Current knowledge of the type specific M antigens of group A streptococci, *J. Immunol.* 89:307–313.

32. Cunningham, M. W., and Beachey, E. H., 1974, Peptic digestion of streptococcal M protein. I. Effects of digestion at suboptimal pH upon the biological and immunochemical properties of purified M protein extracts, *Infect. Immun.* 9:244–248.

33. Beachey, E. H., Campbell, G. L., and Ofek, I., Peptic digestion of streptococcal M protein II. Extraction of M antigen of group A streptococci with pepsin, *Infect. Immun.* 9:891–896.

34. Fischetti, V. A., 1977, Streptococcal M protein extracted by nonionic detergent. II. Analysis of the antibody response to the multiple antigenic determinants of the M-protein molecule, *J. Exp. Med.* 146:1108–1123.

35. Cohen, J. O., Gross, H., and Harrell, W. K., 1977, Immunogenicity and characteristics of M protein released by phage associated lysin from group A streptococci types 1 and 23, *J. Med. Microbiol.* 10:179–194.

36. Fischetti, V. A., Zabriskie, J. B., and Gotschlich, E. C., 1974, Physical, chemical, and biological properties of type 6 M-protein extracted with purified streptococcal phage-associated lysin, in: *Streptococcal Disease and the Community* (M. J. Haverkorn, ed.), Excerpta Medica, Amsterdam.

37. Beachey, E. H., Stollerman, G., Chiang, E. Y., Chiang, T. M., Seyer, J. M., and Kang, A. H., 1977, Purification and properties of M protein extracted from group A streptococci

with pepsin covalent structure of the amino terminal region of type 24 M antigen, *J. Exp. Med.* **145**:1469–1483.

38. Beachey, E. H., Seyer, J. M., and Kang, A. H., 1978, Repeating covalent structure of streptococcal M protein, *Proc. Natl. Acad. Sci. U.S.A.* **75**:3163–3167.

39. Beachey, E. H., Seyer, J. M., and Kang, A. H., 1980, Primary structure of protective antigens of type 24 streptococcal M protein, *J. Biol. Chem.* **255**:5284–6289.

40. Manjula, B. N., and Fischetti, V. A., 1980, Studies on group A streptococcal M proteins: Purification of type 5 M-protein and comparison of its amino terminal sequence with two immunologically unrelated M-protein molecules, *J. Immunol.* **124**:261–267.

41. Manjula, B. N., Acharya, A. S., Mische, S. M., Fairwell, T., and Fischetti, V. A., 1984, The complete amino acid sequence of a biologically active 197-residue fragment of M protein from type 5 group A streptococci, *J. Biol. Chem.* **259**:3686–3693.

42. Manjula, B. N., Acharya, A. S., Mische, S. M., Fairwell, T., Trus, B. L., and Fischetti, V. A., 1985, The complete amino acid sequence of streptococcal pep M5 protein: Presence of two structurally distinct domains and implications to biological functions, in: *Recent Advances in Streptococci and Streptococcal Diseases* (Y. Kimura, S. Kotani, and Y. Shiokawa, eds.), Reedbooks, Ltd., Chertsey, England.

43. Manjula, B. N., and Fischetti, V. A., 1980, Tropomyosin-like seven residue periodicity in three immunologically distinct streptococcal M proteins and its implication for the antiphagocytic property of the molecule, *J. Exp. Med.* **151**:695–708.

44. Manjula, B. N., and Fischetti, V. A., 1986, Sequence homology of group A streptococcal PepM5 protein with other coiled-coil proteins, *Biochem. Biophys. Res. Commun.* **140**:684–690.

45. Manjula, B. N., Trus, B. L., and Fischetti, V. A., 1985, Presence of two distinct regions in the coiled-coil structure of the streptococcal Pep M5 protein: Relationship to mammalian coiled-coil proteins and implications to its biological properties, *Proc. Natl. Acad. Sci. U.S.A.* **82**:1064–1068.

46. Kehoe, M. A., Poirier, T. P., Beachey, E. H., and Timmis, K. N., 1985, Cloning and genetic analysis of serotype 5 M protein determinant of group A streptococci: Evidence for multiple copies of the M5 determinant in the *Streptococcus pyogenes* genome, *Infect. Immun.* **48**:190–197.

47. Hollingshead, S. K., Fischetti, V. A., and Scott, J. R., 1986, Complete nucleotide sequence of type 6 M protein of the group A streptococcus: Repetitive structure and membrane anchor, *J. Biol. Chem.* **261**:1677–1686.

48. Scott, J. R., and Fischetti, V. A., 1983, Expression of streptococcal M protein in *Escherichia coli, Science* **221**:758–760.

49. Robbins, J. C., Spaner, J. G., Jones, S. J., Simpson, W. J., and Cleary, P. P., 1987, *Streptococcus pyogenes* type 12 M protein regulation by upstream sequences, *J. Bacteriol.* **169**:5633–5640.

50. Mouw, A. R., Beachey, E. H., and Burdett, V., 1988, Molecular evolution of streptococcal M protein: Cloning and nucleotide sequence of type 24 M protein gene and relation to other genes of *Streptococcus pyogenes, J. Bacteriol.* **170**:676–684.

51. Bisno, A. L., Craven, D. E., and McCabe, W. R., 1987, M proteins of group G streptococci isolated from bacteremic human infection, *Infect. Immun.* **55**:753–757.

52. Hollingshead, S. K., Fischetti, V. A., and Scott, J. R., 1987, A highly conserved region present in transcripts encoding heterologous M proteins of group A streptococcus, *Infect. Immun.* **55**:3237–3239.

53. Bessen, D., Jones, K. F., and Fischetti, V. A., 1989, Evidence for two distinct classes of streptococcal M protein and their relationship to rheumatic fever, *J. Exp. Med.* **169**:269–283.

54. Widdowson, J. P., 1980, The M-associated protein antigens of group A streptococci, in: *Streptococcal Diseases and the Immune Response* (S. E. Read and J. B. Zabriskie, eds.), Academic Press, Inc., New York, pp. 125–147.

55. Widdowson, J. P., Maxted, W. R., and Pinney, A. M., 1976, An M-associated protein antigen (MAP) of group A streptococci, *J. Hyg.* **69**:553–570.

56. Bisno, A. L., 1980, The concept of rheumatogenic and nonrheumatogenic group A streptococci, in: *Streptococcal Diseases and the Immune Response* (S. E. Read and J. B. Zabriskie, eds.), Academic Press, Inc., New York, pp. 789–803.

57. Homer, C., and Shulman, S. T., 1991, Clinical aspects of acute rheumatic fever, *J. Rheumatol.* **18**:2–13.

58. Ayoub, E. M., 1989, Acute rheumatic fever, in: *Moss' Heart Disease in Infants, Children and Adolescents*, 4th Ed. (F. H. Adams, G. C. Emmanouilides, and T. A. Riemenschneider, eds.), Williams and Wilkins, Baltimore, pp. 692–704.

59. Agarwal, B. L., 1981, Rheumatic heart disease unabated in developing countries, *Lancet* **2**:910–911.

60. Kaplan, E. L., 1991, The resurgence of group A streptococcal infections and their sequelae, *Eur. J. Clin. Microbiol. Infect. Dis.* **10**:55–57.

61. Veasy, L. G., Wiedmeier, S. E., and Orsmond, G. S., 1987, Resurgence of acute rheumatic fever in the intermountain area of the United States, *N. Engl. J. Med.* **316**:421–427.

62. Ayoub, E. M., 1992, Resurgence of rheumatic fever in the United States, *Post Grad. Med.* **92**:133–142.

63. Kaplan, E. L., Johnson, D. R., and Cleary, P. P., 1989, Group A streptococcal serotypes isolated from patients and sibling contacts during the resurgence of rheumatic fever in the United States in the mid-1980's, *J. Infect. Dis.* **159**:101–103.

64. Stollerman, G. H., 1982, Global changes in group A streptococcal diseases and strategies for their prevention, in: *Advances in Internal Medicine*, Vol. 27 (G. H. Stollerman *et al.*, eds.), Year Book Publishers, Chicago, p. 373.

65. Ferretti, J. J., Huang, T., and Hynes, W. L., 1991, Extracellular product genes of group A streptococci, in: *Genetics and Molecular Biology of Streptococci, Lactococci and Enterococci*, American Society for Microbiology, Washington, D.C., pp. 201–205.

66. Hauser, A. R., Goshorn, S. C., Kaplan, E. L., Stevens, D. L., and Schlievert, P. M., 1991, Molecular analysis of the streptococcal pyrogenic exotoxins, in: *Genetics and Molecular Biology of Streptococci, Lactococci and Enterococci*, American Society for Microbiology, Washington, D.C., pp. 195–200.

67. Yu, C., and Ferretti, J. J., 1991, Frequency of the erythrogenic toxin B and C genes (*spe* B and *spe* C) among clinical isolates of group A streptococci, *Infect. Immun.* **59**:211–215.

68. Yu, C., and Ferretti, J. J., 1989, Molecular epidemiologic analysis of the type A streptococcal exotoxin (erythrogenic toxin) gene (*spe* A) in clinical *Streptococcus pyogenes* strains, *Infect. Immun.* **57**:3715–3719.

69. Alouf, J. E., 1980, Streptococcal toxins (streptolysin O, streptolysin S, erythrogenic toxin), *Pharmacol. Ther.* **11**:661–717.

70. Spanier, J. G., Jones, S. J. C., and Cleary, P. P., 1984, Small DNA deletions creating avirulence in *Streptococcus pyogenes*, *Science* **225**:935–938.

71. Cleary, P. P., LaPenta, D., Heath, D., Haanes, E. J., and Chen, C., 1991, A virulence regulon in *Streptococcus pyogenes*, in: *Genetics and Molecular Biology of Streptococci, Lactococci and Enterococci* (G. M. Dunny, P. P. Cleary, and L. L. McKay, eds.), American Society for Microbiology, Washington, D.C., pp. 147–151.

72. Caparon, M. G., and Scott, J. R., 1987, Identification of a gene that regulates expression

of M protein, the major virulence determinant of group A streptococci, *Proc. Natl. Acad. Sci. U.S.A.* **84**:8677–8681.

73. Chen, C., and Cleary, P. P., 1989, Cloning and expression of the streptococcal C5a peptidase gene in *Escherichia coli*; Linkage to the type 12 M protein gene, *Infect. Immun.* **57**:1740–1745.

74. Cleary, P., and Heath, K., 1990, Type II immunoglobulin receptor and its gene, in: *Bacterial Immunoglobulin-Binding Protein*, Vol. I (M. Boyle, ed.), Academic Press, New York, pp. 83–99.

75. Barnett, L. A., and Cunningham, M. W., 1990, A new heart crossreactive antigen in *Streptococcus pyogenes* is not M protein, *J. Infect. Dis.* **162**:875–882.

76. Kil, K.-S., Cunningham, M. W., and Barnett, L. A., 1994, Cloning and sequence analysis of a gene encoding a 60 kilodalton myosin crossreactive antigen of *Streptococcus pyogenes* reveals its homology with class II major histocompatibility antigens, *Infect. Immun.* **62**:2440–2449.

77. Cunningham, M. W., McCormack, J. M., Fenderson, P. G., Ho, M. K., Beachey, E. H., and Dale, J. B., 1989, Human and murine antibodies crossreactive with streptococcal M protein and myosin recognize the sequence GLN-LYS-SER-LYS-GLN in M protein, *J. Immunol.* **143**:2677–2683.

78. Quinn, A., Fischetti, V., and Cunningham, M., 1996, Identification of M protein epitopes in acute rheumatic fever: Connection between the class I M protein epitope and myosin, submitted.

79. McCormack, J. M., Crossley, C. A., Ayoub, E. M., Harley, J. B., and Cunningham, M. W., 1993, Poststreptococcal anti-myosin antibody marker associated with systemic lupus erythematosus and Sjogren's syndrome, *J. Infect. Dis.* **168**:915–921.

80. Stollerman, G. H., 1991, Rheumatogenic streptococci and autoimmunity, *Clin. Immunol. Immunopathol.* **61**:131–142.

81. Cheadle, W. B., 1989, Harvean lectures on the various manifestations of the rheumatic state as exemplified in childhood and early life, *Lancet* **1**:821–832.

82. Wilson, M. G., Schweitzer, M. D., and Lubschez, R., 1943, The familial epidemiology of rheumatic fever, *J. Pediatr.* **22**:468–492.

83. Taranta, A., Torosdag, S., and Metrakos, J. D., 1959, Rheumatic fever in monozygotic and dizygotic twins, *Circulation* **20**:778–792.

84. Glynn, L. E., and Halborrow, E. J., 1961, Relationship between blood groups secretion status and susceptibility to rheumatic fever, *Arthritis Rheum.* **4**:203–207.

85. Ayoub, E. A., 1986, Association of class II human histocompatibility leukocyte antigens with rheumatic fever, *J. Clin. Invest.* **77**:2019–2026.

86. Masharaj, B., Hammond, M. G., and Appadoo, B., 1987, HLA-A, B, DR and DQ antigens in black patients with severe chronic rheumatic heart disease, *Circulation* **76**:259–261.

87. Vakil (Elliott), M., Briles, D. E., and Kearney, J. F., 1991, Antigen-independent selection of T15 idiotype during B cell ontogeny in mice, *Dev. Immunol.* **1**:203–212.

88. Elliott, M., and Kearney, J. F., 1992, Idiotypic regulation of development of the B cell repertoire, *Ann. N.Y. Acad. Sci.* **651**:336–345.

89. Patarroyo, M. E., Winchester, R. J., Vejerano, A., Gibofsky, A., Chalem, F., Zabriskie, J. B., and Kinkel, H. G., 1979, Association of a B-cell alloantigen with susceptibility to rheumatic fever, *Nature* **278**:173–174.

90. Khanna, A. K., Buskirk, D. R., Williams Jr., R. C., Gibofsky, A., 1989, Presence of a non-HLA B cell antigen in rheumatic fever patients and their families as defined by a monoclonal antibody, *J. Clin. Invest.* **83**:1710–1716.

91. Cavelti, P. A., 1945, Autoantibodies in rheumatic fever, *Proc. Soc. Exp. Biol. Med.* **60**:379–381.
92. Kaplan, M. H., and Frengley, J. D., 1969, Autoimmunity to the heart in cardiac disease. Current concepts of the relation of autoimmunity to rheumatic fever, postcardiotomy and post infarction syndromes and cardiomyopathies, *Am. J. Cardiol.* **24**:459–473.
93. Zabriskie, J. B., Hsu, K. C., and Seegal, B. C., 1970, Heart-reactive antibody associated with rheumatic fever: Characterization and diagnostic significance, *Clin. Exp. Immunol.* **7**:147–159.
94. Kaplan, M. H., Bolande, R., Rakita, L., and Blair, J., 1964, Presence of bound immunoglobulins and complement in the myocardium in acute rheumatic fever. Associated with cardiac failure, *N. Engl. J. Med.* **271**:637–645.
95. Kaplan, M. H., and Meyeserian, M., 1962, An immunological cross reaction between group A streptococcal cells and human heart tissue, *Lancet* **1**:706–710.
96. Kaplan, M. H., 1963, Immunologic relation of streptococcal and tissue antigens. I. Properties of an antigen in certain strains of group A streptococci exhibiting an immunologic cross reaction with human heart tissue, *J. Immunol.* **90**:595–606.
97. Kaplan, M. H., and Suchy, M. L., 1964, Immunologic relation of streptococcal and tissue antigens. II. Cross reactions of antisera to mammalian heart tissue with a cell wall constituent of certain strains of group A streptococci, *J. Exp. Med.* **119**:643–650.
98. Kaplan, M. H., and Svec, K. H., 1964, Immunologic relation of streptococcal and tissue antigens. III. Presence in human sera of streptococcal antibody cross reactive with heart tissue. Association with streptococcal infection, rheumatic fever, and glomerulonephritis, *J. Exp. Med.* **119**:651–666.
99. Zabriskie, J. B., and Freimer, E. H., 1966, An immunological relationship between the group A streptococcus and mammalian muscle, *J. Exp. Med.* **124**:661–678.
100. Zabriskie, J. B., 1967, Mimetic relationships between group A streptococci and mammalian tissues, *Adv. Immunol.* **7**:147–188.
101. Beachey, E. H., and Stollerman, G. H., 1973, Mediation of cytotoxic effects of streptococcal M protein by non-type-specific antibody in human sera, *J. Clin. Invest.* **52**:2563–2570.
102. Widdowson, J. P., Maxted, W. R., and Pinney, A. M., 1971, An M-associated protein antigen (MAP) of group A streptococci, *J. Hyg.* **69**:553–564.
103. van de Rijn, I., Zabriskie, J. B., and McCarty, M., 1977, Group A streptococcal antigens cross-reactive with myocardium. Purification of heart-reactive antibody and isolation and characteristics of the streptococcal antigen, *J. Exp. Med.* **146**:579–599.
104. Goldstein, I., Halpern, B., and Robert, L., 1967, Immunological relationship between streptococcus A polysaccharide and the structural glycoproteins of heart valve, *Nature* **213**:44–47.
105. Lyampert, I. M., Vvedenskaya, O. I., and Danilova, T. A., 1966, Study on streptococcus group A antigens common with heart tissue elements, *Immunology* **11**:313–320.
106. Lyampert, I. M., Beletskrya, L. V., and Ugryumova, G. A., 1968, The reaction of heart and other organ extracts with the sera of animals immunized with group A streptococci, *Immunology* **15**:845–854.
107. Sandson, J., Hamerman, D., and Janis, R., 1968, Immunologic and chemical similarities between the streptococcus and human connective tissue, *Trans. Assoc. Am. Physicians* **81**:249–257.
108. Cunningham, M. W., and Russell, S. M., 1983, Study of heart reactive antibody in antisera and hybridoma culture fluids against group A streptococci, *Infect. Immun.* **42**:531.

109. Cunningham, M. W., Graves, D. C., and Krisher, K., 1984, Murine monoclonal antibodies reactive with human heart and group A streptococcal membrane antigens, *Infect. Immun.* **46**:34–41.

110. Krisher, K., and Cunningham, M. W., 1985, Myosin: A link between streptococci and heart, *Science* **227**:413–415.

111. Dale, J. B., and Beachey, E. H., 1986, Epitopes of streptococcal M proteins shared with cardiac myosin, *J. Exp. Med.* **162**:583–591.

112. Cunningham, M. W., and Swerlick, R. A., 1986, Polyspecificity of antistreptococcal murine monoclonal antibodies and their implications in autoimmunity, *J. Exp. Med.* **164**:998–1012.

113. Cunningham, M. W., Krisher, K. K., Swerlick, R. A, Barnett, L. A., and Guderian, P. F., 1988, Molecular mimicry: Streptococci and myosin, in: *Vaccines, New Concepts and Developments* (H. Kohler and P. T. LoVerde, eds.), John Wiley and Sons, New York, pp. 413–423.

114. Barnett, L. A., Ferretti, J. J., and Cunningham, M. W., 1992, A 60 kilodalton acute rheumatic fever associated antigen of *Streptococcus pyogenes*, in: *New Perspectives on Streptococci and Streptococcal Infections* (G. Orefici, ed.), Gustav Fischer, Stuttgart, Jena, New York, pp. 216–218.

115. Cunningham, M. W., Hall, N. K., Krisher, K. K., and Spanier, A. M., 1986, A study of anti-group A streptococcal monoclonal antibodies cross reactive with myosin, *J. Immunol.* **136**:293–298.

116. Cunningham, M. W., McCormack, J. M., Talaber, L. R., Harley, J. B., Ayoub, E. M., Muneer, R. S., Chun, and Reddy, D. V., 1988, Human monoclonal antibodies reactive with antigens of the group A streptococcus and human heart, *J. Immunol.* **141**:2760–2766.

117. Dell, V. A., 1991, Autoimmune determinants of rheumatic carditis: Localization of epitopes in human cardiac myosin. *Eur. Heart J.* **12**(Suppl. D):158–162.

118. Carroll, P., Stafford, D., Schwartz, R. S., and Stollar, B. D., 1985, Murine monoclonal anti-DNA antibodies bind to endogenous bacteria, *J. Immunol.* **135**:1086.

119. Adrezejewski Jr., C., Rauch, J., Lafer, E., Stollar, B. D., and Schwartz, R. S., 1980, Antigen binding diversity and idiotypic cross-reactions among hybridoma autoantibodies to DNA, *J. Immunol.* **126**:226–231.

120. Lafer, E. M., Rauch, J., Andrzejewski Jr., C., Mudd, D., Furie, B., Schwartz, R. S., and Stollar, B. D., 1981, Polyspecific monoclonal lupus autoantibodies reactive with both polynucleotides and phospholipids, *J. Exp. Med.* **153**:897–909.

121. Dighiero, G., Lymberi, P., Mazie, J. C., Rouyre, S., Butler-Browne, G. S., Whalen, R. G., and Avrameas, S., 1993, Murine hybridomas secreting natural monoclonal antibodies reacting with self antigens, *J. Immunol.* **131**:2267–2272.

122. Avrameas, S., Dighiero, G., Lymberi, P., and Guilbert, B., 1983, Studies on natural antibodies and autoantibodies, *Ann. Immunol.* **134**:103–113.

123. Bona, C. A., 1988, V genes encoding autoantibodies: Molecular and phenotypic characteristics, *Annu. Rev. Immunol.* **6**:327–358.

124. Chen, P. P., Liu, M., Sinha, S., and Carson, D. A., 1988, A 16/6 idiotype-positive anti-DNA antibody is encoded by a conserved VH gene with no somatic mutation, *Arthritis Rheum.* **31**:1429.

125. Quinn, A., Adderson, E. E., Shackelford, P. G., Carroll, W. L., and Cunningham, M. W., 1994, Autoantibody germline gene segment encodes V_H and V_L regions of a human anti-streptococcal Mab recognizing streptococcal M protein and human cardiac myosin epitopes, *J. Immunol.* **154**:4203–4212.

126. Schroeder, H. W., and Wang, J. Y., 1990, Preferential utilization of conserved immuno-globulin heavy chain variable region gene segment during human fetal life, *Proc. Natl. Acad. Sci. U.S.A.* **87**:6146–6150.
127. Cunningham, M. W., Antone, S. M., Gulizia, J. M., McManus, B. M., Fischetti, V. A., and Gauntt, C. J., 1992, Cytotoxic and viral neutralizing antibodies crossreact with strepto-coccal M protein, enteroviruses and human cardiac myosin, *Proc. Natl. Acad. Sci. U.S.A.* **89**:1320–1324.
128. Oldstone, M. B. A., 1987, Molecular mimicry and autoimmune disease, *Cell* **50**:819–820.
129. Oldstone, M. B. A., 1989, Overview: Infectious agents as etiological triggers of auto-immune disease, *Curr. Top. Microbiol. Immunol.* **145**:1–3.
130. Shikhman, A. R., Greenspan, N. S., and Cunningham, M. W., 1993, A subset of mouse monoclonal antibodies crossreactive with cytoskeletal proteins and group A strep-tococcal M proteins recognizes N-acetyl-β-D-glucosamine, *J. Immunol.* **151**:3902–3914.
131. Shikhman, A. R., and Cunningham, M. W., 1994, Immunological mimicry between N-acetyl-β-D-glucosamine and cytokeratin peptides, *J. Immunol.* **152**:4375.
132. Shikhman, A. R., Greenspan, N. S., and Cunningham, M. W., 1994, Cytokeratin peptide SFGSGFGGGY mimics N-acetyl-β-D-glucosamine in reaction with antibodies and lectins, and induces *in vivo* anti-carbohydrate antibody response, *J. Immunol.* **153**:5593–5606.
133. Kabat, A., Nickerson, K. G., Liao, J., Gorssbard, L., Ossarman, E. F., Glickman, E., Chess, L., Rabbins, J. B., Schneerson, R., and Yang, Y., 1986, A human monoclonal macroglobulin with specificity for a(2,8)-linked poly-N-acetylneuraminic acid, the capsular polysaccharide of group B meningococci and *E. coli* Ki, which cross-reacts with polynucleotides and denatured DNA, *J. Exp. Med.* **164**:642–654.
134. Cunningham, M. W., 1993, Molecular mimicry: Bacterial antigen mimicry, in: *The Molecular Pathology of Autoimmunity* (C. A. Bona, K. Siminovitch, A. N. Theofilopoulos, and M. Zanetti, eds.), Harwood Academic Publishers, New York, pp. 245–256.
135. McLachlan, A. D., and Stewart, M., 1975, Tropomyosin coiled-coil interactions evi-dence for an unstaggered structure, *J. Mol. Biol.* **98**:393–411.
136. Sargent, S. J., Beachey, E. H., Corbett, C. E., and Dale, J. B., 1987, Sequence of protective epitopes of streptococcal M proteins shared with cardiac sarcolemmal membranes, *J. Immunol.* **139**:1285–1290.
137. Bronze, M. S., Beachey, E. H., and Dale, J. B., 1988, Protective and heart-crossreactive epitopes located within the N-terminus of type 19 streptococcal M protein, *J. Exp. Med.* **167**:1849–1859.
138. Dale, J. B., and Beachey, E. H., 1986, Sequence of myosin cross-reactive epitopes of the streptococcal M protein, *J. Exp. Med.* **164**:1785–1790.
139. Cunningham, M. W., Antone, S. M., Gulizia, J. M., McManus, B. A., and Gauntt, C. J., 1993, α-Helical coiled-coil molecules: A role in autoimmunity against the heart, *Clin. Immunol. Immunopathol.* **68**:118–123.
140. Huber, S. A., and Cunningham, M. W., 1996, Streptococcal M protein peptide with similarity to myosin induces CD4+ T cell dependent myocarditis in MRL/H mice and induces tolerance against coxsackie viral myocarditis, *J. Immunol.*, in press.
141. Antone, S. M., Smart, M., Dale, J. B., and Cunningham, M. W., 1996, Dominant cardiac myosin crossreactive T cell epitopes of group A streptococcal M5 protein, submitted.
142. Antone, S. M., and Cunningham, M. W., 1996, Cytotoxic antistreptococcal monoclonal antibody recognizes laminin, submitted.
143. Kraus, W., Seyer, M., and Beachey, E. H., 1989, Vimentin-cross-reactive epitope of type 12 streptococcal M protein, *Infect. Immun.* **57**:2457–2461.

144. Beck, K., Hunter, I., and Engel, J., 1990, Structure and function of laminin: Anatomy of a multidomain glycoprotein, *FASEB J.* 4:148–160.
145. Dudding, B. A., and Ayoub, E. M., 1968, Persistence of streptococcal group A antibody in patients with rheumatic valvular disease, *J. Exp. Med.* 129:1080–1098.
146. McCarty, M., 1956, Variation in the group specific carbohydrate of group A streptococci, II. Studies on the chemical basis for serological specificity of the carbohydrates, *J. Exp. Med.* 104:629.
147. McCarty, M., 1964, Missing links in the streptococcal chain leading to rheumatic fever: The T. Duckett Jones Memorial Lecture, *Circulation* 29:488–493.
148. Fung, J. C., Wicher, K., and McCarty, M., 1982, Immunochemical analysis of streptococcal group A, B, and C carbohydrates with emphasis on group A, *Infect. Immun.* 37:209–215.
149. Sharif, M., Rook, G., Wilkinson, L. S., Worral, J. G., and Edwards, J. C. W., 1990, Terminal N-acetylglucosamine in chronic synovitis, *Br. J. Rheumatol.* 29:25–31.
150. Read, S. E., and Zabriskie, J. B., 1976, Immunological concepts in rheumatic fever pathogenesis, in: *Textbook of Immunopathology* (P. A. Miescher and H. J. Muller-Eberhard, eds.), Grune and Stratton, New York, p. 471.
151. Beachey, E. H., Alberti, H., and Stollerman, G. H., 1969, Delayed hypersensitivity to purified streptococcal M protein in guinea pigs and in man, *J. Immunol.* 102:42–52.
152. Read, S. E., Zabriskie, J. B., Fischetti, V. A., Utermohlen, V., and Falk, R., 1974, Cellular reactivity studies to streptococcal antigens in patients with streptococcal infections and their sequelae, *J. Clin. Invest.* 54:439–450.
153. Read, S. E., Zabriskie, J. B., Fischetti, V. A., Utermohlen, V., and Falk, R., 1974, Cellular reactivity studies to streptococcal antigens—migration inhibition studies in patients with streptococcal infections and rheumatic fever, *J. Clin. Invest.* 54:439–450.
154. Reid, H. G. M., Read, S. E., Poon-King, T., and Zabriskie, J. B., 1980, Lymphocyte response to streptococcal antigens in rheumatic fever patients in Trinidad, in: *Streptococcal Diseases and the Immune Response*, Academic Press, Inc., New York, pp. 681–693.
155. Gray, E. D., Wannamaker, L. M., Ayoub, E. M., Kholy, E. M., and Abdin, Z. H., 1981, Cellular immune responses to extracellular streptococcal products in rheumatic heart disease, *J. Clin. Invest.* 68:665–671.
156. Gibofsky, A., Williams, R. C., and Zabriskie, J. B., 1987, Immunological aspects of acute rheumatic fever, *Balliere's Clin. Immunol. Allergy* 1:577–590.
157. Hutto, J. H., and Ayoub, E. M., 1980, Cytotoxicity of lymphocytes from patients with rheumatic carditis to cardiac cells in vitro, in: *Streptococcal Diseases and the Immune Response*, Academic Press, Inc., New York, pp. 733–738.
158. Yang, L. C., Soprey, R. R., Wittner, M. K., and Fox, E. N., 1977, Streptococcal induced cell mediated immune destruction of cardiac myofibers in vitro, *J. Exp. Med.* 146:344–360.
159. Friedman, I., Laufer, A., Ron, N., and Davies, A. M., 1971, Experimental myocarditis: *In vitro* and *in vivo* studies of lymphocytes sensitized to heart extracts and group A streptococci, *Immunology* 20:225–232.
160. Dale, J. B., and Beachey, E. H., 1987, Human cytotoxic T lymphocytes evoked by group A streptococcal M proteins, *J. Exp. Med.* 166:1825–1835.
161. Robinson, J. H., Atherton, M. C., Goodacre, J. A., Pinkney, M., Weightman, H., and Kehoe, M. A., 1991, Mapping T-cell epitopes in group A streptococcal type 5 M protein, *Infect. Immun.* 59:4324–4331.
162. Pruksakorn, S., Galbraith, A., Houghten, R. A., and Good, M. F., 1992, Conserved T and B cell epitopes on the M protein of group A streptococci. Induction of bactericidal antibodies, *J. Immunol.* 149:2729–2735.

163. Robinson, J. H., Case, M. C., and Kehoe, M. A., 1993. Characterization of a conserved helper T-cell epitope from group A streptococcal M proteins, *Infect. Immun.* **61**:1062–1068.

164. White, H., Herman, A., Pullen, A. M., Kubo, R., Kappler, J. W., and Marrack, P., 1989, The Vβ specific superantigen staphylococcal enterotoxin B: Stimulation of mature T cells and clonal deletion in neonatal mice, *Cell* **56**:27–35.

165. Drake, C. G., and Kotzin, B. L., Superantigens: Biology, immunology and potential role in disease, *J. Clin. Immunol.* **12**:149–159.

166. Abe, J., Forrester, J., Nakahara, T., Lafferty, J. A., Kotzin, B. L., and Leung, D. Y. M., 1991, Selective stimulation of human T cells with streptococcal erythogenic toxins A and B, *J. Immunol.* **146**:3747–3750.

167. Tomai, M., Kotb, M., Majumdar, G., and Beachey, E. H., 1990, Superantigenicity of streptococcal M protein *J. Exp. Med.* **172**:359–362.

168. Tomai, M., Aelion, J. A., Dockter, M. E., Majumdar, G., Spinella, D. G., and Kotb, M., 1991, T cell receptor V gene usage by human T cells stimulated with the superantigen streptococcal M protein, *J. Exp. Med.* **174**:285–288.

169. Chow, L. H., Yuling, Y., Linder, J., and McManus, B. M., 1989, Phenotypic analysis of infiltrating cells in human myocarditis, *Arch. Pathol. Lab. Med.* **113**:1357–1362.

170. Murphy, G. E., 1960, Nature of rheumatic heart disease with special reference to myocardial disease and heart failure, *Medicine* **39**:289–384.

171. Wagner, B. M., 1960, Studies in rheumatic fever. III. Histochemical reactivity of the Aschoff body, *Ann. N.Y. Acad. Sci.* **86**:992–1008.

172. Marboe, C. C., Knowles, D. M., Weiss, M. B., Ursell, P. C., and Fenoglio Jr., J. J., 1985, Monoclonal antibody identification of mononuclear cells in endomyocardial biopsy specimens from a patient with rheumatic carditis, *Hum. Pathol.* **13**:332–338.

173. Stollerman, G. H., 1975, Pathology of rheumatic fever, in: *Rheumatic Fever and Streptococcal Infection*, Grune and Stratton, New York, p. 135.

174. Springer, T. A., 1994, Traffic signals for lymphocyte recirculation and leukocyte emigration: The multistep paradigm, *Cell* **76**:301–314.

175. Gulizia, J. M., Cunningham, M. W., and McManus, B. A, 1991, Immunoreactivity of antistreptococcal monoclonal antibodies to human heart valves: Evidence for multiple cross-reactive epitopes, *Am. J. Pathol.* **138**:285–301.

176. Yamamura, M., Wang, X.-H., Ohman, J. D., Uyemura, K., Rea, T. H., Bloom, B. R., and Modlin, R. L., 1992, Cytokine patterns of immunologically mediated tissue damage, *J. Immunol.* **149**:1470–1475.

177. Modlin, R. L., and Nutman, T. B., 1993, Type of cytokines and negative immune regulation in human infections, *Curr. Opin. Immunol.* **5**:511–517.

178. Shimizu, Y., Newmann, W., Tanka, Y., and Shaw, S., 1992, Lymphocyte interactions with endothelial cells, *Immunol. Today* **13**:106–112.

179. Guilherme, L., Chuna-Neto, E., Coelho, V., Snitcowsky, R., Pomerantzeff, P. M. A., Assis, R. V., Pedra, F., Neuman J., Goldberg, A., Patarrayo, M. E., Pileggi, F., and Kalil, J., 1995, Human heart-infiltrating T cell clones from rheumatic heart disease patients recognize both streptococcal and cardiac proteins, *Circulation* **92**:415–420.

180. Toyama-Sorimachi, N., Miyake, K., and Miyasaka, M., 1993, Activation of CD44 induces ICAM-1/LFA-1 independent, Ca++, Mg++ independent adhesion pathway in lymphocyte-endothelial cell interaction, *Eur. J. Immunol.* **23**:439–446.

181. Murphy, G. E., and Swift, H. F., 1949, Induction of cardiac lesions closely resembling those of rheumatic fever in rabbits, following repeated skin infections with group A streptococci, *J. Exp. Med.* **89**:687–698.

182. Murphy, G. E., and Swift, H. F., 1950, The induction of rheumatic-like cardiac lesions in rabbits by repeated focal injections with group A streptococci. Comparison with the cardiac lesions of serum disease, *J. Exp. Med.* 91:485–498.
183. Kirschner, L., and Howie, J. B., 1952, Rheumatic-like lesions in the heart of the rabbit experimentally induced by repeated inoculation with haemolytic streptococci, *J. Pathol. Bacteriol.* 64:367–377.
184. Cromartie, W. J., and Craddock, J. G., 1966, Rheumatic-like cardiac lesions in mice, *Science* 154:285–287.
185. Glynn, L. E., and Holborow, E. J., 1961, Relation between blood groups, secretory status and susceptibility to rheumatic fever, *Arthritic Rheum.* 4:203–207.
186. Schwab, J. H., 1962, Analysis of the experimental lesion of connective tissue produced by a complex of C polysaccharide from group A streptococci. I. In vivo reaction between tissue and toxin, *J. Exp. Med.* 116:17–28.
187. Schwab, J. H., 1964, Analysis of the experimental lesion of connective tissue produced by a complex of C polysaccharide from group A streptococci. II. Influence of age and hypersensitivity, *J. Exp. Med.* 119:401–408.
188. Schwab, J. H., 1965, Biological properties of streptococcal cell wall particles. I. Determinants of the chronic nodular lesion of connective tissue, *J. Bacteriol.* 90:1405–1411.
189. Cromartie, W., Craddock, J., Schwab, J., Anderle, S., and Yang, C., 1977, Arthritis in rats after systemic injection of streptococcal cells or cell walls, *J. Exp. Med.* 146:1585–1602.
190. Eisenberg, R., Fox, A., Greenblatt, J., Anderle, S., Cromartie, W., and Schwab, J., 1982, Measurement of bacterial cell wall in tissues by solid phase radioimmunoassay. Correlation of distribution and persistence with experimental arthritis in rats, *Infect. Immun.* 38:127–135.
191. Unny, S. K., and Middlebrooks, B. L., 1983, Streptococcal rheumatic carditis, *Microbiol. Rev.* 47:97–120.
192. Wells, A., Parajasegaram, G., Baldwin, M., Yang, C., Hammer, M., and Fox, A., 1986, Uveitis and arthritis induced by systemic injection of streptococcal cell walls, *Invest. Ophthalmol. Vis. Sci.* 27:921–925.
193. Fox, A., Brown, R., Anderle, S., Chetty, C., Cromartie, W., Gooder, H., and Schwab, J., 1982, Arthropathic properties related to the molecular weight of peptidoglycan-polysaccharide polymers of streptococcal cell walls, *Infect. Immun.* 35:1003–1010.
194. Fox, A., Schallinger, L., and Kirkland, J., 1985, Sedimentation field flow fractionation of bacterial cell wall fragments, *J. Microbiol. Methods* 3:273.
195. Schwab, J., Allen, J., Anderle, S., Dalldorf, F., Eisenberg, R., and Cromartie, W., 1982, Relationship of complement to experimental arthritis induced in rats with streptococcal cells walls, *Immunology* 46:83–88.
196. Greenblatt, J., Hunter, N., and Schwab, J., 1980, Antibody response to streptococcal cell wall antigens associated with experimental arthritis in rats, *Clin. Exp. Immunol.* 42:450–457.
197. Hunter, N., Anderle, S., Brown, R., Dalldorf, F., Clark, R., Cromartie, W., and Schwab, J., 1980, Cell-mediated response during experimental arthritis induced in rats with streptococcal cell walls, *Clin. Exp. Immunol.* 42:441–449.
198. Lerner, M. P., Nordquist, R. E., Donoso, L. A., and Cunningham, M. W., 1996, Immunological mimicry between retinal S antigen and group A streptococcal M protein, *Autoimmunity*, in press.
199. Kodama, M., Matsumoto, Y., Fujiwara, M., Masani, F., Izumi, T., and Shibata, A., 1991, A novel experimental model of giant cell myocarditis induced in rats by immunization with cardiac myosin fraction, *Clin. Immunol. Immunopathol.* 57:250–262.

200. Stollerman, G. H., 1993, The global impact of penicillin, *Mount Sinai J. Med.* **60**:112–119.
201. Bisno, A. L., 1990, Nonsuppurative poststreptococcal sequelae: Rheumatic fever and glomerulonephritis, in: *Principles and Practice of Infectious Diseases* (G. L. Mandell, R. G., Douglas, and J. E. Bennett, eds.), Churchill Livingstone, New York, pp. 1528–1539.
202. Kaplan, E. L., and Johnson, D. R., 1988, Eradication of group A streptococci from the upper respiratory tract by amoxicillin with clavulanate after oral penicillin V treatment failure, *J. Pediatr.* **113**:400–403.
203. Pichichero, M. E., 1991, The rising incidence of penicillin treatment failures in group A streptococcal tonsillopharyngitis: An emerging role for the cephalosporins, *Pediatr. Infect. Dis. J.* **10**:S50–S55.
204. Wittner, M. K., and Fox, E. N., 1977, Homologous and heterologous protection of mice with group A streptococcal M protein vaccines, *Infect. Immun.* **15**:104–108.
205. Fox, E. N., 1974, M proteins of group A streptococci, *Bacteriol. Rev.* **38**:57–86.
206. Beachey, E. H., Stollerman, G. H., Johnson, R. H., Ofek, I., and Bisno, A. L., 1979, Human immune response to immunization with structurally defined polypeptide fragment of streptococcal M protein, *J. Exp. Med.* **150**:862–877.
207. Beachey, E. H., Seyer, J. M., Dale, J. B., Simpson, W. A, and Kang, A. H., 1981, Type-specific protective immunity evoked by synthetic peptide of *Streptococcus pyogenes* M protein, *Nature* **292**:457–459.
208. Beachey, E. H., Gras-Masse, H., Tarter, A., Jolivet, M., Audibert, F., Chedid, L., and Seyer, J., 1986, Opsonic antibodies evoked by hybrid peptide copies of types 5 and 24 streptococcal M proteins synthesized in tandem, *J. Exp. Med.* **163**:1451–1458.
209. Dale, J. B., Chiang, E. Y., and Lederer, J. W., 1993, Recombinant tetravalent group A streptococcal M protein vaccine, *J. Immunol.* **151**:2188–2194.
210. Bessen, D., and Fischetti, V. A., 1988, Influence of intranasal immunization with synthetic peptides corresponding to conserved epitopes of M protein on mucosal colonization by group A streptococci, *Infect. Immun.* **56**:2666–2672.
211. Bronze, M., McKinsey, D., Corbett, C., Beachey, E. H., and Dale, J. B., 1988, Protective immunity evoked by locally administered group A streptococcal vaccines in mice, *J. Immunol.* **141**:2767.
212. Fischetti, V. A., Hodges, W. M., and Hruby, D. E., 1989, Protection against streptococcal pharyngeal colonization with a vaccinia: M protein recombinant, *Science* **244**:1487–1490.
213. Pozzi, G., Oggioni, M. R., Manganelli, R., and Fischetti, V. A., 1992, Expression of M6 protein gene of *Streptococcus pyogenes* in *Streptococcus gordonii* after chromosomal integration and transcriptional fusion, *Res. Microbiol.* **143**:449–457.
214. Medaglini, D., Pozzi, G., King, T. P., and Fischetti, V. A, 1995, Mucosal and systemic responses to a recombinant protein expressed on the surface of the oral commensal bacterium *Streptococcus gordonii* after oral colonization, *Proc. Natl. Acad. Sci. U.S.A.* **92**:6868–6872.
215. Miller, L., Gray, L., Beachey, E., and Kehoe, M., 1988, Antigenic variation among group A streptococcal M proteins, *J. Biol. Chem.* **263**:5668–5673.

3

HLA-B27, Enteric Bacteria, and Ankylosing Spondylitis

JOHN S. SULLIVAN and ANDREW F. GECZY

1. INTRODUCTION

Ankylosing spondylitis (AS) is an inflammatory arthropathy of the axial skeleton and large peripheral joints whose characteristic pathological feature is its tendency to show bony ankylosis around affected joints. Patients with the disease complain of chronic lower back pain and stiffness and a restriction of chest expansion. Other respiratory complications and even cardiovascular disease may sometimes develop.

Although many theories have been advanced as to the cause of AS,[1] the close association between the major histocompatibility complex (MHC) class I antigen HLA-B27 and AS probably provides a clue to understanding the pathogenesis of the disease. In most racial groups studied, more than 85% of individuals who present with AS are also HLA-B27-positive (HLA-B27+), but this antigen is found in no more than 8–10% of most healthy controls.[2] The strength of the HLA-B27 and AS association is striking,[2] but it is important to note that 1) the association is not absolute; 2) the occurrence of other diseases (psoriasis, inflammatory bowel disease, and Reiter's syndrome) may influence susceptibility to AS, and 3) although HLA-B27+ relatives of AS probands are more likely to develop the disease, the incidence of AS in HLA-B27+ individuals in the

JOHN S. SULLIVAN and ANDREW F. GECZY • NSW Red Cross Blood Transfusion Service, Sydney 2000 NSW, Australia.

Microorganisms and Autoimmune Diseases, edited by Herman Friedman *et al.* Plenum Press, New York, 1996.

general population is less than 2%. Thus, apart from genetic predisposition, other factors, possibly environmental, play an important role in the development of the disease.

2. ONE- AND TWO-GENE THEORIES

The strength of the HLA-B27–AS association has resulted in the development of two main theories which might explain this connection. The "one-gene" theory[3] proposes that the HLA-B27 antigen itself is directly involved in the disease process, while the "two-gene" theory[4,5] suggests that HLA-B27 is merely a marker for an as yet unidentified, closely linked disease gene(s). The discordance for AS in some pairs of identical twins, the lack of evidence of independent segregation of the HLA-B27 antigen from AS in a family (except where other diseases such as psoriasis or inflammatory bowel disease also occur), and finally the failure to detect such a disease susceptibility gene have cast doubts on the relevance of the two-gene theory in explaining the HLA-B27–AS correlation. Most studies have concentrated on the one-gene theory as it is more readily addressed from an experimental point of view, but in situations where this approach has not proven satisfactory, the two-gene theory has been supported by default rather than by the availability of convincing data.[6] We have chosen not to embrace the one-gene or the two-gene theories, as neither satisfactorily explains our observations and as these terms are often imprecise and misleading. As we will explain later, we prefer to interpret our data in terms of cross-reactivity between an HLA-B27-associated structure and certain bacteria. Such cross-reactivity is proposed to occur at both the humoral and cellular level and this interpretation does not of course rule out either the one-gene or two-gene theory.

The concept that a unique HLA-B27 antigen might distinguish HLA-B27+ normal individuals from those with AS has received much attention, and both cellular and biochemical techniques have revealed the complex heterogeneity of the HLA-B27 molecule.[7-13] At least eight distinct subtypes of the B27 molecule have been identified, but no single subtype has yet been reliably correlated with disease expression, although B27(B*2703) is thought not to be an associated risk factor for the seronegative arthropathies.[14] Furthermore, direct comparison between B27+AS+ and B27+AS− individuals has failed to reveal specific B27-related differences at either the phenotypic[15] or genotypic[16] levels.

Another version of the one-gene theory is the molecular mimicry theory, which proposes that the B27 molecule possesses epitopes that cross-react directly with antigen(s) expressed by a number of bacteria. On exposure to the relevant bacterium, the B27+ individual generates an antibody response that is incapable of eliminating the organism, and its persistence then leads to the continued production of an antibody which partially cross-reacts with host cells. Evidence favoring such cross-reactivity has come from a number of laboratories,[17-20] and a number of different bacterially derived cell-surface molecules have been identified which appear to cross-react with the B27 molecule. However the relevance of this mechanism *in vivo* has recently been questioned.[21] Recently, Scofield *et al.*[22] have shown that unique among HLA-B molecules, the hypervariable regions of HLA-B27 unexpectedly share short peptide sequences with proteins from a number of gram-negative bacteria. With respect to the role of a class I antigen in the presentation of peptides, Madden and associates[23] recently determined the sequence of endogenous peptides found in the binding cleft of crystallized B27, and Ohno[24] expanded on these sequences to develop a B27 binding motif. This motif includes an invariant arginine in position 2 of a nonapeptide. The B27 molecule contains a nonapeptide sequence that fits the motif at positions 168–172 within the third hypervariable region. Scofield's group[22] found that proteins from gram-negative enteric organisms contain this binding motif significantly more often than proteins from other organisms.

   We have developed a fundamentally different theory of cross-reactivity on the basis of our work on the B27-AS connection. Our results suggest that the cells of B27+AS+ individuals (but not of B27+ normal individuals) display a cell-surface antigenic structure which cross-reacts with a determinant expressed by a wide range of enteric bacteria. Although it is generally assumed[25] that our results fall into the one-gene theory category, we have interpreted our data in terms of cross-reactivity between a B27-associated structure and a limited number of enteric bacteria.[26]

2.1. Debates and Controversies Surrounding These Theories

   The nature and extent of the specific cross-reactivity between certain enteric bacteria and the tissues of B27+AS+ individuals has been the subject of intense and spirited debate.[6,27,28] The skepticism surrounding these studies has stemmed from a number of nonconfirmatory reports which cast doubt on the validity of these findings.[29] However, the first confirmatory work by Archer *et al.* was based on cells from AS patients from a London population, but we supplied the reagents and the assays were performed in our laboratory in Sydney.[30] A more objective and comprehensive approach, initiated by a Dutch group,[31] involved the collection of cells from 26 Dutch AS patients and 19 healthy controls and the shipment of these cryopreserved cells under code to Sydney where they were tested for the HLA-B27-associated determinant. The reassuring outcome of this series of studies was that the specific cytotoxicity of a range of antibacterial sera for B27+AS+ cells would be reproduced in a blind trial.[31] In a follow-up blind study,[32] the ability of cross-reactive antisera to distinguish between the cells of Dutch patients with AS and normal controls was investigated. Of the 45 samples tested, 29 were fresh peripheral blood mononuclear cells (PBMC), while 16 were cryopreserved PBMC. No false positives were identified, but there was one false negative among the 45 samples, and the negative sample was confirmed after the recoded cryopreserved cell sample from this patient was retested. It was concluded from this exercise that the "cross-reactive" antisera raised in Sydney gave good discrimination between patients and normals.[32] The possible reasons for the failure of some other laboratories to confirm these observations still remain unknown.

3. DESCRIPTION OF HLA-B27-ASSOCIATED DISEASE MARKERS IN THE PATHOGENESIS OF ANKYLOSING SPONDYLITIS

The basis of our research work has been to analyze the nature of the observed cross-reactivity between B27+AS+ cells and enteric organisms and to elucidate the possible relevance of this cross-reactivity to the pathogenesis of AS. So far, strains of *Klebsiella, Escherichia coli, Salmonella, Shigella, Staphylococcus,* and *Clostridium* have been identified which express a factor that cross-reacts with B27+AS+ cells.[33,34] Antisera to these strains raised in rabbits specifically lyse B27+AS+ lymphocytes in a complement-dependent chromium release assay while having no effect on cells from B27+ or B27− normal individuals. These organisms have been found to shed a cross-reactive factor (the "modifying" factor; MF) into their culture media. This factor modifies B27+ normal cells so that they become serologically similar to B27+AS+ cells and are lysed by the specific antisera in the chromium release assay.[35] Results which are consistent with the modification of HLA-B27 by a putative arthritogenic peptide have recently been reported by Wang and associates.[36] This group identified three anti-HLA-B27 monoclonal antibodies (B27.M1, B27.M2, and Ye-2) which did not react with "empty" HLA-B27 but which recognized HLA-B27 only in the context of certain peptides, either added exogenously or expressed endogenously. These findings raise the possibility that these peptides represent arthritogenic factors which play a role in the pathogenesis of the spondyloarthropathies. It is possible that our antisera, which discriminate between B27+AS+ and B27+AS− individuals, recognize foreign peptides in the context of HLA-B27.

With reference to our previous work, MF can be isolated from the outer membrane of cross-reactive organisms, where it is associated with a specific protein component with an isoelectric point of 5.5 and an approximate molecular weight of 30,000 kDa.[37,38] The activity of MF seems to be associated with a subunit of this protein, since trypsin treatment does not inactivate it; the modifying activity seems to be associated with a lower molecular weight fraction.[37] The significance of this finding is still to be determined. It is tempting to suggest that an MF may be a precursor of the arthritogenic peptide(s), proposed by other workers. The biochemical properties of the modifying factor are common to all other cross-reactive organisms investigated so far[33] and also to the cross-reactive determinant expressed on B27+AS+ lymphoblastoid cell lines.[39] This similarity suggests that there must be a common genetic sequence in the cross-reactive organisms and B27+AS+ cell lines.

Studies have suggested that the gene coding for the MF in *Klebsiella* K43-BTS1 and other cross-reactive organisms is carried on an extrachromosomal element or plasmid.[40] Cesium chloride–purified plasmid preparations from cross-reactive organisms can transform previously non-cross-reactive bacteria so that they permanently acquire the genetic element coding for MF (unpublished results). Transconjugation experiments produce similar results.[40] The indication that the MF gene resides on a mobile genetic element raises interesting possibilities with regard to triggering the expression of the factor.

3.1. Relevance of MF and MF-Producing Organisms in the Pathogenesis of AS

The *in vitro* modification of B27+AS− cells by MF has been found to be dependent on a large variety of metabolic events such as protein synthesis and prostaglandin metabolism.[41,42] It is not yet known whether specific metabolic events are directly involved in the modification of B27+AS− cells or whether the only requirement is a metabolically competent cell. It seems, though, that the *in vitro* modification process involves more than just the passive attachment of MF to the cell membrane of B27+AS− cells.[41,42]

The essential difference between *in vitro* modified B27+AS− and

B27+AS+ cells is that in the former, the expression of the cross-reactive factor is only transient, since the cells lose their susceptibility to lysis by cross-reactive antisera after washing and resuspension in fresh media in the absence of additional MF after 8–12 hr.[43] However, B27+AS+ cell lines continuously express the cross-reactive determinant in culture in the complete absence of exogenous MF.[44] In order to reconcile these facts, we have proposed that the B27+AS+ cells have permanently acquired the genetic element that codes for the cross-reactive determinant, whereas the modification of B27+AS− cells *in vitro* is mediated by the product of this genetic element, i.e., MF.[45] The permanent modification of B27+AS+ cells could have occurred through an interaction between a member of the bowel flora carrying the plasmid coding for MF and a B27+ susceptible cell, resulting in the transfer of genetic material from the bacterium to the human cells. This idea represents a novel pathogenic mechanism, and although unconventional, it is fully consistent with our results. The verification of this controversial hypothesis remains a major research priority.

This postulate is supported by data demonstrating the presence and maintenance of specific cross-reactive bacteria in the bowel flora of all B27+AS+ patients tested,[34,46] and it is likely that the majority of bowel organisms from B27+AS+ patients express the MF determinant. Repeated sampling of a number of B27+AS+ patients revealed that the cross-reactive organisms persist over long periods.[34] In contrast, cross-reactive organisms were isolated from only one of 35 B27−AS− and one of 20 B27+AS− individuals.[34] The intriguing fact that the cross-reactive organisms are stably maintained in B27+AS+ patients remains unexplained but may suggest that the infectious process in AS may result from a subtle, ongoing pathogenetic process rather than an acute infectious episode typical of many other seronegative arthropathies.

4. EVIDENCE FOR HLA-B27-RESTRICTED CYTOTOXIC T LYMPHOCYTES IN THE PATHOGENESIS OF ANKYLOSING SPONDYLITIS AND OTHER SERONEGATIVE ARTHROPATHIES

One possible consequence of the modification of B27-positive cells *in vitro* may be the production of B27-restricted cytotoxic T lymphocytes (CTL), which could kill target cells bearing the B27-associated modified determinant. About ten years ago we reported that CTL, raised by stimulating the PBMC of an HLA-B27-positive clinically normal individual

(B27+AS−) with the PBMC of an HLA-identical sibling suffering from AS (B27+AS+), will specifically lyse B27+AS+ PBMC but not PBMC from HLA-B27+ (B27+AS−) or HLA-B27− (B27−AS−) normal individuals.[47] Moreover, the "disease-specific" CTL will lyse B27+AS− PBMC that have been modified *in vitro* with culture filtrate from one of our arthritogenic bacteria (e.g., *Klebsiella, Salmonella, Shigella*). The CTL of similar specificity can also be raised by immunizing B27+AS− PBMC with autologous cells modified *in vitro* by certain arthritogenic antigens.[47] Furthermore, we have also isolated CTL of similar specificity from the peripheral blood of B27+AS+ patients.[48]

In a recent study Herman and coworkers[49] tested a panel of 354 CD8+ T-lymphocyte clones (TLCs) derived from the synovial fluid of four patients with reactive arthritis and two patients with AS. In one patient with *Yersinia*-induced reactive arthritis, two TLCs were identified that specifically killed *Yersinia*-infected B27-positive target cells. In another patient with *Salmonella*-induced reactive arthritis, one B27-restricted CD8+ TLC that recognized both *Salmonella* and *Yersinia* was identified. Further, in five of the six patients autoreactive CTLs were found and showed B27-restricted killing of uninfected cell lines. These findings, together with our previous demonstration of B27-restricted CTL directed against a B27-modified determinant, should provide new insights into the nature of the arthritogenic determinant recognized by T cells as well as the significance of this recognition as part of an early event in the initiation of the spondyloarthropathies. While many questions still remain unanswered, such as why many HLA-B27-positive individuals do not develop arthritis following infection with arthritogenic bacteria, our CTL data and those of Hermann and associates[49] may represent an important clue in understanding the pathogenic mechanisms in these B27-associated diseases.

If the effector mechanisms suggested by our studies are significant in the pathogenesis of AS, then certain cell types in the affected sacroiliac region may present cytotoxic effector cells a more "recognizable" structure than cells in the vicinity of the peripheral joints. Another possibility is that the activity or the density of CTL, or both, may be greater in the sacroiliac region. The demonstration that neutrophils and monocytes from B27+AS+ individuals function as targets for these CTL (Sullivan and Geczy, unpublished results) raises the possibility that such cells might have some pathological relevance, as exaggerated inflammatory episodes are thought to play a major role in the development of AS.

Ileocolonoscopic studies of the distal small bowel have suggested that subclinical abnormalities are extremely common in AS.[50] This observation is relevant to our results on fecal carriage, as it could be proposed that the carriage of these nonpathogenic cross-reactive bacteria might initiate an inflammatory reaction at the level of the small bowel and that this subclinical damage may reflect an early pathological lesion in AS. Indeed, it is possible that these inflamed regions of the bowel may facilitate the transfer of either bacterial products or DNA into the circulation, thereby allowing them to elicit an effect in the susceptible individual. More extensive work both on the nature of the subclinical bowel abnormalities and on their possible relationship to gut microflora is required to adequately explore such a possibility.

5. CONCLUSIONS AND FUTURE DIRECTIONS

The significance of gastrointestinal infection and of the serological cross-reactivity between certain organisms and B27+AS+ cells in the clinical manifestations of AS has yet to be elucidated. Fundamentally different theories of cross-reactivity as advocated by the Ebringers and our group will remain theories unless specific pathological processes can be related to either. Our own "altered self" theory of cross-reactivity provides a possible disease mechanism in the generation of specific CTL which is amenable to experimental verification. It is noteworthy that the specific CTL activity is essentially identical to the pattern of reactivity seen with the cross-reactive antibacterial sera. The next step in this part of the work is to clone CTL against HLA-B27-associated determinants with the expectation that cloned CTL will provide us with a more reliable typing reagent than the rabbit antisera that we have been using for the serological detection of these determinants. As well as typing for B27-associated determinants on the cells of AS patients, CTL should enable us to detect putative disease-associated structures on the cells of normal HLA-B27-positive family members of AS patients who harbor cross-reactive bowel organisms but who do not express the serologically defined cell-surface modification. The PBMC of these family members will also be used as stimulator cells, with the PBMC of other family members acting as responder cells. The object of these *in vitro* stimulations would be to induce CTL to serologically "latent" B27-associated determinants. It is anticipated that these CTL may uncover, on the cells of clinically normal individuals, hitherto undetected B27-associated determinants that may precede the serologically detectable antigens; that is, these determinants may be an early clinical manifestation of disease. Ultimately, molecular biological techniques should contribute significantly to clarifying the mechanisms controlling disease susceptibility in AS. Moreover, such techniques will simultaneously clarify the precise role of HLA-B27 and environmental triggering agents in the etiology of AS.

REFERENCES

1. Keat, A., 1986, Is spondylitis caused by Klebsiella? *Immunol. Today* **7**:144–149.
2. Tiwari, J. L., and Terasaki, P. I., 1985, *HLA and Disease Associations*, Springer-Verlag, New York, pp. 85–98,
3. Ebringer, A., 1978, The link between genes and disease, *New Scientist* **9**:865–868.
4. McDevitt, H. O., and Bodmer, W. F., 1974, HLA, immune response genes and disease, *Lancet* **i**:1269–1275.
5. Benacerraf, B., and McDevitt, H. O., 1972, Histocompatibility-linked immune response genes. A new class of genes that controls the formation of specific immune responses has been identified, *Science* **175**:273–279.
6. Georgopoulous, K., Carson-Dick, W., Goodacre, J. A., and Pain, R. H., 1985, A reinvestigation of the cross-reactivity between Klebsiella and HLA-B27 in the aetiology of ankylosing spondylitis, *Clin. Exp. Immunol.* **62**:662–671.
7. Breuning, M., Lucas, C. J., Breuer, B. S., Engelsma, M. Y., de Lange, G. G., Dekker, A. J., Biddison, W. E., and Ivanyi, P., 1982, Subtypes of HLA-B27 detected by cytotoxic T lymphocytes and their role in self-recognition, *Hum. Immunol.* **5**:259–268.
8. Toubert, A., Gomard, E., Grumet, F. C., Amor, B., Muller, J. Y., and Levy, J. P., 1984, Identification of several functional subgroups of HLA-B27 by restriction of the activity of antiviral T killer lymphocytes, *Immunogenetics* **20**:513–525.
9. Aparicio, P., Vega, M. A., and Lopez de Castro, J. A., 1985, One allogeneic cytolytic T-lymphocyte clone distinguishes three different HLA-B27 subtypes: Identification of amino acid residues influencing the specificity and avidity of recognition, *J. Immunol.* **135**:3074–3081.
10. Molders, H. H., Breuning, M. H., Ivanyi, P., and Ploegh, H. L., 1983, Biochemical analysis of variant HLA-B27 antigens, *Hum. Immunol.* **6**:111–117.
11. Choo, S. Y., Seyfreid, C., Hansen, J. A., and Nepom, G. T., 1986, Tryptic peptide mapping identifies structural heterogeneity among six variants of HLA-B27, *Immunogenetics* **23**: 409–412.
12. DeWaal, L. P., Krom, F. E. J. M., Breur-Vriesendorp, B. S., Engelfriet, C. P., Lopez de Castro, J. A., and Ivanyi, P., 1987, Conventional alloantisera can recognise the same HLA-B27 polymorphism as detected by cytotoxic T lymphocytes, *Hum. Immunol.* **20**: 265–271.
13. Breur-Vriesendorp, B. S., Dekker-Saeys, A. J., and Ivanyi, P. P., 1987, Distribution of HLA-B27 subtypes in patients with ankylosing spondylitis: The disease is associated with a common determinant of the various B27 molecules, *Ann. Rheum. Dis.* **46**:353–356.
14. Hill, A. V., Allsopp, C. E., Kwiatkowski, D., Anstey, N. M., Greenwood, B. M., and McMichael, A. J., 1991, HLA class I typing by PCR: HLA-B27 and an African B27 subtype, *Lancet* **337**:640–642.
15. Karr, R. W., Habin, Y., and Schwartz, B. D., 1982, Structural identity of human histocompatibility leukocyte antigen-B27 molecules from patients with ankylosing spondylitis and normal individuals, *J. Clin. Invest.* **69**:443–450.
16. Coppin, H. L., and McDevitt, H. O., 1986, Absence of polymorphism between HLA-B27

genomic exon sequences isolated from normal donors and ankylosing spondylitis patients, *J. Immunol.* 137:2168–2172.

17. Ebringer, A., Baines, M., and Ptaszynska, T., 1985, Spondyloarthritis, uveitis, HLA-B27 and *Klebsiella, Immunol. Rev.* 86:101–116.

18. Van Bohemen, C. G., Grumet, F. C., and Zanen, H. C., 1984, Identification of HLA-B27 M1 and M2 cross-reactive antigens in *Klebsiella, Shigella* and *Yersinia, Immunology* 52: 607–610.

19. Ogasawara, M., Koon, D. H., and Yu, D. T. Y., 1986, Mimicry of human histocompatibility HLA-B27 antigens by *Klebsiella pneumoniae, Infect. Immun.* 51:901–908.

20. Schwimmbeck, P. L., Yu, D. T. Y., and Oldstone, M. A., 1987, Autoantibodies to HLA-B27 in the sera of HLA-B27 patients with ankylosing spondylitis and Reiter's syndrome, *J. Exp. Med.* 1166:173–181.

21. Kapasi, K., Chui, B., and Inman, R. D., 1992, HLA-B27/microbial mimicry: An *in vivo* analysis, *Immunology* 77:456–461.

22. Scofield, R. H., Warren, W. L., Koelsch, G., and Harley, J. B., 1993, A hypothesis for the HLA-B27 immune dysregulation in spondyloarthropathy: Contributions from enteric organisms, B27 structure, peptides bound by B27, and convergent evolution, *Proc. Natl. Acad. Sci. U.S.A.* 90:9330–9334.

23. Madden, D. R., Gorga, J. C., Strominger, J. L., and Wiley, D. C., 1991, The structure of HLA-B27 reveals nonamer self-peptides bound in an extended conformation, *Nature* 353:321–325.

24. Ohno, S., 1992, How cytotoxic T-cells manage to discriminate nonself from self at the nonapeptide level, *Proc. Natl. Acad. Sci. U.S.A.* 89:4643–4647.

25. Sheldon, P., 1985, Specific cell-mediated responses to bacterial antigens and clinical correlations in reactive arthritis, Reiter's syndrome and ankylosing spondylitis, *Immunol. Rev.* 86:5–25.

26. Geczy, A. F., Prendergast, J. K., Sullivan, J. S., Upfold, L. I., McGuigan, L. E., Bashir, H. V., Prendergast, M., and Edmonds, J. P., 1987, HLA-B27, molecular mimicry, and ankylosing spondylitis: Popular misconceptions, *Ann. Rheum. Dis.* 46:171–172.

27. Beaulieu, A. D., Rousseau, F., Israel-Assayag, E., and Roy, R., 1983, *Klebsiella* related antigens in ankylosing spondylitis, *J. Rheumatol.* 10:102–105.

28. Kinsella, T. D., Fritzler, M. J., and McNeil, D. J., 1983, Ankylosing spondylitis. A disease in search of microbes, *J. Rheumatol.* 10:1–4.

29. Benjamin, R., and Parham, P., 1990, Guilt by association: HLA-B27 and ankylosing spondylitis, *Immunol. Today* 11:137–142.

30. Archer, J. R., Stubbs, M. M., Currey, H. L. F., and Geczy, A. F., 1985, Antiserum to Klebsiella K43 BTS1 specifically lyses lymphocytes of HLA-B27-positive patients with ankylosing spondylitis from a London population, *Lancet* i:344–345.

31. Van Rood, J. J., Van Leeuwen, A., Ivanyi, P., Cats, A., Breur-Vriesendorp, B. S., Dekker-Saeys, A. J., Kijlstra, A., and van Kregten, E., 1985, Blind confirmation of Geczy factor in ankylosing spondylitis, *Lancet* ii:943–944.

32. Geczy, A. F., Van Leeuwen, A., Van Rood, J. J., Ivanyi, P., Breur, B. S., and Cats, A., 1986, Blind confirmation in Leiden of Geczy factor on the cells of Dutch patients with ankylosing spondylitis, *Hum. Immunol.* 17:239–245.

33. Prendergast, J. K., Sullivan, J. S., Geczy, A., Upfold, L. I., Edmonds, J. P., Bashir, H. V., and Reiss-Levy, E., 1983, Possible role of enteric organisms in the pathogenesis of ankylosing spondylitis and other seronegative arthropathies, *Infect. Immun.* 41:935–941.

34. Prendergast, J. K., McGuigan, L. E., Geczy, A. F., Kwong, T. S. L., and Edmonds, J. P., 1984, Persistence of HLA-B27 cross-reactive bacteria in bowel flora of patients with ankylosing spondylitis, *Infect. Immun.* 46:686–689.

35. Geczy, A. F., Alexander, K., Bashir, H. V., and Edmonds, J., 1980, A factor(s) in *Klebsiella* culture filtrates specifically modifies an HLA-B27-associated cell-surface component, *Nature* 283:782–784.

36. Wang, J., Yu, D. T. Y., Fukazawa, T., Kellner, H., Wen, J., Cheng, X.-K., Roth, G., Williams, K. M., and Raybourne, R. B., 1994, A monoclonal antibody that recognises HLA-B27 in the context of peptides, *J. Immunol.* 152:1197–1205.

37. Sullivan, J., Upfold, L., Geczy, A. F., Bashir, H. V., and Edmonds, J. P., 1982, Immuno-chemical characterisation of *Klebsiella* antigens which specifically modify an HLA-B27-associated cell-surface component, *Hum. Immunol.* 5:295–307.

38. Upfold, L. I., Sullivan, J. S., and Geczy, A. F., 1986, Biochemical studies on a factor isolated from *Klebsiella* K43-BTS1 that cross-reacts with cells from HLA-B27-positive patients with ankylosing spondylitis, *Hum. Immunol.* 17:224–238.

39. Orban, P., Sullivan, J. S., Geczy, A. F., Upfold, L. I., Coulitis, N., and Bashir, H. V., 1983, A factor shed by lymphoblastoid cell lines of HLA-B27 positive patients with ankylosing spondylitis, specifically modifies the cells of HLA-B27-positive normal individuals, *Clin. Exp. Immunol.* 53:10–16.

40. Cameron, F. H., Russell, P. J., Sullivan, J., and Geczy, A. F., 1983, Is a *Klebsiella* plasmid involved in the aetiology of ankylosing spondylitis in HLA-B27-positive individuals? *Mol. Immunol.* 20:563–566.

41. Sullivan, J. S., and Geczy, A. F., 1985, The modification of HLA-B27-positive of HLA-B27-positive lymphocytes by the culture filtrate of *Klebsiella* K43 BTS1 is a metabolically active process, *Clin. Exp. Immunol.* 62:672–677.

42. Sullivan, J. S., and Geczy, A. F., 1986, The modification of HLA-B27-positive lymphocytes by the culture filtrate of Klebsiella K43 BTS 1: Influence of metabolic inhibitors, *Int. Rev. Clin. Sci.* 14:1151–1152.

43. Geczy, A. F., Alexander, K., Bashir, H. V., Edmonds, J. P., Upfold, L., and Sullivan, J., 1983, HLA-B27, *Klebsiella* and ankylosing spondylitis: Biological and chemical studies, *Immunol. Rev.* 70:23–50.

44. Alexander, K., Edwards, C., Misko, I. S., Geczy, A. F., Bashir, H. V., and Edmonds, J. P., 1981, The distribution of a specific HLA-B27-associated cell surface component on the tissues of patients with ankylosing spondylitis, *Clin. Exp. Immunol.* 45:158–164.

45. Sullivan, J. S., Prendergast, J. K., and Geczy, A. F., 1983, The aetiology of ankylosing spondylitis: Does a plasmid trigger the disease in genetically susceptible individuals? *Hum. Immunol.* 6:185–187.

46. McGuigan, L. E., Prendergast, J. K., Geczy, A. F., Edmonds, J. P., and Bashir, H. V., 1986, Significance of non-pathogenic cross-reactive bowel flora in patients with ankylosing spondylitis, *Ann. Rheum. Dis.* 45:566–571.

47. Geczy, A. F., McGuigan, L. E., Sullivan, J. S., and Edmonds, J. P., 1986, Cytotoxic T lymphocytes against disease-associated determinant(s) in ankylosing spondylitis, *J. Exp. Med.* 164:932–937.

48. Edwards, C., Sullivan, J., Geczy, A., McGuigan, L., and Edmonds, J., 1988, *In vivo* CTL derived from HLA-B27 positive patients with ankylosing spondylitis, *Aust. Soc. Immunol.* (abstract):179.

49. Hermann, E., Yu, D. T. Y., Meyer Zum Buschenfelde, K.-H., and Fleischer, B., 1993, HLA-B-27-restricted CD8 T cells derived from synovial fluids of patients with reactive arthritis and ankylosing spondylitis, *Lancet* 342:646–650.

50. Mielants, H., Veys, E. M., Goemaere, S., Cuvelier, C., and DeVos, M., 1993, A prospective study of patients with spondyloarthropathy with special reference to HLA-B27 and to gut history, *J. Rheumatol.* 20:1353–1358.

4

Triggering of Autoimmune Antibody Responses in Syphilis

ROBERT E. BAUGHN

1. INTRODUCTION

Early stimulation of the humoral and cellular immune response is clearly one of the hallmarks of natural and experimental syphilis.[1] The depth and breadth of the humoral immune response in the early stages of disease is represented by antiphospholipid/anticardiolipin antibodies, antitreponemal antibodies (the majority exhibiting extensive cross-reactions with other pathogenic and nonpathogenic treponemes), and a vast array of autoantibodies to blood cells, serum components, and tissue constituents (Table I). Although certain autoimmune diseases may have a microbial etiology, syphilis is unquestionably an infectious disease. It is not, nor does it represent, an autoimmune state. The varied autoimmune antibody responses associated with or accompanying infection do, however, appear to be triggered as secondary reactions. Were it not for the fact that *Treponema pallidum* subsp. *pallidum* is capable of inciting immunopathologic mechanisms, it would be easy to discount the presence of autoantibodies as little more than the end result of an infection-related immune injury. In fact, tissue destruction alone can favor the induction

ROBERT E. BAUGHN • V. A. Medical Center, Houston, Texas 77030-4298.

Microorganisms and Autoimmune Diseases, edited by Herman Friedman *et al.* Plenum Press, New York, 1996.

TABLE I
Types of Autoantibodies
Detected in Natural and
Experimental Syphilis[a]

Antiphospholipid antibodies
Cold autoantibodies to erythrocytes
Cold autoantibodies to lymphocytes
Rheumatoid factors
Cryoglobulins
Antifibronectin antibodies
Anticollagen antibodies
Antilaminin antibodies
Anti–creatine kinase antibodies

[a]See Ref. 2 for additional information.

of autoimmune antibodies if host proteins are sufficiently altered through the enzymatic actions of an offending organism. Although the controversy concerning the nature of the cardiolipin responsible for elicitation of anticardiolipin antibodies reactive in the Wassermann and Venereal Disease Research Laboratories (VDRL) type tests has not yet been resolved, the inciting event may simply be the release of cardiolipin from infected host cells via enzymatic action.[2] As isotype switching (IgM → IgG) and increases in affinity are seen with respect to other autoantibodies in this disease (discussed below), alternative triggering mechanisms must be considered. These include conformational changes in host proteins through attachment and binding, the possible involvement of multiple mechanisms in the evolution of unlinked responses, and flaws in immunoregulatory mechanisms. Several factors or mechanisms appear to impact immunoregulation and incite autoimmune responses;[3] these are listed in Table II. My intent in this chapter is to focus on three of these factors (polyclonal B-cell activation, idiotypic networks, and molecular mimicry), which in the early stages of syphilitic infection may contribute to flawed immunoregulation, rather than deal at length with the various autoantibodies detected. Emphasis on these factors stems from earlier work suggesting that a relationship exists between circulating immune complexes and specific autoimmune responses and that the latter may be elicited by different mechanisms.

There are a number of inherent problems in examining autoimmune events in syphilitic infections and subsequently attempting to determine whether the autoantibodies invoked have any pathogenic

TABLE II
Factors or Mechanisms that Possibly
Affect Immunoregulation
of Autoantibody Responses[a]

Polyclonal B-cell activation	Infectious agents
Intrinsic T-cell defects	Molecular mimicry
Suppressor cells	Heat-shock proteins
Idiotypic networks	Superantigens

[a]See Ref. 3 for additional information.

potential. As this microaerophilic spirochete cannot be propagated *in vitro* under standard cultural conditions, the infection of appropriate animal models with organisms harvested from orchitic testes creates a problem, especially if the preparations used for injection contain host proteins, enzymes, or cells. This problem may be compounded if guinea pigs are injected with treponemes from orchitic rabbit testes. Secondly, just as mice are less than ideal models for syphilitic infections, rabbits and guinea pigs are not the models of choice for examining autoimmunity. Our present knowledge of essentially all of the basic principles of auto-immunity stems from experiments in laboratory mice, including those with intrinsic abnormalities and transgenic and gene knockout mice. Despite these obstacles, one has to rely principally on animal studies. Ethical considerations preclude studying the disease in man over a pro-tracted time course. In addition, the significance of seroepidemiological results for a sexually transmitted disease and for an organ-specific auto-immune disorder are not comparable. With respect to the latter, parame-ters such as clustering within families or communities and geographical distributions may be very important. Although the obstacles related to the animal models are not insuperable, they have made it difficult to identify potentially pathogenic autoantibodies in syphilis. Much of the evidence is indirect or inferred from similar findings in other autoimmune states. Pathogenic autoantibodies appear to result from mechanisms that medi-ate B-cell responses to conventional antigenic stimuli. Thus, they are antigen-driven, influenced by T-cell help, and forced to undergo clonal selection. The prominent features of clonal selection are class switching, increased antibody affinity, and somatic hypermutation of immunoglobu-lin variable region genes as demonstrated in murine and human systemic lupus erythematosus (SLE).[4-6]

2. OVERVIEW OF THE DISEASE

Syphilis is an excellent example of an acute symptomatic infection, which if allowed to progress, will ultimately result in a chronic bacterial disease syndrome.[7] Progression of the disease in humans through several distinct stages, in the absence of treatment, attests to this fact. The appearance of specific antitreponemal antibodies during the first week of infection (primary syphilis) constitutes evidence that immunorecognitive and stimulatory forces are invoked early in the disease. Yet the organisms are not eradicated; *T. pallidum* persists in the tissues and the clinical manifestations of disease progress despite the development of profound humoral and cellular immune responses. The secondary stage is accompanied by recognition of spirochetal antigens by an expanded number of antibodies; such an expansion would not be possible if the treponemes were completely eradicated. This fact, together with the induction of autoimmune antibodies to several endogenous host proteins and the subsequent formation of circulating immune complexes (IC) in both natural and experimental syphilis,[8–10] led us to postulate that the hosts' failure to eradicate the organism is a reflection of abnormalities in immunoregulation. In terms of possible mechanisms, we reasoned that the interaction of immune complexes with Tγ cells might lead to an irreversible loss of Fc receptors for IgG. As in several parasitic infections,[11–14] the chronicity of latent syphilis with persistence of treponemes suggests that the immune system and the organism evolve into a stable, balanced host-parasite relationship. Maintenance and control of this relationship appears to be dependent on immunoregulatory interactions which involve relevant idiotypic relationships (discussed below).

3. IMMUNE COMPLEXES AND AUTOIMMUNE RESPONSES IN SYPHILIS

While secondary syphilis in man is often cited as a classical example of a type III hypersensitivity reaction or an immune complex disorder,[15] no direct evidence, prior to 1980, had implicated a functional role for IC in mediating the lesions seen in this disease. Moreover, the possibility that these immune complexes contained endogenous host antigens, and therefore merely reflected manifestations of the autoimmune phenomena known to accompany syphilitic infection, had not been examined.

Those considerations led us to study systematically the incidence of elevated IC levels in *T. pallidum*–infected rabbits and later to characterize the putative host and treponemal antigens present in purified IC.[8-10] The results of those studies, which demonstrated the presence of treponemal antigens, host antigens, and antibodies to those components in IC from infected rabbits, provided a framework for later experiments aimed at characterizing IC from the sera of patients with secondary syphilis,[16] infants with congenital syphilis,[17] and a second animal model, the guinea pig.[18]

Common threads appeared to be woven into the fabric of purified immune complexes from the sera of all infected species. Demonstration of an 83-kDa fibronectin (Fn) binding protein of treponemal origin, Fn degradation products, and IgG antibodies directed against these components provided us with the first direct link between the induction of autoimmune antibodies and IC formation. In addition, those results suggested that the importance of Fn in syphilis, and particularly that of the RGD (Arg-Gly-Asp) amino acid sequence in the cell-binding domain of Fn, extended beyond that of mediation of cytoadherence of the organism.

Purified complexes from patients with secondary syphilis, infants with congenital syphilis, and animals with experimental infection also were found to contain creatine kinase (CK) and autoantibodies directed against this host component. The possible mechanism leading to the induction of anti–creatine kinase antibodies is still not known. In all likelihood it may simply reflect tissue destruction. Creatine kinase, like cardiolipin, appears to be released from affected mitochondria. Unlike Fn, specific *T. pallidum* ligands for this host protein have never been identified.

Intrigued by the fact that different control mechanisms might be responsible for both the induction and regulation of various humoral immune responses expressed during syphilitic infection, we embarked on quantitatively assessing specific antibody responses with respect to isotype profiles. Those studies led first to the discovery that secondary syphilis was characterized by coordinate, restricted expression of IgG1 and IgG3 responses to *T. pallidum* antigens.[19] Later, using similar approaches, autoimmune responses were quantified with respect to isotype profile in the hope of determining the extent of coordination between class and subclass responses to host components and those invoked against *T. pallidum*.[20] Anti-Fn and anti-CK responses were shown to be predominantly IgG1 and IgG3, much like the responses to treponemal antigens. This finding was interpreted as possibly reflecting basic functional and/or mechanistic linkages, as two of the most plausible explanations involved immaturity of the immune response or polyclonal B-cell activation.

That study also raised the question as to whether multiple mechanisms were involved in the evolution of the various autoimmune responses associated with syphilis. An inverse relationship was noted between anti-Fn IgG and antitreponemal IgG. This led us to postulate that
Fn, unlike CK and cardiolipin (as in the VDRL antigen) with its high
binding affinity for the 83-kDa antigen, might be presented differently
from a normally sequestered antigen which would be freed only after
tissue damage. In addition, anti-VDRL reactivity was predominantly IgM
with all of the detectable IgG confined to IgG1. Thus, the anticardiolipin
responses did not appear to be linked to specific antitreponemal responses or to the anti-CK and anti-Fn responses.

4. POLYCLONAL B-CELL ACTIVATION

Secondary syphilis in man and the intravenous (IV) or disseminated
rabbit model, much like the acute stage of parasitic infections, such as
African and American trypanosomiasis and malaria, are associated with
polyclonal B-cell activation and immune complex–like symptomatology.[7]
As mentioned above, this triad of events preceding latency constitutes
evidence that normal immune functions are skewed in favor of the
organism. Until recently, much of the evidence for polyclonal B-cell
activation was essentially indirect, despite the fact that hypergammaglobulinemia in syphilis had been well established.[21] As described below,
the results of our epitope mapping studies with the 15- and 17-kDa
lipoproteins of *T. pallidum*, designated TpN15 and TpN17, respectively,
strongly supports the existence of polyclonal B-cell activation throughout
the first few weeks of a syphilitic infection. Alternatively, the hypergammaglobulinemia might simply reflect the symmetric expansion of complementary B-lymphocyte clones within the idiotypic network or the unregu-
lated proliferation of a relatively small number of clones. Of the 15
patients with secondary syphilis who were examined in one of our earlier
studies,[16] three exhibited IgG levels > 3 SD above the normal mean and
four others had IgG levels > 2 SD above the normal mean. Although IgA
and IgM levels in four of the patients were above normal range values,
those slight elevations were not statistically significant. For 50 patients
with secondary syphilis who were examined since that time, the trend
continues (Baughn, unpublished results); of these, 34 have exhibited
significantly elevated IgG levels using a direct ELISA for quantitation. In
the IV rabbit model, concomitant elevations in circulating immunoglobulins, especially IgG, appear to be consistent with the hyperimmunoglobu-

linemia that develops during natural disease in man.[9] Slight to moderate increases in IgM levels were noted at days 14 and 21 postinfection. Expressed as a percent increase over preinfection baseline values, increases of 19.6 ± 12% and 16.3 ± 9% were seen at each time point, respectively. The IgG levels were significantly ($p < 0.05$) elevated on days 28 and 35, represented by increases of 35.7 ± 17% and 29.8 ± 13%, respectively. Again, even though direct ELISAs are currently used in my laboratory for quantitation of rabbit Ig levels, those general trends have not changed over the past 15 years.

Based on inferences from human and murine SLE and chronic parasitic infection, polyclonal B-cell activation usually results in autoantibodies with no apparent relevance to the underlying disorder. In addition, the resulting autoantibodies are predominantly IgM and thus not pathogenic. Almost all of the autoantibodies detected in primary syphilis and in the first few weeks of experimental syphilis meet this criteria. However, the mechanisms leading to polyclonal B-cell activation in syphilis are not known. Presumably, those unidentified activators would trigger lymphocytes to express their total V-gene repertoire, which, in turn, would activate self-reactive B cells to produce antibodies to self antigens.[22] In direct contrast, other investigators[23] have argued that the serologic response to infection develops slowly, since patients may be serononreactive many weeks after exposure. This opposing view is difficult to reconcile, except that there are many variables that possibly impact such a conclusion, such as the initial inoculum size and the sensitivity and type of test used. While only 80 to 86% of patients with primary syphilis are reactive in conventional, nontreponemal tests,[1] both the sensitivity and the specificity are much greater if the VDRL antigen is used in an ELISA.[24]

In early 1992, we began a series of B-cell epitope scanning studies on two of the smaller treponemal lipoproteins with known sequences, namely TpN15 and TpN17. These two proteins were chosen because of their potential diagnostic importance in congenital and primary syphilis; they are described in detail elsewhere.[25] Using previously described procedures,[26] our strategies for these mapping studies consisted of initially synthesizing overlapping peptides (decapeptides and octapeptides) covering the entire sequences of each lipoprotein. These "general net" syntheses were carried out using commercially available kits; peptides were synthesized in duplicate with Fmoc chemistry according to the software and manuals supplied with the kits. The peptides were then tested repeatedly in indirect ELISA with sera and immunoglobulin fractions from normal and syphilitic sera (both animal and human) to identify those B-cell stimulants that might be suitable for diagnostic purposes, in addition to those with the potential to elicit polyspecific antibodies capable of cross-reacting with host proteins by virtue of sequence homology (molecular mimicry). The second and third phases of our strategies consisted of resynthesizing "windows" of 5-, 6-, 7-, and 8-mers to define the boundaries of those epitopes of interest and then synthesizing peptides with alanine replacements in each position so as to determine which amino acid residues were critical for antibody binding.

Although the results obtained in subsequent antibody binding studies with these synthetic peptides failed to identify any epitopes on TpN15 and TpN17 that might be directly involved in molecular mimicry as it relates to the induction of autoimmune events, they did provide us with several important insights.[27] First, the results with both of these relatively small proteins confirmed that, in general, several linear or continuous epitopes were involved in polyclonal response. Figure 1 illustrates the fact that IgM antibodies to multiple epitopes of TpN15 were

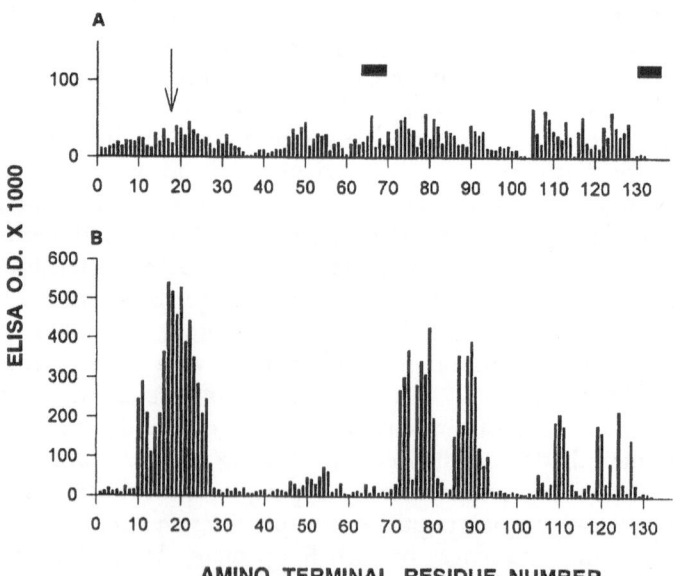

FIGURE 1. Reactivities of normal preinfection (A) and syphilitic 7 days after infection (B) rabbit IgM with 134 overlapping octapeptides covering the entire sequence of the 15-kDa lipoprotein of *Treponema pallidum* (TpN15). The arrow indicates the consensus signal peptidase II cleavage site which gives rise to a signal peptide and a mature protein consisting of 17 and 124 residues, respectively. The black bars represent the two predicted regions of B-cell determinants using the algorithm of Chou and Fasman.[36]

present within the first week of an IV infection in rabbits. These IgM responses were not restricted to only those overlapping peptides within the mature protein; responses were also seen against those within the signal peptide (first 17 peptides in Fig. 1). Specific IgM antibodies to several individual peptide motifs of TpN17 also were detectable within the first week of an experimental animal infection. Similar response patterns were seen with the TpN15 and TpN17 synthetic peptides and individual IgM fractions from the sera of patients with primary syphilis. Basically, these results constitute compelling evidence that polyclonal B-cell activation occurs early in syphilitic infection. With time, IgM responses appeared more narrowly restricted and peaked between 14 and 21 days postinfection. Maturation and isotype switching occurred at approximately the same time. Between days 21 and 56, IgG reactivity was principally directed against four B-cell epitopes within two regions of TpN15 spanning amino acid positions 49–62 and 70–99.

Identification of B-cell epitopes on those lipoproteins with predictive algorithms would have completely missed the mark. In fact, the immunodominant B-cell epitopes of TpN15 and TpN17 identified in our mapping studies did not even overlap with the predicted windows. Conserved sequences were identified which also must exist on avirulent treponemes, as extensive reactions were seen with the majority of the immunodominant B-cell epitopes and antisera to T. phagedenis biotype Reiter, T. refringens, and T. vincentii. With the various sets of synthetic peptides, testing for IgM reactivity required fractionation so as to avoid competing reactions with higher-affinity IgGs present in both animal and human sera.

Aside from the extensive cross-reactions seen when peptides were tested with antisera to avirulent organisms, immunization of animals with synthetic peptides corresponding to the immunodominant motifs or mimetics of those motifs resulted in polyspecific antibodies capable of recognizing treponemal proteins other than TpN15 and TpN17. As shown in Table III, affinity-purified IgGs directed against single peptide or mimetic motifs of TpN15 exhibited polyspecific responses in that they were capable of recognizing treponemal proteins other than the 15-kDa lipoprotein on immunoblots.

In all likelihood, the B-lymphocyte-stimulating activity is due to the lipoprotein nature of many of the membrane-associated polypeptides of T. pallidum.[25] Amino-terminal synthetic lipopeptide analogs of spirochetal antigens have been investigated as immunomodulators.[28] Although they were shown to be potent macrophage activators, inducing interleukins IL-1β, IL-6, and IL-12, their role in polyclonal B-cell activation was not examined. Even though the cytokine IL-12 tends to shift

TABLE III
Polyspecific Nature of Antibodies Raised
in Rabbits against Individual Peptide Motifs

Immunizing motif	Reactions with antigens[a] of *T. pallidum* on immunoblot
53MVQVVYD59	94, 83, 60, 47, 44.5
73DYARVMY79	94, 83, 60, 47, 44.5, 42, 37, 15
77VMYASSG83	94, 83, 60, 47, 44.5, 42, 37, 15
87EAAFREL93	94, 83, 60, 47, 44.5

[a]Weights are expressed in kilodaltons (kDa).

cellular responses toward Th1 cells, it would seem that in early syphilis, secretion of IL-12 alone is not sufficient to retard the profound antibody induction by Th2 cells.

5. IDIOTYPIC NETWORKS

The network theory of immune regulation, first proposed by Jerne,[29] suggests that idiotype (Id)–anti-Id reactions control the response of a host to an antigen via a positive (enhancing) or negative (suppressing) feedback mechanism. Despite numerous studies which had implicated Id networks in regulation of the immune response to a variety of antigens, infectious agents, and tumors, the role of anti-Id responses in syphilitic infection had not been explored. This seemed unfortunate, as anti-Id responses undoubtedly affect the induction and downregulation of both humoral and cellular responses to *T. pallidum*. More importantly, as network controls also regulate the formation of autoantibodies, we deemed it imperative to examine the possibility that the autoimmune responses seen in syphilis might result from flaws in immunoregulatory mechanisms, principally involving the Id network. Drawing upon hypotheses concerned with functional Id networks, we reasoned that autoantibodies to host proteins, particularly to Fn, might arise from perturbations in the idiotypic network. Conceivably, as diagrammed in Fig. 2, anti-Fn antibodies might in fact be antiidiotypic (anti-Id) antibodies to antitreponemal antibodies, specifically those directed against the 83-kDa, Fn-binding protein or receptor of *T. pallidum*. In addition, it seems plausible that anti-Id antibodies might suppress efficient immune responses to *T. pallidum* by binding to and blocking idiotypes from reacting with the antigen binding sites on this or other treponemal proteins.

A   *Treponema pallidum* - FIBRONECTIN INTERACTIONS:

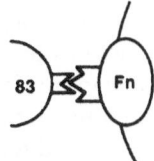

83-kDa protein of
T. pallidum **(83) binds**
fibronectin (Fn)

B   INDUCTION/GENERATION OF FIRST ANTIBODY (Ab-1) DIRECTED
AGAINST THE 83-kDa PROTEIN OF *T. pallidum:*

anti-83 binds the
83-kDa protein

C   INDUCTION/GENERATION OF ANTI-ANTI-83 OR ANTI-Id (Ab-2 $_\beta$)
DIRECTED AGAINST THE Id OF Ab-1:

anti-Id binds the
Id of Ab-1

D   THE Ab-2$_\beta$ COULD BLOCK THE Ab-1 FROM REACTING WITH THE
83-kDa PROTEIN

OR

Ab-2$_\beta$ COULD BIND TO FIBRONECTIN AND MIMIC AUTOIMMUNE ANTIBODY

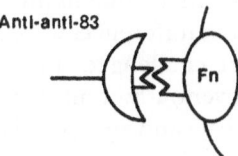

FIGURE 2. Figurative explanation of one possible mechanism involving the Id network in
secondary syphilis. (From Ref. 29.)

Our initial Id–anti-Id study[30] was aimed at determining whether naturally occurring autologous anti-Id (autoanti-Id) antibodies to *T. pallidum*-specific $F(ab')_2$ fragments could be demonstrated at any time during the first 6 months of experimental syphilis or within a comparable period after reinfection. Such experiments were considered prerequisites to undertaking studies designed to assess the role of anti-Id in resistance to reinfection in this model and the impact of the Id network on auto-immune and *T. pallidum*-specific responses in human infection. In that study, IgGs from serial bleeds were obtained one week to 5 months after infection of rabbits with *T. pallidum* and examined for autologous anti-Id antibodies. The capacities of IgGs to bind radiolabeled ^{125}I-$F(ab')_2$ of anti-*T. pallidum* IgG prepared from earlier bleeds were determined using radioimmunoassays. Autoanti-Id IgGs were found to be demonstrable between 60 and 180 days postinfection. Following reinfection, 90 days after the initial infection, autoanti-Id IgGs peaked 6 to 8 weeks later, even though they were first noted as early as 2 weeks postreinfection. In the reinfection model the autoanti-Id continued to persist throughout the remaining 3 months of the study. Since autoanti-Id antibodies to anti-Fn antibodies also were demonstrable in reinfected animals, those findings raised questions regarding the possible role of the anti-Id in modulating the Id expression of both *T. pallidum*-specific and autoimmune humoral responses. Overall, the success achieved in that study fulfilled the first criterion of the network theory in that autoanti-Id antibodies were shown to be an integral part of the normal immune response of rabbits experi-mentally infected with *T. pallidum.*

Subsequent attempts were made to specifically address the hypoth-esis that anti-Id to anti-83-kDa antibodies might mimic autoantibodies to the amino acid sequence (Arg-Gly-Asp) *RGD* site of Fn.[7] Rabbits were immunized with either affinity-purified 83-kDa antigen or the synthetic Fn-7 peptide, KYG*RGD*S and subsequently challenged with *T. pallidum.* Compared with nonimmunized and sham-immunized control rabbits, accelerated lesion development was noted in the rabbits immunized with the 83-kDa antigen. Surprisingly, profound differences were noted in the animals immunized with Fn-7 and later challenged. In those animals challenged with 4×10^7 organisms intravenously, a minimal number of lesions (< 20) developed with delayed onset and an atypical appearance. Unlike the control animals, immune complexes were not detectable in serum samples obtained from these animals between the third and tenth week following infection. In the Fn-7-immunized rabbits challenged in-tradermally at four sites with 10^6 *T. pallidum,* lesion development also was delayed and atypical since lesions were smaller, were minimally ulcerated, and healed rapidly. Anti-Id to anti-RGD F(ab')$_2$ were demonstrable within 2 weeks following infection of Fn-7-immunized rabbits. In addition, binding reactions between Id and anti-Id were significantly inhibited by free Fn and RGD-containing peptides. Chemical cross-linking experiments indicated that anti-Id also reacted with the 83-kDa antigen. We also found that anti-Ids from these *T. pallidum*–challenged animals were capable of suppressing RGD-induced lymphocyte proliferation. Compared with non-challenged Fn-7-immunized rabbits, the induction and development of anti-Ids in the challenged animals was accelerated.

Collectively, those results suggested that downregulation of a specific autoimmune response could affect directly the progression of syphilis, at least in an experimental model. Such findings were not without precedent. Patients with systemic lupus erythematosus[31] in early remission had been shown to develop detectable serum levels of anti-DNA anti-Ids, which are not present during active phases of disease. In contrast, a reciprocal arrangement of DNA antibody (Id) and autoanti-Id seems to exist in patients with severe SLE, implying failure of the network. Even though these studies directly implicate network interactions in syphilis in downregulating autoreactive clones, there is little evidence that they contribute to the elicitation of cross-reactive autoantibodies.

This in turn raised several intriguing questions. Are generalized lesions, such as those present in secondary syphilis and those best exemplified by the disseminated rabbit model, indicative of the network failure? Network failure, in this context, would be best described as the host's inability to downregulate polyspecific antibodies to specific epitopes, that is, those antibodies having the potential to cross-react with host proteins and initiate damage. Is lesion resolution associated with restoration of network function? If so, does this represent an attempt to bring the varied autoimmune responses in syphilitic infection under control and restore balance/self-tolerance? Are the occasional relapses which are seen following clinical recovery or the establishment of latency also indicative of network failure? Is "chancre immunity" in the rabbit a valid model of protection or does it perhaps more accurately reflect restoration of self-tolerance? In other words, are polyspecific cross-reactive antibodies major players in lesion formation and progression?

6. MOLECULAR MIMICRY

According to the hypothesis advanced to describe molecular mimicry in a variety of bacterial and viral diseases,[32] autoimmune responses

may develop as the result of structural similarities between self determinants on host antigens and antigenic determinants (epitopes) of the
organism, which in turn elicit adverse host reactions. The relationship
between infection with specific "rheumatogenic" strains of group A
streptococci, the development of a vigorous humoral response to antigens in the bacterial capsule, cell wall, and cell membrane, and the
subsequent development of rheumatic fever and carditis has been extensively studied and well reviewed.[33] Rheumatic fever is prototypic of diseases that cause tissue injury because of an unfortunate biologic accident
resulting from molecular mimicry.

Although such a mechanism had been described in another spirochetal infection, Lyme disease,[34] its role in syphilitic infections had not
been examined. To address this possibility, as mentioned above, our
research in early 1992 was directed at comprehensively mapping B-cell
epitopes on lipoproteins of T. pallidum. Following the mapping studies of
the TpN15 and TpN17, which failed to identify any linear sequences with
the potential to elicit autoimmune responses, our efforts were redirected
at the immunodominant 47-kDa (TpN47) lipoprotein. In sharp contrast,
the epitope mapping studies of this protein led to the early identification
of linear amino acid motifs which we believe are unquestionably ideal
paradigms of molecular mimicry in syphilis.

Our strategy for epitope mapping of TpN47, because of its size,
differed slightly from that taken with the smaller lipoproteins. As this
lipoprotein has 434 residues, our plan of attack consisted of first synthesizing 12-mers with an offset of four amino acids and an overlap of eight
amino acids so as to identify regions of B-cell reactivity and then at a later
time at confirming our findings by resynthesizing overlapping 8-mers
with an offset of 1 within those reactive regions for retesting of sera and Ig
fractions. Using this approach, 106 synthetic 12-mer peptides plus six
control peptides were synthesized in duplicate for epitope scanning
purposes. The completed peptides obtained in this manner were then
subjected to repeated antibody reactivity testing by the modified indirect
ELISA.

Figure 3 shows the results obtained in a representative epitope
scanning study with this set of noncleavable peptides. Altogether, these
peptides were tested in 37 ELISA using IgM fractions and sera (for IgG
reactivity) from syphilitic animals and patients (both pooled and individual samples).

Based on composite results from those individual ELISA we resynthesized overlapping octapeptides (offset = 1) within the immunodominant
regions identified in the above studies. Figures 4 and 5 show representa-

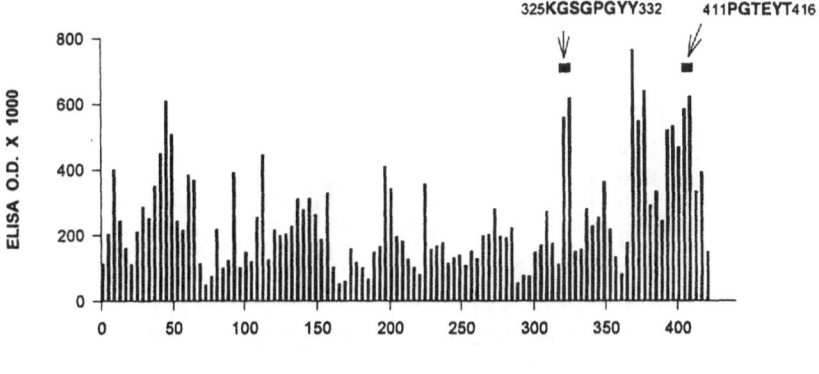

FIGURE 3. Epitope scan of 12-mer peptides (overlap = 8; offset = 4) encompassing the entire sequence of TpN47. Pooled IgG from six rabbits 38 days after IV infection with *T. pallidum* (1:400 dilution). The bifurcated arrows indicate in each case the pairs of 12-mers containing the designated sequences or motifs.

tive results obtained within two of the nine regions. As an adjunct to these epitope scanning studies, the immunoreactive sequences within these regions were then entered into the Hitachi DNASIS/PROSIS Analysis System for homology searches in the NBRF-PIR and SWISS-PROT databases. The modified results of those searches as shown in Figs. 6 and 7 seemed to directly implicate the 411PGTEYT416 and 325KGSGPGYY332 motifs of TpN47 as potential molecular mimics.

To further explore the importance of the 411PGTEYT416 motif as a probable culprit or instigator in the induction of polyspecific cross-reactive antibodies, peptides were resynthesized with alanine replacements in each of the positions throughout an 8–amino acid stretch, FTPGTEYT (amino acid positions 409 through 416 of the sequence as it occurs in TpN47). The results of six representative experiments with those alanine-substituted peptides are presented in Table IV. Basically, they suggest that within PGTEYT, the phe(F), glu(E), tyr(Y), and second thr(T) residues are critical for antibody binding and that the identification of antibody-binding contact residues is an important step toward understanding immune recognition at the molecular level.

Preliminary efforts have also been directed at examining the ontogeny or maturation of autoantibody responses to this culprit motif in syphilitic rabbits. As illustrated in Fig. 8, IgM antibodies to this motif peak 14 days after IV infection and are basically short-lived. In contrast, the IgG

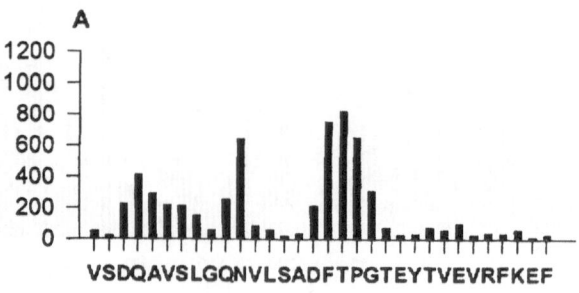

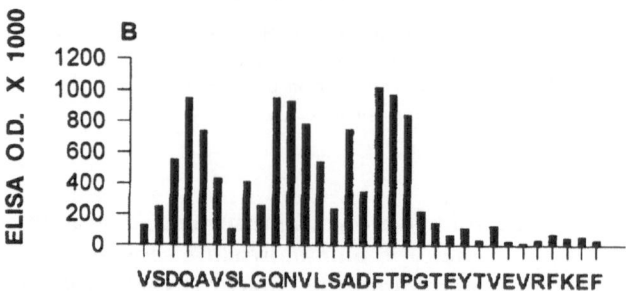

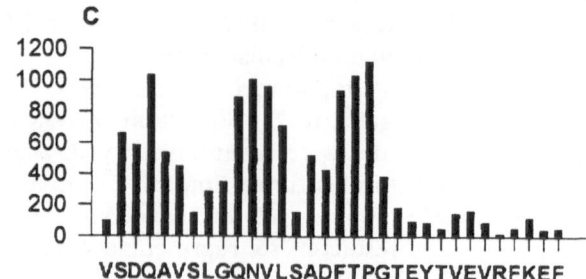

AMINO ACID RESIDUE

FIGURE 4. Epitope scan of octapeptides (overlap = 7; offset = 1) encompassing amino acids 393 through 424. (A) Pooled IgM from six rabbits 14 days after IV infection with *T. pallidum* (1:100 dilution). (B) Pooled IgG from six rabbits 35 days after IV infection with *T. pallidum* (1:400 dilution). (C) Pooled IgG from six patients with secondary syphilis (1:200 dilution).

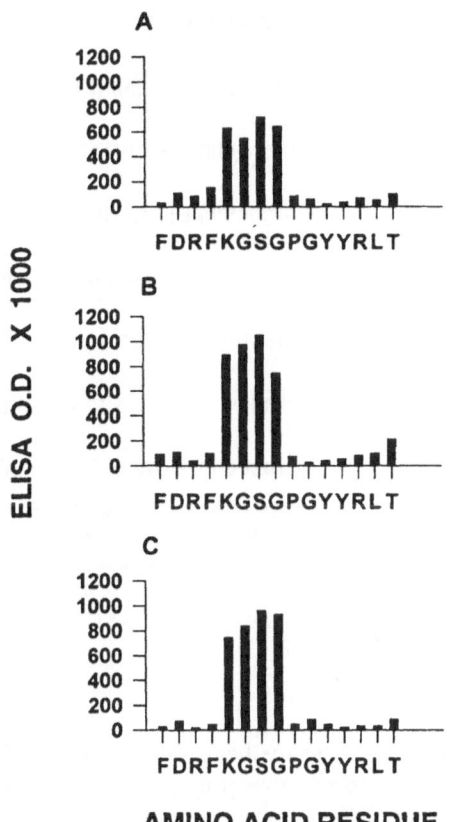

FIGURE 5. Epitope scan of octapeptides (overlap = 7; offset = 1) encompassing amino acids 320 through 335. (A) Pooled IgM from six rabbits 14 days after IV infection with *T. pallidum* (1:100 dilution). (B) Pooled IgG from six rabbits 35 days after IV infection with *T. pallidum* (1:400 dilution). (C) Pooled IgG from six patients with secondary syphilis (1:200 dilution).

response to the motif is biphasic. The first peak is at day 14 with lesser activity noted at day 21, the time at which the disseminated rash is prominent. The major IgG response peaks between days 35 and 56 postintravenous infection. Antibodies at 90 or 120 days postinfection had declined to baseline levels. More importantly, animals reinfected after 6 months and bled 1 to 4 weeks later failed to exhibit a typical anamestic response; the magnitude and duration of those responses were less than those seen following the initial infection. These results suggest that

411 **PGTEYT** 416   TpN47(A43852)
: : : : : :
1844 **PGTEYT**1849   Fibronectin, Bovine (FNBO)
: : : : : :
1965 **PGTEYT**1970   Fibronectin, Precursor, Human (FNHU)
: : . : : :
101 **PGSEYT**106   Fibronectin ED-A, Human, Fragment (S03917)
: : . : : :
85 **PGSEYT**90   Fibronectin, Guinea pig, Fragment (A40790)
: : : : :
867 **PGTEYK**872   Type XIV Collagen Alpha 1 Chain, (A45974)
: : : : :
582 **PGTEYK**587   Collagen Alpha 1[XIV] Chain (S22916)
: : : : :
472 **PGTEYV**477   Fibronectin, Bovine, Fragments (A23292)

FIGURE 6. Selected results of homology searches using the SWISS-PROT and NBRF-PIR databases and the PGTEYT motif. The above examples were selected from a total of 52 entries showing > 83.3% homology in five or more amino acid positions. Amino acid sequence identity with TpN47 is denoted by a colon above the residue, whereas sequence similarity is denoted by a dot. Nonidentity is denoted by the absence of a colon or a dot. The NBRF-PIR ascension numbers are in parentheses.

control mechanisms, possibly involving the Id network, develop late in infection to limit these cross-reactive, polyspecific antibodies with immunopathogenic potential.

The PGTEYT motif, as designated in Fig. 3, is located in the carboxyl-terminal region of the lipoprotein. Most of the reactivity against linear epitopes of this immunodominant lipoprotein is, in fact, located within this region. In disseminated syphilis, whether in humans or rabbits, the antibodies against this motif are predominantly IgG. Our hypothesis is that with the antigen-driven IgM switch to IgG, those IgG antibodies with greater binding affinity recognize and bind to the variant motifs on fibronectin and collagen that mimic the triggering epitope. Such events

→

FIGURE 7. Selected results of homology searches using the SWISS-PROT and NBRF-PIR databases and the KGSGPGYY motif. The above examples were selected from a total of 436 entries showing > 50% homology in five or more amino acid positions. Amino acid sequence identity with TpN47 is denoted by a colon above the residue and sequence similarity is denoted by a dot. Nonidentity is denoted by the absence of a colon or a dot. The NBRF-PIR ascension numbers are in parentheses.

325 **KGSGPGYY**332   TpN47 (A43852)
　　．．：：：：：
123 **SAAGPGYY**130   Ig Heavy Chain V Region, Human (S31589)
　　：：：．．：：
53 **PGSGSAYY**60   Ig Heavy Chain V-D-J Region Alpha CD4 (S19964)
　　：：：．．：：
53 **PGSGSNYY**60   Ig Heavy Chain V-D-J Region Alpha CD4 (S19969 & S19966)
　　：：　：：：：
389 **KGLGPGYL**396   Collagen Alpha 4(IV) Chain, Rabbit (A45137)
　　：：　：：：：
238 **KGFGPGYL**245   Collagen Alpha 4(IV) Chain, Human (S36854 and S28777)
　　：：：．：：
54 **HHSGPTYY**61   Ig Heavy Chain V Region, VH4DJH6 (S19668 & S24445)
　　：：：．．：：
33 **GGSGTTYY**40   Ig Heavy Chain V-A1 Region (H30517)
　　：：．：　：：
1813 **EGSNPPYY**1820   Proteoglycan 24K Core Protein Precurs (A60979)
　　．．．：：．：：
309 **RACGPDYY**316   Epidermal Growth Factor Receptor (A36325 & A43818)
　　．：：：：
180 **CICDPGYY**187   Adhesion Molecule LECAM-1, Rat (S23936)
　　．：：：：
1592 **ELCAPGYY**1599   Heparan Sulfate Proteoglycan HSPG2/PE (A38096)
　　：：．　．．：：
94 **KGKVSAYY**101   Ig Heavy Chains V-III Regions (H3HUTL)
　　：：：　．：：
53 **PGSGNTYY**60   Ig Heavy Chain V Region, Clone 26F.1 (PH1165)
　　．：：：：
915 **GTCDPGYY**922   Laminin Chain B2 Precursor, Mouse (MMMSB2)
　　：：：　．：：
53 **PGSGNTYY**60   Ig Heavy Chain V Region, Clone 202.10 (PH1000)
　　．：：：：
914 **LKVAPGYY**921   Integrin Beta-4 Chain Precursor (JN0786)
　　：：：　：：
436 **VFSGPRYY**443   Neutrophil Collagenase, EC 3.4.24.34 (KCHUN)
　　：：：　：：
243 **VFSGPRYY**250   Procollagenase, Human (S11026)
　　：：　．：：：
4 **GGSFSGYY**11   Ig Mu Chain, Human Fragment (S37454)
　　：：　．：：：
26 **GGSFSGYY**33   Ig Heavy Chain V Region, Human (S26806, S26805, S14474)
　　：：：．：：
502 **YGSGSGYG**509   Keratin K5, 58K Type II, Epidermal (A29904)
　　：：：：：
81 **PPSGPGYP**88   MAC-2 Antigen, Human (A36071)
　　：：：．：：
86 **GGSGAGYG**93   Keratin 3, Type 1, Cytoskeletal (KRXL)

TABLE IV
Results of Six Different ELISAs Using the Pins
from the Alanine Replacement Synthesis of the Parent Motif FTPGTEYT[a]

Parent motif FTPGTEYT	Pooled rabbit anti-human FN (1:200)	Pooled CS IgM (1:200)	Pooled 2⁰ human IgM (1:200)	Pooled 2⁰ human IgG (1:200)	Pooled 35-day syphilitic rabbit IgG (1:400)	Pooled rabbit anti-human type IV collagen (1:200)
ATPGTEYT	0.230[b] (↓ 64.9%)[c]	0.157 (↓ 77.3%)	0.429 (↓ 16.9%)	0.131 (↓ 82.5%)	0.142 (↓ 71.0%)	0.236 (↓ 52.4%)
FAPGTEYT	0.723 (↑ 10.0%)	0.755 (↑ 9.1%)	0.573 (↑ 11.0%)	0.762 No change	0.439 No change	0.486 No change
FTAGTEYT	0.269 (↓ 59.0%)	0.166 (↓ 76.0%)	0.194 (↓ 62.4%)	0.181 (↓ 75.8%)	0.126 (↓ 74.3%)	0.000 (↓ 100%)
FTPATEYT	0.670 No change	0.563 (↓ 18.6%)	0.580 (↑ 12.4%)	0.717 No change	0.504 No change	0.491 No change
FTPGAEYT	0.691 (↑ 5.3%)	0.592 (↓ 14.5%)	0.573 (↑ 11.0%)	0.749 No change	0.506 No change	0.485 No change
FTPGTAYT	0.340 (↓ 48.2%)	0.184 (↓ 73.4%)	0.202 (↓ 60.9%)	0.160 (↓ 78.6%)	0.169 (↓ 65.6%)	0.047 (↓ 90.5%)
FTPGTEAT	0.254 (↓ 61.3%)	0.294 (↓ 57.5%)	0.174 (↓ 66.3%)	0.160 (↓ 78.6%)	0.151 (↓ 69.2%)	0.054 (↓ 89.1%)
FTPGTEYA	0.277 (↓ 57.8%)	0.266 (↓ 67.3%)	0.186 (↓ 64.0%)	0.274 (↓ 63.4%)	0.158 (↓ 67.8%)	0.050 (↓ 89.8%)

[a] Normal human sera and pre-immunization normal rabbit sera yielded O.D.'s of < 0.060 in all assays.
[b] ELISA O.D.; Numbers in bold represent significant reductions.
[c] Percent increase or decrease associated with the amino acid replacement. Significant values are in bold. No change = < ±5% change.

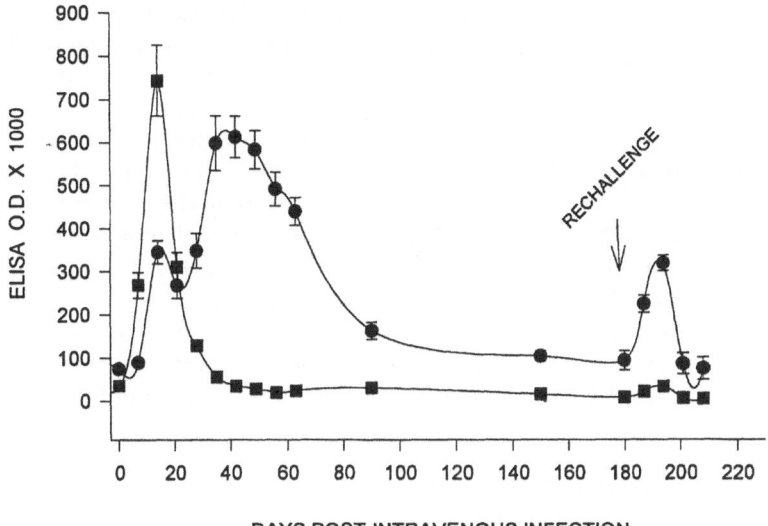

FIGURE 8. Maturation of the IgM (●) and IgG (■) responses against the PGTEYT motif in rabbits with experimental syphilis. The synthetic peptide, ac**PGTEYT**GSGSK, was chemically coupled to bovine serum albumin. For a solid-phase ELISA, 20 μg/well of this material was then used to coat Immulon 2 plates. Pooled sera from six animals at each time point were assayed in triplicate at 1:100 dilutions.

might in turn set off a cascade so that additional epitopes on those host proteins become targets for further antigen presentation and processing and the subsequent loss of tolerance. It is tempting to speculate that those molecular mimics on host proteins are actually the selecting antigens for clonal expansion in syphilis, although this may be difficult if not impossible to prove. Nonetheless, we do not feel that these paradigms of molecular mimicry merely represent ineluctable statistics.

Clearly, several rather important questions remain unanswered: Why aren't these responses controlled earlier in the course of a syphilitic infection? How are these responses eventually controlled? How do the control mechanisms relate to the development of the so-called chancre immunity? Is it possible that responses to immunodominant epitope(s) within the carboxyl-terminal portion mask epitopes at the amino-terminal portion that are perhaps capable of stimulating protective responses? Such a phenomenon has been observed in *Trypanosoma cruzi* infections,[35] leading to the speculation that organisms may evade lethal host immune responses through the strategic placement of immunodominant epitopes on surface-associated proteins. In this regard, we have demonstrated that high-affinity antibodies against one epitope can inhibit binding of antibodies directed against two other epitopes when these three sites are linked in tandem on a 22-mer synthetic peptide of TpN15.[27]

Despite these gaps in our understanding, the importance of these findings would seem to relate directly to vaccine development. Over the past few years there has been considerable interest in the use of recombinant-derived polypeptides of *T. pallidum* as potential immunogens. If certain motifs on treponemal antigens ultimately lead to the induction of deleterious antibodies or cellular responses by virtue of their molecular mimicry, it is imperative to ensure that those sequences are not included so as to preclude immunopathological consequences. Otherwise, the testing of such a potential vaccine candidate would be moot.

7. CONCLUDING REMARKS

Of necessity this chapter has been limited to a brief discussion of those mechanisms/factors that affect immunoregulation in early syphilitic infection and induce the production of certain autoantibodies. All of these humoral responses appear to be secondary reactions, despite the fact that multiple mechanisms may be involved. As such, the autoreactive phenomena in this disease are contributory but not the sole reason for the immunopathogenesis of syphilis.

Our focus also has been limited to early events in syphilitic infection and then to only a few of the varied autoantibody responses. It is possible that the Id network and molecular mimicry both affect immunoregulatory mechanisms in later stages (tertiary and neurosyphilis) of the disease. The flagella of spirochetes are quite similar in structure, with a central viable region together with conserved regions in their amino- and carboxy-terminal ends.[28] In addition, N-terminal sequence similarity has been documented between the flagella of *T. pallidum* and *Borrelia burgdorferi*, and species with the genus *Borrelia* contain cross-reactive epitopes. Monoclonal antibodies directed against *B. burgdorferi* flagellin have been shown to cross-react with antigens on myelinated nerve fibers, neurons in the central nervous system, cardiac muscle, and synovial cells. The prevalence of autoimmune antibodies to host cell determinants, specifically myelin, cardiolipin, galactocerebrosides, and myelin basic proteins, in patients with Lyme neuroborreliosis raises questions as to whether similar autoantibodies are characteristic of neurosyphilis, and if so, what mechanisms are involved.

Immunoregulatory mechanisms in this infectious disease are unquestionably complex,[7] involving interleukins, the generation of T-helper and T-suppressor cells, and other factors such as antigen clearance and localization. It would seem that the missing component in our discussion relates to the inherent defenses against autoimmunization, such as T-cell clonal deletion and inactivation.[3] Whether these defenses are intact in the individual or animal infected with *T. pallidum* is neither known nor understood.

REFERENCES

1. Baughn, R. E., and Musher, D. M., 1995, Immune responses to spirochetes, in: *Clinical Immunology. Principles and Practice* (R. R. Rich, T. A. Fleisher, B. D. Schwartz, W. T. Shearer, and W. Strober, eds.), Mosby, St. Louis, MO, pp. 519–524.
2. Wicher, K., and Wicher, V., 1990, Autoimmunity in syphilis, in: *Organ-specific Autoimmunity* (P. E. Bigazzi, G. Wick, and K. Wicher, eds.), Marcel Dekker, Inc., New York, pp. 108–124.
3. Schwartz, R. S., 1995, Mechanisms of autoimmunity, in: *Clinical Immunology. Principles and Practice* (R. R. Rich, T. A. Fleisher, B. D. Schwartz, W. T. Shearer, and W. Strober, eds.), Mosby, St. Louis, MO, pp. 1053–1061.
4. Diamond, B., Katz, J. B., Paul, E., Aranow, C., Lustgarten, D., and Scharff, M. D., 1992, The role of somatic mutation in the pathogenic anti-DNA response, *Annu. Rev. Immunol.* **10**:731.
5. Tillman, D. M., Jou, N.-T., and Marion, T. N., 1992, Both IgM and IgG anti-DNA antibody are the products of clonally selective B cell stimulation in (NZB X NZW)F1 mice, *J. Exp. Med.* **176**:761.
6. VanEs, J. H., Meyling, F. H. J., wan de Akker, W. R. M., 1991, Somatic mutations in the variable regions of a human IgG anti-double-stranded DNA autoantibody suggest a role for antigen in the induction of systemic lupus erythematosus, *J. Exp. Med.* **173**:461.
7. Baughn, R. E., and Musher, D. M., 1992, Evidence that autologous idiotypic regulation of anti-arginine-glycine-aspartic acid autoantibodies may influence development and progression of syphilitic lesions in infected rabbits, *Infect. Immun.* **60**:3861.
8. Baughn, R. E., Tung, K. S. K., and Musher, D. M., 1980, Detection of circulating immune complexes in the sera of rabbits with experimental syphilis: Possible role in immunoregulation, *Infect. Immun.* **29**:575.
9. Baughn, R. E., Adams, C. B., and Musher, D. M., 1982, Immune complexes in experimental syphilis: A methodologic evaluation, *Sex. Transm. Dis.* **9**:170.
10. Baughn, R. E., 1983, Immunoregulatory effects in experimental syphilis, in: *Pathogenesis and Immunology of Treponemal Infection* (R. F. Schell and D. M. Musher, eds.), Marcel Dekker, Inc., New York, pp. 271–295.
11. Colley, D. G., 1987, Dynamics of the human immune response to schistosomes, in: *Bailliere's Clinical Tropical Medicine and Communicable Diseases* (A. A. F. Mahmoud, ed.), vol. 2, no. 2, *Schistosomiasis*, W. B. Saunders, Eastbourne, England, pp. 315–332.
12. Sher, A., and Colley, D. G., 1989, Immunoparasitology, in: *Fundamental Immunology*, 2nd Ed. (W. E. Paul, ed.), Raven Press, New York, pp. 957–983.

13. Lambert, P. H., Goldman, M., Rose, L. M., 1982, Idiotypic interactions and immuno-pathology, *Ann. Inst. Pasteur Immunol.* **133C**:239–243.

14. Rose, L. M., Goldman, M., and Lambert, P. H., 1982, Simultaneous induction of an idiotype, corresponding anti-idiotypic antibodies, and immune complexes during African trypanosomiasis in mice, *J. Immunol.* **128**:79.

15. Gilliand, B., and Mannik, M., 1983, Immune complex diseases, in: *Harrison's Principles of Internal Medicine*, 10th Ed. (R. G. Peterdorf, R. D. Adams, E. Braunwald, K. J. Isselbacher, J. B. Martin, and J. D. Wilson, eds.), McGraw-Hill, New York, pp. 378–382.

16. Baughn, R. E., McNeely, M. C., Jorizzo, J. L., and Musher, D. M., 1986, Characterization of the antigenic determinants and host components in immune complexes from patients with secondary syphilis. *J. Immunol.* **136**:1406.

17. Dobson, S. R. M., Taber, L. H., and Baughn, R. E., 1988, Characterization of the components in circulating immune complexes from infants with congenital syphilis, *J. Infect. Dis.* **158**:940.

18. Baughn, R. E., Wicher, V., Jakubowski, A., and Wicher, K., 1987, Humoral response in *Treponema pallidum*-infected pigs. II. Circulating immune complexes and autoimmune responses, *J. Immunol.* **138**:4435.

19. Baughn, R. E., Jorizzo, J. L., Adams, C. B., and Musher, D. M., 1988, Ig class and IgG subclass responses to *Treponema pallidum* in patients with syphilis, *J. Clin. Immunol.* **8**:128.

20. Versalovic, J., Nash, Z. D., Carinhas, R., Musher, D. M., and Baughn, R. E., 1990, Immunoglobulin class and subclass restriction of autoimmune responses in secondary syphilis, *Clin. Exp. Immunol.* **80**:381.

21. van der Sluis, J. J., and Boer, M., 1980, Uneven distribution of antitreponemal antibody activity in differing immunoglobulin G fractions from patients with early syphilis, *Infect. Immun.* **29**:837.

22. Primi, D., Smith, C. I. E., Hammarstrom, L., Lundquist, P. G., and Moller, G., 1977, Evidence for the existence of self-reactive human B lymphocytes, *Clin. Exp. Immunol.* **29**:316.

23. Lukehart, S. A., and Holmes, K. K., 1994, Syphilis, in: *Harrison's Principles of Internal Medicine*, 13th Ed. (K. J. Isselbacher, E. Braunwald, J. Wilson, J. B. Martin, A. S. Fauci, and K. L. Kasper, eds.), McGraw-Hill, New York, pp. 726–737.

24. Pedersen, N. S., Orum, O., and Mouritsen, S., 1987, Enzyme-linked immunosorbent assay for detection of antibodies to the venereal disease research laboratory (VDRL) antigen in syphilis, *J. Clin. Microbiol.* **25**:1711.

25. Norris, S. J., and the *Treponema pallidum* polypeptide research group, 1993, Polypeptides of *Treponema pallidum*: Progress toward understanding their structural, functional and immunologic roles, *Microbiol. Rev.* **57**:750.

26. Geysen, M. H., Rodda, S. J., Mason, T. J., Tribbick, G., and Schoofs, P. G., 1987, Strategies for epitope analysis using peptide synthesis, *J. Immunol. Methods* **102**:259.

27. Baughn, R. E., Demecs, M., and Musher, D. M., 1996, Epitope mapping of B-cell determinants on the 15-kilodalton lipoprotein of *Treponema pallidum* with synthetic peptides, in press.

28. Radolf, J. D., Arndt, L. L., Akins, D. R., Curetty, L. L., Levi, M. E., Shen, Y., Davis, L. S., and Norgard, M. V., 1995, Treponema pallidum and Borrelia burgdorferi lipoproteins and synthetic lipopeptides activate monocytes/macrophages, *J. Immunol.* **154**:2866.

29. Jerne, N. K., 1974, Towards a network theory of the immune system, *Ann. Immunol. Inst. Pasteur* **125c**:373.

30. Baughn, R. E., 1990, Demonstration and immunochemical characterization of natural, autologous anti-idiotypic antibodies throughout the course of experimental syphilis, *Infect. Immun.* **58**:766.

31. Zouali, M., and Eyquem, A., 1983, Idiotypic/anti-idiotypic interactions in systemic lupus erythematosus: Demonstration of oscillating levels of anti-DNA autoantibodies and reciprocal antiidiotypic activity in a single patient, *Ann. Immunol. Paris* **134C:**377.

32. Oldstone, M. B. A., 1989, Molecular mimicry as a mechanism for the cause and as a probe uncovering etiologic agent(s) of autoimmune disease, *Curr. Top. Microbiol. Immunol.* **145:**127.

33. Stollerman, G. H. (ed.), 1992, Rheumatic fever and other rheumatic diseases of the heart, in: *Heart Disease, A Textbook of Cardiovascular Medicine,* 4th Ed., Vol. 1 (E. Braunwald, ed.), W. B. Saunders, Philadelphia, pp. 1721–1741.

34. Szczepanski, A., and Benach, J. L., 1991, Lyme borreliosis: Host responses to Borrelia burgdorferi, *Microbiol. Rev.* **55:**21.

35. Wrightsman, R. A., Dawson, B. D., Fouts, D. L., and Manning, J. E., 1994, Identification of immunodominant epitopes in Trypansoma cruzi trypomastigote surface antigen-1 protein that mask protective epitopes, *J. Immunol.* **153:**3148.

36. Chou, P. Y., and Fasman, G. D., 1978, Prediction of the secondary structure of proteins from their amino acid sequence, *Adv. Enzymol.* **47:**45.

5

Viruses and Diabetes Mellitus

THOMAS DYRBERG, PETER MACKAY,
BIRGITTE MICHELSEN, JACOB PETERSEN,
ALLAN KARLSEN, and VAGN BONNEVIE

1. INTRODUCTION

Insulin-dependent (type I) diabetes appears on a background of inherited predisposition localized mainly to the major histocompatibility complex (MHC)[1] and is believed to be mediated by mononuclear leukocytes which infiltrate the islets of Langerhans (insulitis) and selectively destroy the insulin-producing β cells.[2] The triggering event is unknown, but since the concordance rate in human monozygotic twins is only about 50%,[3] it has been suggested that environmental viruses might play a role. This notion finds support in the seasonal variability of disease incidence[4] and from the observation made in the 1920s that cyclical peaks of diabetes incidence paralleled previous outbreaks of mumps.[5] Since then, case reports of infections with rubella[6] or Coxsackievirus[7] and subsequent development of diabetes have supported the possibility that infectious microorganisms are involved. Further evidence was provided by Yoon *et*

THOMAS DYRBERG and PETER MACKAY • Diabetes Immunology Department, Novo Nordisk A/S, 2880 Bagsvaerd, Denmark.   BIRGITTE MICHELSEN and JACOB PETERSEN • Hagedorn Research Institute, 2820 Gentofte, Denmark.   ALLAN KARLSEN • Steno Diabetes Center, 2820 Gentofte, Denmark.   VAGN BONNEVIE • Department of Medical Microbiology, University of Odense, 5000 Odense C, Denmark.

Microorganisms and Autoimmune Diseases, edited by Herman Friedman *et al.* Plenum Press, New York, 1996.

al. who isolated from the pancreas of a child with acute-onset diabetes a Coxsackievirus that induced diabetes when inoculated into mice.[8]

In murine model systems, β-cell tropic strains of encephalomyocarditis virus (EMCV),[9] mengovirus,[10] and Coxsackievirus[11] produce diabetes upon inoculation *in vivo*. There is no evidence for autoimmune cell-mediated β-cell toxicity induced by these fast viruses which seem to infect and lyse the host β cells directly. The diabetogenicity of the EMCV D strain, compared to the nondiabetogenic B strain, may be related to the lower interferon (IFN) levels induced by the former strain[12] and to its lower IFN sensitivity,[13] by which it escapes clearance and infects and lyses β cells within days of inoculation.[14]

More consistent with the long prodromal period in human diabetes[15] are the reports that persistent infection of rodents with slow viruses like reovirus type 1[16] and rubella virus[17] is associated with autoimmune phenomena such as insulitis and islet cell autoantibodies. In the spontaneously diabetic non-obese diabetic (NOD) mouse, an endogenous type C retroviral product is found in a subpopulation of β cells,[18,19] but its possible role in diabetes has not been clarified. In humans, case reports of individuals with congenital rubella syndrome have linked rubella virus with insulitis and diabetes,[20,21] but this association has not been confirmed by others.[22] *In vitro* infection of rat and human β cells with reovirus has been found to upregulate MHC class I expression on the cell surface,[23] thus making them potentially better targets for cytotoxic T cells. This observation is of interest since residual β cells in newly diagnosed patients hyperexpress class I MHC.[24]

All the viruses mentioned above can be detected in the β cells of inoculated mice or, in the case of NOD mice, as endogenous particles. In those mouse models of virus-induced diabetes where there is evidence of cell-mediated β-cell destruction it is not known whether viral or self antigens are recognized by the islet-infiltrating lymphocytes or whether an autoimmune component is involved. Classical tissue-specific autoimmune disease results from loss of tolerance to self antigens and immune reactivity to the cells that elaborate them, and although little is known about how self-tolerance is broken, virus infections may in some cases be involved.

In human type I diabetes patients, the presence of insulitis at diagnosis,[2] the *in vitro* T-cell reactivity to β-cell antigens,[25] the ability of cyclosporin treatment to suppress disease,[26] and the failure to detect genomic sequences from candidate viruses in diabetic pancreata[27,28] together argue against viruses as direct effectors of β-cell destruction. However, current evidence does not preclude that viral infection can initiate or enhance autoimmune reactivity in some individuals. Our recent finding that the interferon inducer polyinosinic:polycytidylic acid (poly I:C) precipitates diabetes in diabetes-prone (DP) BB rats that have remained normoglycemic throughout the age period during which spontaneous disease normally occurs[29] suggests that even an agent that increases cytokine production in an antigen-nonspecific manner can reactivate an autoimmune state which is otherwise quiescent.

In the following sections we will review some of the evidence for virus involvement in autoimmune diabetes and some mechanisms by which these viruses may exert their effect.

2. VIRUSES AND AUTOIMMUNE DIABETES

2.1. Virus Infection Prevents Autoimmune Diabetes

Inadvertent exposure of an NOD mouse colony to mouse hepatitis virus (MHV) resulted in a decreased diabetes incidence in both sexes.[30] In female mice the incidence dropped from about 80% prior to MHV exposure to 25% after but rose dramatically to 95% in MHV seronegative female progeny of new founders delivered by cesarean section and maintained in a specific pathogen–free environment. Experimental infection with lymphocytic choriomeningitis virus (LCMV),[31] lactate dehydrogenase virus (LDV),[32] or EMCV[33] has also been found to prevent diabetes in NOD mice.

Protection of NOD mice by LCMV was associated with the selective infection of a small population of CD4+ T cells that may have closely delineated a diabetogenic subset, since immunity to a panel of standard CD4+ T cell–dependent antigens was not found to be affected.[34] In other mouse strains, LCMV infection reportedly leads to polyclonal activation of as much as half of the CD8+ T-cell compartment, only a minority of which is LCMV-specific,[35] and other effects include modulation of adhesion molecules and downregulation of the CD8 accessory structure. In LCMV-infected NOD mice any of these effects on CD4+ and CD8+ T cells may play a role in the preventive effect of virus on diabetes.

While LCMV and LDV are known to affect T-cell immunity[36] and antigen-presenting cell function,[37] respectively, the mechanisms by which EMCV and MHV prevent diabetes are less well understood but may include increased production of lymphokines which correct NOD suppressor cell deficiency[38] and/or induce dormancy in the effector pathway.[39] The suppression of spontaneous diabetes in NOD mice by EMCV is particularly interesting since it was the β-cell tropic D strain which had this effect.[33] After initial transient hyperglycemia was observed in a few inoculated mice, all EMCV-infected NOD mice were completely protected against disease, with no evidence of general T-cell suppression. In view of the fact that treatment with poly I:C also prevents diabetes in NOD mice[40] the possibility must be considered that IFNs induced by EMCV and MHV interfere with endogenous viruses such as the retrovirus detected in NOD mouse β cells (see below).

In the BB rat we have shown, as in NOD mice, that infection with LCMV prevents insulitis and diabetes,[41] possibly by a transient depletion of T cells which in DP BB rats were found to be uniquely susceptible to binding and replication of LCMV, as compared to cells from diabetes-resistant (DR) BB rats.[42] In both DP BB rat and NOD mouse colonies, the incidence of spontaneous diabetes is decreased if environmental pathogens are not strictly eliminated,[43,33] and in both animal models early immune stimulation with complete Freund's adjuvant (CFA) prevents diabetes,[44,45] possibly by inducing natural, antigen–nonspecific suppressor cells.[46]

2.2. Virus Infection Promotes Autoimmune Diabetes

Currently, the best direct evidence for viral involvement in autoimmune diabetes comes from the DR BB rat substrain where inoculation of viral antibody–free (VAF) rats with Kilham's rat virus (KRV) was found to reproducibly induce insulitis and diabetes.[47] The KRV antigens were not detectable in the islets but were abundant in peripancreatic lymph nodes, cervical lymph nodes, and spleen, suggesting that diabetes was caused by self-reactive inflammatory cells that appeared in the islets and not by viral lysis or immunity to membrane-bound viral antigens.[48] The KRV did not induce diabetes in non-BB rat strains, nor did it accelerate disease in rats from the DP substrain unless these were first reconstituted with DR spleen cells.[47] Lymphocyte subset distributions in KRV- infected DR rats, including that of the normally diabetes-suppressive RT6+ T-cell subset, did not differ from noninfected control rats. Therefore, it did not appear that virus infection had depleted or increased any particular population of immunocytes, but rather had triggered a genetically programmed pathogenic immune response unique to DR BB rats.

In the murine EMCV model of virus-induced diabetes, contrasting results have been published concerning the role of the immune system. Thus, Yoon *et al.* found no signs of immunity to β cells in mice made diabetic by infection with the EMCV D variant.[9] Immunosuppression by treatment with antilymphocyte serum or by thymectomy did not decrease diabetes incidence in infected mice, nor could the disease be adoptively transferred into irradiated recipients by spleen cells. Conversely, a study by Jansen *et al.* showed that immunosuppression by X-irradiation prior to infection with EMCV M variant decreased disease incidence, even in the face of increased virus multiplication.[49] This study, as well as others where immune phenomena were invoked by the observation that treatment with anti–T cell antibody reduced the incidence of disease induced by EMCV M variant,[50,51] suffer from the fact that no insulitis, and thus no direct sign of cell-mediated immunity, was evident in those mice made diabetic by EMCV infection alone. It is unclear what importance the depletion of distal T cells has in these studies, one of which did show the appearance of T cells in EMCV-infected mice which were cytotoxic to normal β cells *in vitro.*[51]

In the murine model of reovirus-induced diabetes, a virus strain that had been passaged *in vitro* in murine β cells induced transient glucose intolerance upon inoculation, but no overt diabetes.[16] This glucose intolerance appeared to be immune-mediated, since it was accompanied by lymphocytic islet inflammation and islet cell autoantibodies and could be prevented by antilymphocyte serum or cyclophosphamide.[52] Although immune reactivity in this model may contribute to β-cell pathology, it is not known whether islet-infiltrating immunocytes recognize virally modified or normal β-cell antigens or whether repeated virus inoculation would induce overt diabetes.

As mentioned in the introduction, endogenous retroviral particles are detected in the β cells of NOD mice.[18] These particles were found less frequently in the related but non-diabetes-prone NON strain,[19] they were present only in endocrine β cells, and their frequency could be correlated with the occurrence of both insulitis and diabetes.[53] Cyclophosphamide treatment of normal NOD mice accelerates diabetes, and in the latter study[53] this agent enhanced both the expression in β cells of retroviral particles and the intensity of insulitis. Since cyclophosphamide inhibits DNA synthesis it may be speculated that it enhances retrovirus expression by inhibiting the synthesis of a factor that otherwise suppresses expression of retroviral particles. The type C retrovirus has been partially cloned, and results indicate that the virus is replication-defective and that it differs from other known retroviruses.[54]

Treatment of genetically susceptible mouse strains with multiple low doses of streptozotocin (STZ) induces insulitis and diabetes. In a study by Appel *et al.* it was shown that lymphocytic infiltration in the islets of STZ-treated CD-1 mice was preceded by the appearance of type A and type C murine leukemia virus particles within the β cells.[55] Multiple STZ-induced diabetes is dependent on cellular immunity in susceptible mouse strains[56] and since β-cell necrosis in the study of Appel *et al.* was observed only after the development of insulitis the authors speculated that STZ-induced leukemia virus products had caused structural changes in the β cells which were then recognized by the immune system. Because STZ-induced diabetes in our experience can only be adoptively transferred into hosts that have been primed with STZ,[57] it is possible that structural changes may occur in the β cells after STZ exposure, but whether these are caused by STZ itself or by an induced virus is not known.

3. INTERFERON AND DIABETES

Interferons are a group of proteins, classified in mouse and man into IFN-α, IFN-β (type I IFN), and IFN-γ (type II IFN), that serve as antigen-nonspecific defense against viral infection[58] and complement the later antigen-specific T-cell and B-cell responses. IFN-α and IFN-β can be produced in virus-infected tissue cells, while IFN-γ is a product of acti-vated immune system T and NK cells.[59,60] However, all three classes can be produced by blood mononuclear cells, including B cells[61] or macro-phages,[62] depending on the stimulus, and target cells bear IFN receptors on their surface that discriminate between type I and II IFNs.[63] Interferon increases the cytotoxic potential of NK cells,[64] T cells,[65] and macro-phages,[66] cell types that are found in the insulitis lesion and which are cytotoxic to β cells in BB rats[67,68] and/or in NOD mice.[69,70] Treatment with IFN has been found to accelerate spontaneous autoimmune disease in a murine lupus model,[71] an effect that mimics the enhancement obtained with LCMV infection in the same model.[72]

Evidence for IFN involvement in type I diabetes comes from the observation that antibodies to both IFN-α and IFN-γ prevent spontaneous disease in BB rats[73,74] and from the observation that the IFN inducer poly I:C accelerates disease in this model.[73] Upon poly I:C administration, the time of diabetes onset in DP BB rats was found to be inversely correlated with both the poly I:C dose and the serum level of IFN-α.[75] At a given poly I:C dose, serum IFN-α levels were higher in non-diabetes-prone Wistar control rats than in (T-lymphopenic) DP BB rats, but Wistar rats never became diabetic, showing that poly I:C–induced lymphokines cannot precipitate autoimmune diabetes in a normal immunological milieu.

In the DR BB rat, the immunological milieu includes both diabeto-genic and suppressive T-cell populations, the latter of which keeps the former in check.[76] Thomas *et al.* found that removal of the suppressive RT6+ T-cell subset by injection of RT6-specific antibody resulted in diabetes, but only if the animals were maintained in an environment where viral pathogens, including KRV, were present.[7] If VAF DR rats from a more strictly controlled environment were subjected to a combined treatment of anti-RT6 antibody and poly I:C, 94% became diabetic versus 0% by RT6 lymphocyte depletion alone and 22% by poly I:C alone.

These data show first that upon perturbation of the immune system (depletion of regulatory cells), the environment has a profound effect on diabetes incidence in this model, and second, that the effect can be mimicked by treatment with a synthetic IFN inducer and therefore may be related to immune activation and increased cytokine production. Because KRV was present in the non-VAF environment that made DR rats susceptible to anti RT6-induced diabetes,[77] and because this virus is diabetogenic in VAF BB DR rats upon inoculation (Section 2.2),[47] one may speculate that it activates local immunocytes during its occupation of lymphoid tissues, upon which a diabetogenic population migrates into the islets. Whether or not the T-cell response to KRV infection is clonally limited is unknown, but both the poly I:C data and the finding that the diabetes-promoting effect of environmental pathogens in the RT6-depleted DR rat can be reproduced by inducing peritoneal inflammation with a suspension of sterile feces[78] suggest that diabetogenic T cells can be activated by antigen-nonspecific means.

In the NOD mouse, IFN also seems to play a role in disease pathogenesis, since anti-IFN-γ antibody prevents diabetes[79] and since IFN-γ is expressed by T cells in the islet infiltrate.[80] As noted above, however, poly I:C prevents diabetes in this model, possibly because regulatory lymphokines other than IFN are induced which are important in NOD mice. It is not currently understood why poly I:C has opposite effects on diabetes development in NOD mice and BB rats.

In human type I diabetes, a role for IFN-α and IFN-γ was indicated by the finding by Foulis *et al.* that IFN-α was present in the residual β cells of newly diagnosed diabetes patients[81] and that periinsular and infiltrating lymphocytes stained positive for IFN-γ.[82] Nearly all remaining β cells hyperexpressed MHC class I and some had aberrant expression of class II. Expression of β-cell IFN-α was unique to type I diabetic pancreas in that it was absent in type II (maturity onset) pancreatic islets and in chronic pancreatitis specimens,[81] the latter observation suggesting that IFN-α induction in type I patients was not the result of general pancreatic inflammation. All these observations are compatible with viral infection of β cells, and indeed, when studying control pancreata, these authors only found similar β-cell IFN-α expression in a few individuals with neonatal acute Coxsackie B or congenital rubella virus infection. On this basis Foulis has speculated that β-cell IFN-α expression represents a link between virus infection and autoimmune diabetes and proposed a model where a latent virus infection induced β-cell IFN-α synthesis and hyper-expression of MHC class I.[83] The combined effects of virus infection and IFN-γ secreted by infiltrating immunocytes could lead to the aberrant MHC class II expression observed by Bottazzo *et al.* on the β cells of acute-onset human diabetics.[84] Aberrant MHC class II expression on endocrine cells has been suggested as a mechanism by which autoantigens can be presented in a context not previously encountered by T cells and to which they are not tolerant.[85]

Our finding that basal 2′,5′-oligoadenylate synthetase activity is enhanced in peripheral blood mononuclear cells from diabetic subjects is compatible with a latent virus infection.[86]

3.1. Interferon Effects on Beta Cells

Consistent with the hypothesis that IFN synthesis in human diabetic β cells is a response to a viral infection is the observation that IFN reduces viral replication and cell damage in mouse islet cells infected with EMCV *in vitro*.[87] The effect is likely mediated at least in part by the activity of the antiviral 2′,5′-A synthetase pathway which is upregulated in rat β cells by IFN-α treatment *in vitro*.[88]

In addition to antiviral activity, IFN has several effects on β cells that may be conducive to destruction by islet-invading effector cells. Thus, IFN-γ upregulates MHC class I expression on rodent and human β cells,[89–91] and while there was no evidence of MHC class II induction in these studies, other investigators found that IFN-γ also induced the latter restriction elements on mouse and rat β cells.[92,93] The discrepancies may be due to species differences in the recombinant IFNs and the target cells used.

In human β cells, a combination of IFN-γ and either tumor necrosis factor (TNF) or lymphotoxin was found to induce *de novo* expression of class II molecules.[94] Other investigators found this combination to inhibit insulin release[95] and ultimately to cause cell destruction in normal rodent islets.[96] A later study in BB rats showed that IFN-γ sensitized islet cells to the *in vitro* cytotoxic action of effector lymphocytes and also to the toxic effect of IL-1 and TNF in combination.[97] In addition to its MHC-inducing and sensitizing functions, IFN-γ induces intercellular adhesion molecules (ICAM-1) on human islets,[98] further improving their potential as targets for effector lymphocytes.

While these experiments have addressed the effects of exogenous IFN on β cells, experiments with transgenic mice have investigated the effect of IFN expression by the β cells themselves. In the experiments by Sarvetnick *et al.* transgenic mice expressing a mouse IFN-γ gene under government of the human insulin gene promoter (*ins-IFN-γ* mice) developed insulitis and diabetes,[99] suggesting that IFN is an effective recruiter of immune cells when expressed in islets, even in a non-diabetes-prone mouse. Lymph node cells isolated from *ins-IFN-γ* mice were cytotoxic to nontransgenic, histocompatible islets *in vitro*, supporting the notion that recruited immunocytes recognized normal and not altered β-cell membrane structures. When the *ins-IFN-γ* transgene was backcrossed onto mice of the severe combined immune deficiency (SCID) strain, which cannot produce mature T and B lymphocytes, no β-cell necrosis was observed, indicating that recruited effector cells, and not IFN-γ itself, were responsible for the islet cell destruction.[100]

Since IFN-α was the cytokine type seen by Foulis *et al.* in human β cells, it is of particular interest that transgenic mice expressing the gene product in β cells also develop diabetes,[101] with islet inflammation that includes both CD4+ and CD8+ T cells. Prophylactic treatment with antibody to IFN-α reduced mononuclear cell infiltration and prevented β-cell necrosis and diabetes, suggesting that local production of biologically active IFN was the instigator of islet inflammation in this model.

4. MOLECULAR MIMICRY BETWEEN VIRUSES AND BETA-CELL AUTOANTIGENS

Among the pathogenic mechanisms that have been implicated in autoimmunity, the "molecular mimicry" hypothesis has received considerable interest. The term was originally coined to suggest a mechanism by which microbes could adapt to a given host milieu by mimicking host self proteins, thus escaping immune detection.[102] Subsequently the term has been taken to imply the generation of an immune response cross-reactive between a microbial epitope and a host antigen.[103]

Humoral cross-reactivity between endogenous retroviral particles and insulin has been observed in the NOD mouse.[104] Cellular cross-reactivity between these antigens was not investigated, but the result is likely to have been negative since spontaneous T-cell reactivity to insulin is not seen in NOD mice.[105] However, since immunization with porcine insulin or human insulin B chain,[106] in addition to oral tolerance induction with porcine insulin,[107] has been reported to prevent diabetes in this animal model, insulin peptides and/or microbial peptides with similar epitopes cannot be excluded as contributing antigens in the insulitis lesion.

Human type 1 diabetes is characterized by the presence of islet cell autoantibodies[108,109] and T-cell reactivity against islet cell antigens,[25,110] and among these autoantigens interest has centered on the two isoforms of the GABA-synthesizing enzyme glutamic acid decarboxylase, GAD_{65} and GAD_{67}.[111,112] GAD autoimmunity is suggested to play an important role in the initial pathogenic events leading to clinical diabetes since humoral and cellular anti-GAD reactivity is seen in 70–80% of type 1 diabetes patients at diagnosis[113,114] and can be detected several years before clinical onset.[115] Furthermore, tolerization of NOD mice with GAD_{65} not only protects the animals from developing insulitis and diabetes, but also reduces the spreading of the immune response to other islet cell antigens such as heat-shock protein and carboxypeptidase H.[116–118] Taken together, these data suggest that the initiating target antigen for the β-cell-destructive autoimmune response in diabetes could be GAD_{65} or a structurally related antigen.

A complementary approach to studying immune cross-reactivity to self and foreign antigens, such as that seen in NOD mice between insulin and retroviral particles, has been to search for common amino acid sequences in autoantigens and virus proteins.[103] For example, the acetylcholine receptor alpha chain has been found to share an epitope with a herpes simplex virus glycoprotein,[119] and autoanbitodies to this acetylcholine receptor epitope from patients with myasthenia gravis cross-react with the virus, implicating herpes simplex virus as an etiologic agent in some cases of myasthenia.[119] From studies in experimental allergic encephalomyelitis (EAE) it is known that immunization with a 6–amino acid residue from the N-terminal region of myelin basic protein is sufficient to induce the disease,[120] and since immunization with smaller peptides could not induce EAE, a hexamer may define the minimum sequence identity of interest for molecular mimicry in T cell–dependent autoimmune diseases.

A search for structurally similar epitopes between GAD_{65}/GAD_{67} and viral proteins has revealed numerous homologous sequences of six and seven amino acids (Table I and Fig. 1). Several of the viruses sharing sequences with GAD_{65}/GAD_{67} have been implicated in the pathogenesis of diabetes, notably Coxsackievirus B4[7,8] and cytomegalovirus (CMV).[121] The P2-C protein of Coxsackievirus B4 shares sequence homology with a fragment of GAD_{65}[122,123] (Table I) that has been suggested to represent an immunodominant T-cell epitope because it is recognized by T cells from NOD mice several weeks before manifest diabetes.[110] Thus, it is possible

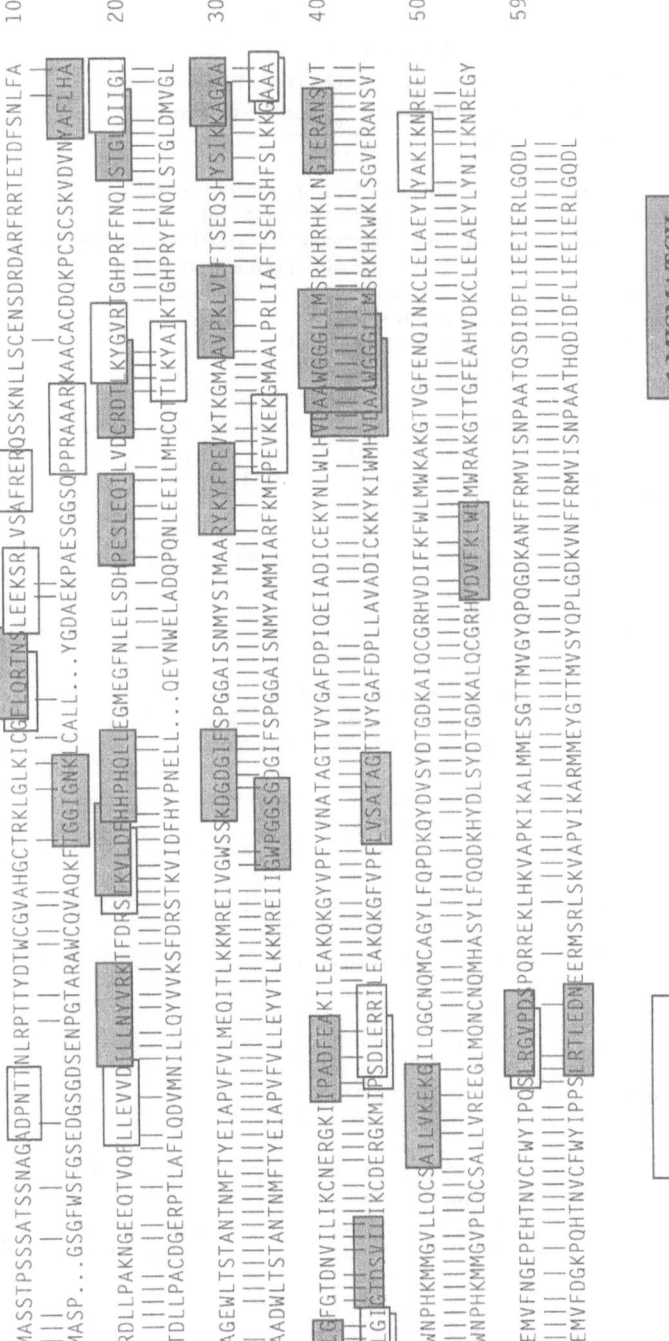

FIGURE 1. The PIR/PSQ protein databases were searched for homologous sequences between virus proteins and the rat islet glutamic acid decarboxylase$_{67}$ (GAD$_{67}$) (top sequence) or human islet glutamic acid decarboxylase$_{65}$ (GAD$_{65}$) (bottom sequence). The viruses containing shared sequences are shown in Table I. The homologous GAD determinants where all residues are identical are shown in open boxes and determinants with one residue difference are shown in shaded boxes.

TABLE I
Shared Epitopes on Glutamic Acid Decarboxylase$_{65}$
and Glutamic Acid Decarboxylase$_{67}$ and Virus Proteins[a]

Glutamic acid decarboxylase$_{65}$	
Visna virus	Transregulatory splicing protein [67–73]
Human rotavirus A	Inner capsid protein VP6 [48–54]
Human herpes virus 1	Immediate-early protein IE175 [250–256], glycoprotein B [53–59], major capsid protein [469–475]
Human herpes virus 2	Immediate-early protein RL2 [422–428]
Human herpesvirus 3	Glycoprotein E [539–545]
Human herpesvirus 4	Probable nuclear antigen [55–61]
Rubella virus	Genome polyprotein [11–17]
Japanese encephalitis virus	Genome polyprotein [261–267]
Vaccinia virus	C9L protein/74K HindIII-C protein [119–126], H4L/H5 protein [521–527]
Coxsackievirus B1 and B4	Genome polyprotein [1134–1141]
Human papillomavirus	E6 protein [25–31]
Human cytomegalovirus	HVLF4 protein [310–316], HXLF3 protein [178–184], hypothetical protein UL88 [411–417]
Influenza A virus	Hemagglutinin precursor [415–412]
Sendai virus	Nucleocapsid protein [477–483]
Measles virus	Nonstructural protein C [168–174]
Variola virus	D7L protein [101–110]
Semliki Forest virus	Structural polyprotein [303–309]
Ross River virus	Structural polyprotein [304–310]
O'nyong-nyong virus	Structural polyprotein [295–301]
Glutamic acid decarboxylase$_{67}$	
Human papillomavirus	L2 protein [211–217], E7 protein [48–54], hypothetical protein L1 [66–72]
Vaccinia virus	Hemagglutinin precursor [17–23]
Hepatitis B virus	Structural protein 2 precursor [453–459]
Human herpesvirus 1	UL7 protein [118–124], UL41 protein [237–243], major capsid protein [469–475]
Human herpesvirus 3	Major capsid protein [867–873], glycoprotein B [539–545]
Human herpesvirus 4	BVRF1 (EC-RF2) protein [266–272], membrane antigen gp220/350 [439–445], nuclear antigen EBNA-3A [279–285]
Human herpesvirus 6	SFLG protein [34–40]
Human immunodeficiency virus	*Pol* polyprotein [954–960]
Human hepatitis A virus	Genome polypeptide [278–284], [1310–1316]
Human adenovirus	Terminal protein [598–604]
Human cytomegalovirus	UL104 protein [217–223], hypothetical protein UL102 [304–310], hypothetical protein UL70 [580–586]
Reovirus type 1	Mu2 protein [300–306]
Reovirus type 3	Sigma 1 protein precursor [265–271]
Reovirus type 1, 2, and 3	Lambda 3 protein [802–808]

TABLE I
(Continued)

Human rotavirus A	Glycoprotein VP7 precursor [17–24]
Yellow fever virus	Genome polyprotein [792–798], [1677–1683]
Bunyamwera virus	Hypothetical 14.7K protein [39–45]
Japanese encephalitis virus	Genome polyprotein [825–831], [1649–1655]
Kunjin virus	Genome polyprotein [822–828]
Fowlpox virus	F4 protein [63–69]
Measles virus	Fusion glycoprotein [121–127], nonstructural protein C [168–174]
Rinderpest virus	Fusion glycoprotein [114–120]
Middelburg virus	Nonstructural protein NS72 [945–951]
Human poliovirus	Genome polyprotein [1618–1624]
Berne virus	Envelope protein [154–161]
Human immunodeficiency virus type 2	*Pol* polyprotein [535–541]
Human adenovirus 7	Terminal protein [284–290]
Human adenovirus 12	Terminal protein [249–255]

aViruses with amino acid sequence similarities to human islet glutamic acid decarboxylase$_{65}$ (GAD$_{65}$) and/or rat islet glutamic acid decarboxylase$_{67}$ (GAD$_{67}$). The specific virus proteins and amino acid positions (within square brackets) are shown for each virus. The homologous sequences in GAD$_{65}$ and GAD$_{67}$ are shown in Fig. 1.

that infection with Coxsackievirus B4 can initiate a T-cell response against a viral protein and that the responding cells may cross-react with GAD$_{65}$.

Cytomegalovirus has also been implicated in human diabetes since viral genomic material was detected in peripheral blood lymphocytes of 22% of recent-onset type 1 diabetes patients compared to only 3% of healthy control subjects.[121] Human CMV contains a total of six sequences mimicking epitopes on the GAD$_{65}$/GAD$_{67}$ molecules (Table I), and T cells that react with any of these may also recognize GAD. However, the majority of the viruses with sequence homology to GAD have not been associated with the development of type 1 diabetes (Table I), and the homologous determinants in GAD appear to be evenly spread throughout the molecules rather than being confined to specific regions (Fig. 1). These and other data[123] indicate that the mere presence of a sequence identity between a viral protein and an autoantigen does not by itself implicate the virus as a pathogenic factor.

5. TRANSGENIC MODELS OF VIRUS-INDUCED DIABETES

Several studies of tissue-specific tolerance have provided insight into how virus infection can lead to islet autoimmunity and diabetes, for example, in transgenic mice carrying genes coding for the LCMV glyco-protein (GP),[124–126] the nuclear protein (NP),[124] or influenza virus hem-agglutinin (HA)[127] under the control of the rat insulin promoter (RIP). These systems permit studies of immunological tolerance to non self antigens expressed in β cells and the occurrence and consequences of tolerance loss subsequent to infection with virus.

Mice expressing a viral sequence on their β cells did not sponta-neously develop diabetes. Infection with LCMV of RIP-GP mice, but not of nontransgenic mice, caused insulitis and persistent hyperglycemia 9–11 days postinfection, whereas infection with vaccinia virus had no ef-fect.[124,126] These findings suggested that LCMV infection could induce diabetes in RIP-GP mice by breaking tolerance to GP on the β cell. However, the situation appeared to be different in transgenic RIP-HA mice, in which neither influenza virus inoculation nor a vaccinia-HA recombinant virus inoculation caused β-cell destruction.[128]

In mice that were made transgenic for genes coding for both RIP-GP and a rearranged T-cell receptor (TCR) specific for the transgenically expressed GP peptide p32-42 (RIP-GP/TCR mice), the onset of diabetes was accelerated due to the larger number of responding T cells.[126] Only activated T cells seemed able to recognize the GP peptide on the β cells, since both RIP-GP and RIP-GP/TCR mice remained unresponsive to GP prior to virus inoculation. Infection with LCMV induced GP-specific cytotoxic T lymphocytes in RIP-GP and RIP-GP/TCR mice, and diabetes could be adoptively transferred by spleen cells from LCMV-infected mice to naive RIP-GP recipient mice.[126] These experiments show that in a model system, antigen-specific cytotoxic T cells generated in response to virus infection can destroy β cells expressing viral antigens.

Somewhat surprisingly, the RIP-GP transgenic mice were not tolerant to the transgenic antigen. Thus, in RIP-GP/TCR mice, transgenic thymo-cytes were positively and not negatively selected, as shown by skewing of thymocytes to the CD8+ population. Furthermore, after *in vitro* stimula-tion, cytotoxic T lymphocytes derived from uninfected RIP-GP/TCR double transgenic mice were able to lyse target cells expressing LCMV GP.[126] Peripheral tolerance to GP in this model was therefore maintained because the viral sequence-specific T cells were not activated in unin-fected RIP-GP mice. In contrast, in RIP-HA transgenic mice both CD4+ and CD8+ T cells (but not B lymphocytes) were tolerant to HA, and inoculation with influenza virus or vaccinia virus expressing HA did not break the tolerance or lead to β-cell destruction.[127]

It is not clear why RIP-HA mice develop tolerance toward the trans-gene product while RIP-GP mice do not. One possibility not investigated in detail in these studies could be differences in the time of expression of the antigens during development, as this has previously been shown to influence the induction of tolerance.[128] Furthermore, although the expression site of the transgene to a large extent will be defined by the insulin promoter, expression in other tissues can occur, for example, in the thymus, and markedly change the time course of disease development.[129] Still other factors may modulate the expression and consequences of the transgene. Thus, approximately 50% of RIP-GP/TNFα double transgenic mice spontaneously became hyperglycemic,[130] data which suggests that local production of TNF-α in the islets attracts and activates GP-specific T cells which lyse the β cells.

Overall, work in transgenic mice expressing viral sequences in the β cells demonstrates that β-cell destruction is influenced not only by the infectious agent, but also by the MHC haplotype, the level and site of expression of cytokines, and the number of self-reactive cytotoxic T lymphocytes generated. In these respects, the transgenic models resemble spontaneous diabetes in man where multiple factors are believed to govern whether islet cell autoimmunity develops and whether this in turn will progress until a sufficient number of β cells are lost to cause clinically manifest diabetes. In conclusion, the virus/β cell transgenic mice provide an experimental model to study tolerance and immunity to β cell–expressed viral antigens. Furthermore, these models may eventually be useful in studies of therapies to intervene in the preclinical disease process.

6. CONCLUDING REMARKS

At present there is little data in humans that unequivocally implicates viruses as being directly responsible for the selective β-cell loss that causes type I diabetes. However, circumstantial evidence does not exclude the possibility that virus infection of the β cells can instigate events, such as lymphocytic cross-reactivity and upregulation of MHC and adhesion molecules, which ultimately promote an autoimmune cascade. Since transplantation of "normal" pancreatic tissue from a nondiabetic twin to a long-term diabetic identical sibling results in rapid β-cell destruction,[131] it appears that the immune system of diabetes patients can recognize unmodified β-cell antigens, even years after clinical onset.

The findings that KRV induces diabetes in DR BB rats and that antigen-nonspecific IFN induction by poly I:C accelerates disease in DP BB rats suggest that immune-activating events distal to the islets can modulate a genetically predisposed immune milieu and precipitate or enhance the occurrence of diabetes. It is possible that environmental viruses have similar effects in diabetes-prone human beings, and this notion should be considered in future research into the possible role of viruses in human type I diabetes mellitus.

REFERENCES

1. Trucco, M., and Dorman, J. S., 1989, Immunogenetics of insulin-dependent diabetes in humans, *Crit. Rev. Immunol.* **9**:201–245.
2. Gepts, W., 1965, Pathology and anatomy of the pancreas in juvenile diabetes mellitus, *Diabetes* **14**:619–633.
3. Barnett, A. H., Eff, C., Leslie, R. D. G., and Pyke, D. A., 1981, Diabetes in identical twins: A study of 200 pairs, *Diabetologia* **20**:87–93.
4. Gamble, D. R., and Taylor, K. W., 1969, Seasonal incidence of diabetes mellitus, *Br. Med. J.* **3**:631–633.
5. Gundersen, E., 1927, Is diabetes of infectious origin? *J. Infect. Dis.* **41**:197–202.
6. Johnson, G. M., and Tudor, R. B., 1970, Diabetes mellitus and congenital rubella infection, *Am. J. Dis. Child* **120**:453–455.
7. King, M. L., Bidwell, D., Shaikh, A., Voller, A., and Banatvala, J. E., 1983, Coxsackie-B-virus specific IgM responses in children with insulin-dependent diabetes mellitus, *Lancet* **i**:1397–1399.
8. Yoon, J.-W., Austin, M., Onodera, T., and Notkins, A. L., 1979, Virus induced diabetes mellitus: Isolation of a virus from the pancreas of a child with diabetic ketoacidosis, *N. Engl. J. Med.* **300**:1173–1179.
9. Yoon, J.-W., McClintock, P. R., Bachurski, C. J., Longstreth, J. D., and Notkins, A. L., 1985, Virus-induced diabetes mellitus. No evidence for immune mechanisms in the destruction of β-cells by the D-variant of encephalomyocarditis virus, *Diabetes* **34**: 922–925.
10. Yoon, J.-W., Morishima, T., McClintock, P. R., Austin, M., and Notkins, A. L., 1984, Virus-induced diabetes mellitus: Mengovirus infects pancreatic beta cells in strains of mice resistant to encaphalomyocarditis virus, *J. Virol.* **50**:684–690.
11. Yoon, J.-W., Onodera, T., and Notkins, A. L., 1978, Virus-induced diabetes mellitus: Beta cell damage and insulin-dependent hyperglycemia in mice infected with Coxsackie virus B4, *J. Exp. Med.* **148**:1068–1080.
12. Yoon, J.-W., McClintock, P. R., Onodera, T., and Notkins, A. L., 1980, Virus-induced diabetes mellitus. XVIII. Inhibition by a non-diabetic variant of encephalomyocarditis (EMC) virus, *J. Exp. Med.* **152**:878–882.
13. Cohen, S. H., Bolton, V., and Jordan, G. W., 1983, Relationship of the interferon-inducing particle phenotype to encephalomyocarditis virus-induced diabetes mellitus, *Infect. Immun.* **42**:605–611.
14. Jordan, G. W., and Cohen, S. H., 1987, Encephalomyocarditis virus-induced diabetes mellitus in mice: Model of viral pathogenesis, *Rev. Infect. Dis.* **9**:917–924.
15. Gorsuch, A. N., Lister, J., Dean, B. M., Spencer, K. M., McNally, J. M., and Bottazzo, G. F., 1981, Evidence for a long prediabetic period in type I (insulin-dependent) diabetes mellitus, *Lancet* **ii**:1363–1365.

16. Onodera, T., Jenson, A. B., Yoon, J.-W., and Notkins, A. L., 1978, Virus-induced diabetes mellitus: Reovirus infection of pancreatic β cells in mice, *Science* 201:529–531.

17. Rayfield, E. J., Kelly, K. J., and Yoon, J.-W., 1986, Rubella virus-induced diabetes in hamsters, *Diabetes* 35:1278–1281.

18. Fujita, H., Fujino, H., Nonaka, K., Tarui, S., and Tochino, H., 1984, Retrovirus-like particles in pancreatic B-cells of NOD (non-obese diabetic) mice, *Biomed. Res.* 5:67–70.

19. Gaskins, H. R., Prohazka, M., Hamaguchi, K., Serreze, D. V., and Leiter, E. H., 1992, Beta cell expression of endogenous xenotropic retrovirus distinguishes diabetes-susceptible NOD/Lt from resistant NON/Lt mice, *J. Clin. Invest.* 90:2220–2227.

20. Menser, M. A., Forrest, J. M., and Bransby, R. D., 1978, Rubella infection and diabetes mellitus, *Lancet* i:57–60.

21. Rubinstein, P., Walker, M. E., Fedun, B., Witt, M. E., Cooper, L. Z., and Ginsberg-Fellner, F., 1982, The HLA system in congenital rubella patients with and without diabetes, *Diabetes* 31:1088–1091.

22. Smithsells, R. W., Sheppard, S., Marshall, W. C., and Peckham, C, 1978, Congenital rubella and diabetes mellitus, *Lancet* i:1048–1049.

23. Campbell, I. L., Harrison, L. C., Ashcroft, R. G., and Jack, I., 1988, Reovirus infection enhances expression of class I MHC proteins on human β-cell and rat RINm5F cells, *Diabetes* 37:362–365.

24. Foulis, A. K., Farquharson, M. A., and Hardman, R., 1987, Aberrant expression of class II major histocompatibility complex molecules by B cells and hyperexpression of class I major histocompatibility complex molecules by insulin containing islets in type I (insulin-dependent) diabetes mellitus, *Diabetologia* 30:333–343.

25. Roep, B. O., Arden, S. D., de Vries, R. R. P., and Hutton, J. C., 1990, T-cell clones from a type-1 diabetes patient respond to insulin secretory granule proteins, *Nature* 345:632–634.

26. Stiller, C., Martel, R., Dupré, J., Gent, M., Jenner, M. R., Keown, P. A., Laupacis, A., Martell, R., Rodger, N. W., von Graffenried, B., and Wolfe, B. M. J., 1984, Effects of cyclosporin immunosuppression in insulin-dependent diabetes mellitus of recent onset, *Science* 223:1362–1367.

27. Foy, C. A., Quirke, P., Williams, D. J., Lewis, F. A., Grant, P. J., Eglin, R., and Bodanski, H. J., 1994, A search for candidate viruses in type I diabetic pancreas using polymerase chain reaction, *Diabet. Med.* 11:564–569.

28. Buesa-Gomez, J., de la Torre, J. C., Dyrberg, T., Landin-Olsson, M., Mauseth, R. S., Lernmark, Å., and Oldstone, M. B. A., 1994, Failure to detect genomic viral sequences in pancreatic tissues from two children with acute-onset diabetes mellitus, *J. Med. Virol.* 42:193–197.

29. MacKay, P., 1995, Adoptive transfer to and from old normoglycemic BB rats, *Diabetologia* 38:145–152.

30. Wilberz, S., Partke, H. J., Dagnaes-Hansen, F., and Herberg, L., 1991, Persistent MHV (mouse hepatitis virus) infection reduces the incidence of diabetes mellitus in non-obese diabetic mice, *Diabetologia* 34:2–5.

31. Oldstone, M. B. A., 1988, Prevention of type I diabetes in nonobese diabetic mice by virus infection, *Science* 239:500–502.

32. Takei, I., Asaba, Y., Kasatani, T., Maruyama, T., Watanabe, K., and Yanagawa, T., 1992, Suppression of development of diabetes in NOD mice by lactate dehydrogenase virus infection, *J. Autoimmun.* 5:665–673.

33. Hermitte, L., Vialettes, B., Naquet, P., Atlan, C., Payan, M. J., and Vague, P., 1990, Paradoxical lessening of autoimmune process in non-obese mice after infection with the diabetogenic variant of encephalomyocarditis virus, *Eur. J. Immunol.* 20:1297–1303.

34. Oldstone, M. B. A., 1990, Viruses as therapeutic agents. I. Treatment of nonobese insulin-dependent diabetes mice with virus prevents insulin-dependent diabetes mellitus while maintaining general immune competence, *J. Exp. Med.* 171:2077–2089.

35. Andersson, A. C., Christensen, J. P., Marker, O., and Thomsen, A. R., 1994, Changes in cell adhesion molecule expression on T cells associated with systemic virus infection, *J. Immunol.* 152:1237–1245.

36. McChesney, M. B., and Oldstone, M. B. A., 1987, Viruses perturb lymphocyte functions: Selected principles characterizing virus-induced immunosuppression, *Ann. Rev. Immunol.* 5:279–304.

37. Isakov, N., Feldman, M., and Segal, S., 1982, Acute infection of mice with lactic dehydrogenase virus (LDV) impairs the antigen-presenting capacity of their macrophages, *Cell. Immunol.* 66:317–332.

38. Hatamori, N., Yokono, K., Nagata, M., Kawase, Y., Hayakawa, M., Akiyama, H., Sakamoto, T., Yonezawa, K., Yaso, S., Shii, K., Doi, K., and Baba, S., 1989, Suppressor T-cell abnormality in NOD mice before onset of diabetes, *Diabetes Res. Clin. Pract.* 6:265–270.

39. Uleato, D., Lacy, P. E., Kipnis, D. M., Kanagawa, O., and Unanue, E. R., 1992, A T-cell dormant state in the autoimmune process of nonobese diabetic mice treated with complete Freund's adjuvant, *Proc. Natl. Acad. Sci. U.S.A.* 89:3927–3931.

40. Serreze, D. V., Hamaguchi, K., and Leiter, E. H., 1989, Immunostimulation circumvents diabetes in NOD/Lt mice, *J. Autoimmun.* 2:759–776.

41. Schwimmbeck, P. L., Dyrberg, T., and Oldstone, M. B. A., 1990, Abrogation of diabetes in BB rats by acute virus infection, *J. Immunol.* 140:3394–3400.

42. Shyp, S., Tishon, A., and Oldstone, M. B. A., 1990, Inhibition of diabetes in BB rats by virus infection II. Effects of viral infection on the immune response to non-viral and viral antigens, *Immunology* 69:501–507.

43. Like, A. A., Guberski, D. L., and Butler, L. 1991, Influence of environmental viral agents on frequency and tempo of diabetes mellitus in BB/Wor rats, *Diabetes* 40:259–262.

44. Sadelain, M. J. W., Qin, H. Y., Lauzon, J., and Singh, B., 1990, Prevention of type I diabetes in NOD mice by adjuvant immunotherapy, *Diabetes* 39:583–589.

45. Sadelain, M. J. W., Qin, H.-Y., Sumoski, W., Parfrey, N., Singh, B., and Rabinovitch, A., 1990, Prevention of diabetes in the BB rat by early immunotherapy using Freund's adjuvant, *J. Autoimmun.* 3:671–680.

46. Quin, H.-Y., Suarez, W. L., Parfrey, N., Power, R., and Rabinovitch, A., 1992, Mechanisms of complete Freund's adjuvant protection against diabetes in BB rats: Induction of non-specific suppressor cells, *Autoimmunity* 12:193–199.

47. Guberski, D. L., Thomas, V. A., Shek, W. R., Like, A. A., Handler, E. S., Rossini, A. A., Wallace, J. E., and Welsh, R. M., 1991, Induction of type I diabetes by Kilham's rat virus in diabetes-resistant BB/Wor rats, *Science* 254:1010–1013.

48. Brown, D. W., Welsh, R. M., and Like, A. A., 1993, Infection of peripancreatic lymph nodes but not islets precedes Kilham rat virus–induced diabetes in BB/Wor rats, *J. Virol.* 67:5873–5878.

49. Jansen, F. K., Müntefering, H., and Schmidt, W. A. K., 1977, Virus induced diabetes and the immune system. I. Suggestion that appearance of diabetes depends on immune reactions, *Diabetologia* 13:545–549.

50. Haynes, M. K., Huber, S. A., and Craighead, J. E., 1987, Helper-inducer T lymphocytes mediate diabetes in EMC-infected BALB/cByJ mice, *Diabetes* 36:877–881.

51. Babu, P. G., Huber, S. A., and Craighead, J. E., 1986, Contrasting features of T

lymphocyte-mediated diabetes in encephalomyocarditis virus–infected BALB/cBy and BALB/cCum mice, *Am. J. Pathol.* **124**:193–198.

52. Onodera, T., Ray, U. R., Melez, K. A., Suzuki, H., Toniolo, A., and Notkins, A. L., 1982, Virus-induced diabetes mellitus: Autoimmunity and polyendocrine disease prevented by immunosuppression, *Nature* **297**:66–68.

53. Suenaga, K., and Yoon, J.-W., 1988, Association of β-cell-specific expression of endogenous retrovirus with development of insulitis and diabetes in NOD mouse, *Diabetes* **37**:1722–1726.

54. Tomita, K., Nakayima, H., Nakagawa, C., Miyagawa, J., Yamamoto, K., Hamaguchi, T., Noguchi, T., Tanaka, T., Hanafusa, T., Kono, N., and Matsuzawa, Y., 1993, Cloning of type C retrovirus in the pancreas of NOD mouse, *Autoimmunity* **15**(Suppl. 1):50.

55. Appel, M. C., Rossini, A. A., Williams, R. M., and Like, A. A., 1978, Viral studies in streptozotocin-induced pancreatic insulitis, *Diabetologia* **15**:327–336.

56. Paik, S. G., Fleischer, N., and Shin, S., 1980, Insulin-dependent diabetes mellitus induced by subdiabetogenic doses of streptozotocin: Obligatory role of cell-mediated autoimmune processes, *Proc. Natl. Acad. Sci. U.S.A.* **77**:6129–6133.

57. MacKay, P., and Gotfredsen, C., 1987, Passive transfer of streptozotocin-induced diabetes in mice, *Diabetes* **36**(Suppl. 1):268.

58. Isaacs, I., and Lindemann, J., 1957, Virus interference. I. The interferon, *Proc. R. Soc. B.* **147**:258–267.

59. Stobo, J., Green, I., Jackson, L., and Baron, S., 1974, Identification of a subpopulation of mouse lymphoid cells required for interferon production after stimulation with mitogens. *J. Immunol.* **112**:1589–1593.

60. Timonen, T., Saksela, E., Virtanen, I., and Cantell, K., 1980, Natural killer cells are responsible for the interferon production induced in human lymphocytes by tumor cell contact, *Eur. J. Immunol.* **10**:422–427.

61. Weigent, D. A., Langford, M. P., Smith, E. M., Blalock, J. E., and Stanton, G. J., 1981, Human B lymphocytes produce leukocyte interferon after interaction with foreign cells, *Infect. Immun.* **32**:508–512.

62. Ito, Y., Aoki, H., Kimura, Y., Takano, M., Shimokata, K., and Maeno, K., 1981, Natural interferon-producing cells in mice, *Infect. Immun.* **31**:519–523.

63. Branca, A. A., and Baglioni, C., 1981, Evidence that type I and type II interferons have different receptors, *Nature* **294**:768–770.

64. Herberman, R. R., Ortaldo, J. R., and Bonnard, G. D., 1979, Augmentation by interferon of human natural and antibody-dependent cell-mediated cytotoxicity, *Nature* **227**:221–223.

65. Lindahl, P., Leary, P., and Gresser, I., 1972, Enhancement by interferon of the specific toxicity of sensitized lymphocytes, *Proc. Natl. Acad. Sci. U.S.A.* **69**:721–725.

66. Schultz, R. M., Papamatheakis, J. D., and Chirigos, M. A., 1977, Interferon: An inducer of macrophage activation by polyanions, *Science* **197**:674–676.

67. MacKay, P., Jacobson, J., and Rabinovitch, A., 1986, Spontaneous diabetes mellitus in the Bio-Breeding/Worcester rat. Evidence in vitro for natural killer cell lysis of islet cells, *J. Clin. Invest.* **77**:916–924.

68. Varsanyi Nagy, M., Chan, E. K., Teruya, M., Forrest, L. E., Likhite, V., and Charles, M. A., 1989, Macrophage-mediated islet cell cytotoxicity in BB rats, *Diabetes* **38**:1329–1331.

69. Nagata, M., Santamaria, P., Kawamura, T., Utsugi, T., and Yoon, J.-W., 1994, Evidence for the role of CD8+ cytotoxic T cells in the destruction of pancreatic β-cells in nonobese diabetic mice, *J. Immunol.* **152**:2042–2050.

70. Maruyama, T., Wanatabe, K., Takei, I., Kasuga, A., Shimada, A., Yanagawa, T., Kasatani,

T., Suzuki, Y., Kataoka, K., Saruta, K., and Habu, S., 1991, Anti-asialo GM1 antibody suppression of cyclophosphamide-induced diabetes in NOD mice, *Diabetes Res.* **17**:16–23.

71. Engleman, E. G., Sonnenfeld, G., Dauphinee, M., Greenspan, J. S., Talal, N., McDevitt, H. O., and Merigan, T. C., 1981, Treatment of NZB/NZW F1 hybrid mice with *Mycobacteria bovis* strain BCG or type II interferon preparations accelerates autoimmune disease, *Arthritis Rheum.* **24**:1396–1402.

72. Tonietti, G., Oldstone, M. B. A., and Dixon, F., 1970, The effect of induced chronic viral infections on the immunologic diseases of New Zealand mice, *J. Exp. Med.* **132**:89–109.

73. Ewel, C., Sobel, D. O., Zeligs, B. J., and Bellanti, J. A. 1992, Poly I:C accelerates development of diabetes mellitus in diabetes-prone BB rat, *Diabetes* **41**:1016–1021.

74. Nicoletti, F., Meroni, P. L., Landolfo, S., Gariglio, M., Guzzardi, S., Barcellini, W., Lunetta, M., Mughini, L., and Zanussi, C., 1990, Prevention of diabetes in BB/Wor rats treated with monoclonal antibodies to interferon-γ, *Lancet* **336**:319.

75. Sobel, D. O., Ewel, C. H., Zeligs, B., Abbassi, V., Rossio, J., and Bellanti, J. A., 1994, Poly I:C induction of α-interferon in the diabetes-prone BB and normal Wistar rats. Dose-response relationships, *Diabetes* **43**:518–522.

76. Greiner, D. L., Mordes, J. P., Handler, E. S., Angelillo, M., Nakamura, N., and Rossini, A. A., 1987, Depletion of RT6.1+ T lymphocytes induces diabetes in resistant Biobreeding/Worcester (BB/W) rats, *J. Exp. Med.* **166**:461–475.

77. Thomas, V. A., Woda, B. A., Handler, E. S., Greiner, D. L., Mordes, J. P., and Rossini, A. A., 1991, Altered expression of diabetes in BB/Wor rats by exposure to viral pathogens, *Diabetes* **40**:255–258.

78. Like, A. A., 1990, Depletion of RT6.1+ T lymphocytes alone is insufficient to induce diabetes in diabetes-resistant BB/Wor rats, *Am. J. Pathol.* **136**:565–574.

79. Debray-Sachs, M., Carnaud, C., Boitard, C., Cohen, H., Gresser, I., Bedossa, P., and Bach, J. F., 1991, Prevention of diabetes in NOD mice treated with antibody to murine IFN-γ, *J. Autoimmun.* **4**:237–248.

80. Toyoda, H., Formby, B., Magalong, D., Redford, A., Chan, E., Takei, S., and Charles, M. A., 1994, In situ islet cytokine gene expression during development of type I diabetes in the non-obese diabetic mouse, *Immunol. Lett.* **39**:283–288.

81. Foulis, A. K., Farquharson, M. A., and Meager, A., 1987, Immunoreactive α-interferon in insulin-secreting β cells in type I diabetes mellitus, *Lancet* **ii**:1423–1427.

82. Foulis, A. K., McGill, M., and Farquharson, M. A., 1991, Insulitis in type I (insulin-dependent) diabetes mellitus in man. Macrophages, lymphocytes, and interferon-γ containing cells, *J. Pathol.* **165**:97–103.

83. Foulis, A. K., 1990, Does viral infection initiate autoimmunity in type I diabetes? *J. Autoimmun.* **3**(Suppl.):21–26.

84. Bottazzo, G. F., Dean, B. M., McNally, J. M., MacKay, E. H., Swift, P. G. F., and Gamble, D. R., 1985, In situ characterization of autoimmune phenomena and expression of HLA molecules in the pancreas in diabetic insulitis, *N. Engl. J. Med.* **313**:353–360.

85. Bottazzo, G. F., Pujoll-Borrell, R., Hanafusa, T., and Feldman, M., 1983, Role of aberrant HLA-DR expression and antigen presentation in induction of endocrine autoimmunity, *Lancet* **ii**:1115–1118.

86. Bennevie-Nielsen, V., Larsen, M. L., Frifelt, J. J., Michelsen, B., and Lernmark, Å., 1989, Association of type I diabetes and attenuated response of 2′,5′-oligoadenylate synthetase to yellow fever, *Diabetes* **38**:1636–1642.

87. Wilson, G. L., Bellomo, S. C., and Craighead, J. E., 1983, Effect of interferon on encephalomyocarditis virus infection of cultured mouse pancreatic B cells, *Diabetologia* **24**:38–41.

88. Bonnevie-Nielsen, V., Gerdes, A. M., Fleckner, J., Petersen, J. S., Michelsen, B., and Dyrberg, T., 1991, Interferon stimulates the expression of 2′,5′-oligoadenylate synthetase and MHC class 1 antigens in insulin-producing cells, *J. Interferon Res.* 11: 255–260.

89. Campbell, I. L., Bizilj, K., Colman, P. G., Tuch, B. E., and Harrison, L. C., 1986, Interferon-γ induces the expression of HLA-A,B,C but not HLA-DR on human pancreatic β-cells, *J. Clin. Endocrinol. Metab.* 35:1220–1223.

90. Campbell, I. L., Wong, G. H. W., Schrader, J. W., and Harrison, L. C., 1985, Interferon-γ enhances the expression of the major histocompatibility complex class I antigens on mouse pancreatic beta cells, *Diabetes* 34:1205–1209.

91. Campbell, I. L., Harrison, L. C., Colman, P. G., Papaioannou, J., and Ashcroft, R. G., 1986, Expression of class I MHC proteins on RIN-m5F cells is increased by interferon-γ and lymphokine-conditioned medium, *Diabetes* 35:1225–1228.

92. Leiter, E. H., Christianson, G. J., Serreze, D. V., Ting, A. T., and Worthen, S. M., 1989, MHC antigen induction by interferon-γ on cultured mouse pancreatic β cells and macrophages, *J. Exp. Med.* 170:1243–1262.

93. Varey, A. M., Lydyard, P. M., Dean, B. M., van der Meide, P., Baird, J. D., and Cooke, A., 1988, Interferon-γ induces class II MHC antigens on RINm5F cells, *Diabetes* 37: 209–212.

94. Pujoll-Borrell, R., Todd, I., Doshi, M., Bottazzo, G. F., Sutton, R., Gray, D., Adolf, G. R., and Feldman, M., 1987, HLA class II induction in human islet cells by interferon-γ plus tumor necrosis factor or lymphotoxin, *Nature* 326:304–306.

95. Campbell, I. L., Iscaro, A., and Harrison, L. C., 1988, IFN-γ and tumor necrosis factor-α: Cytotoxicity to murine islets of Langerhans, *J. Immunol.* 141:2325–2329.

96. Pukel, C., Baquerizo, H., and Rabinovitch, A., 1988, Destruction of rat islet cell monolayers by cytokines: Synergistic interactions of interferon-γ, tumor necrosis factor, lymphotoxin and interleukin-1, *Diabetes* 37:133–136.

97. Baquerizo, H., and Rabinovitch, A., 1990, Interferon-gamma sensitizes rat pancreatic islet cells to lysis by cytokines and cytotoxic cells, *J. Autoimmun.* 3(Suppl.):123–130.

98. Campbell, I. L., Cutri, A., Wilkinson, D., Boyd, A. W., and Harrison, L. C., 1989, Intercellular adhesion molecule 1 is induced on isolated endocrine islet cells by cytokines but not by reovirus infection, *Proc. Natl. Acad. Sci. U.S.A.* 86:4282–4286.

99. Sarvetnik, N., Liggitt, D., Pitts, S. L., Hansen, S. E., and Stewart, T. A., 1988, Insulin-dependent diabetes mellitus induced in transgenic mice by ectopic expression of class II MHC and interferon-gamma, *Cell* 52:773–782.

100. Sarvetnick, N., Shizuru, J., Liggitt, D., Martin, L., McIntyre, B., Gregory, A., Parslow, T., and Stewart, T., 1990, Loss of pancreatic islet tolerance induced by β-cell expression of interferon-γ, *Nature* 346:844–847.

101. Stewart, T. A., Hultgren, B., Huang, X., Pitts-Meek, S., Hully, J., and MacLachlan, N. J., 1993, Induction of type 1 diabetes by interferon-α in transgenic mice, *Science* 260:1942–1946.

102. Damian, R. T., 1989, Molecular mimicry: Parasite evasion and host defence, *Curr. Top. Microbiol. Immunol.* 145:101–115.

103. Oldstone, M. B. A., 1987, Molecular mimicry and autoimmune disease, *Cell* 50:819–820.

104. Serreze, D. L., Leiter, E. H., Kuff, E. L., Jardieu, P., and Ishizaka, K., 1988, Molecular mimicry between insulin and retroviral antigen p73. Development of cross-reactive autoantibodies in sera from NOD and c57BL/KsJ *db/db* mice, *Diabetes* 37:351–358.

105. Hurtenbach, U., and Maurer, C., 1989, Type I diabetes in NOD mice is not associated with insulin-specific, autoreactive T cells, *J. Autoimmun.* 2:151–161.

106. Muir, A., Luchetta, R., Song, H.-Y., Peck, A., Krischer, J., and Maclaren, N., 1993, Insulin immunization protects NOD mice from diabetes, *Autoimmunity* 15(Suppl. 1):58.
107. Zhang, J., Davidson, L., Eisenbarth, G., and Weiner, H. L., 1991, Suppression of diabetes in nonobese mice by oral administration of porcine insulin, *Proc. Natl. Acad. Sci. U.S.A.* 88:10252–10256.
108. Bottazzo, G. F., Florin-Christensen, A., and Doniach, D., 1974, Islet cell antibodies in diabetes mellitus with autoimmune polyendocrine deficiencies, *Lancet* ii:1279–1283.
109. Bækkeskov, S., Aanstoot, H.-J., Christgau, S., Reetz, A., Solimena, M., Cascalho, M., Folli, F., Richter-Olesen, H., and De-Camilli, P., 1990, Identification of the 64K auto-antigen in insulin dependent diabetes as the GABA-synthesizing enzyme glutamic acid decarboxylase, *Nature* 347:151–156.
110. Atkinson, M. A., Kaufmann, D. L., Campbell, I. L., Gibbs, K. A., Shah, S. C., Bu, D.-F., Erlander, M. G., Tobin, A. J., and Maclaren, N. K., 1992, Response of peripheral-blood mononuclear cells to glutamate decarboxylase in insulin-dependent diabetes, *Lancet* 339:458–459.
111. Karlsen, A. E., Hagopian, W. A., Petersen, J. S., Boel, E., Dyrberg, T., Grubin, C. E., Michelsen, B. K., Madsen, O. D., and Lernmark, Å., 1992, Recombinant glutamic acid decarboxylase (representing the single isoform expressed in human islets) detects IDDM-associated 64,000-Mr autoantibodies, *Diabetes* 41:1355–1359.
112. Michelsen, B. K., Petersen, J. S., Boel, E., Møldrup, A., Dyrberg, T., and Madsen, O. D., 1991, Cloning, characterization and autoimmune recognition of rat islet glutamic acid decarboxylase in insulin-dependent diabetes mellitus, *Proc. Natl. Acad. Sci. U.S.A.* 88:8754–8758.
113. Hagopian, W. A., Karlsen, A. E., Gottsäter, A., Landin-Olsson, M., Grubin, C. E., Sundkvist, G., Petersen, J. S., Boel, E., Dyrberg, T., and Lernmark, Å., 1993, Quantitative assay using recombinant human islet glutamic acid decarboxylase (GAD65) shows that 64K autoantibody positivity at onset predicts diabetes type, *J. Clin. Invest.* 91:368–374.
114. Petersen, J., Hejnæs, K. R., Moody, A., Karlsen, A. E., Marshall, M. O., Høier-Madsen, M., Boel, E., Michelsen, B., and Dyrberg, T., 1994, Detection of GAD65 antibodies in diabetes and other autoimmune diseases using a simple radioligand assay, *Diabetes* 43:459–467.
115. Bækkeskov, S., Landin, M., Kristensen, J. K., Srikanta, J., Bruining, G. J., Mandrup-Poulsen, T., de Beaufort, C. D., Soeldner, J. S., Eisenbarth, G., Lindgren, F., Sundquist, G., and Lernmark, Å., 1987, Antibodies to a 64,000 Mr human islet cell antigen precede the clinical onset of insulin dependent diabetes, *J. Clin. Invest.* 79:926–934.
116. Kaufman, D. L., Clare-Salzler, M., Tian, J., Forsthuber, T., Ting, G. S. P., Robinson, P., Atkinson, M. A., Sercarz, E. E., Tobin, A. J., and Lehmann, P. V., 1994, Spontaneous loss of T-cell tolerance to glutamic acid decarboxylase in murine insulin-dependent diabetes, *Nature* 366:69–72.
117. Tisch, R., Yang, X.-D., Singer, S. M., Liblau, R. S., Fugger, L., and McDevitt, H. O., 1994, Immune response to glutamic acid decarboxylase correlates with insulitis in non-obese diabetic mice, *Nature* 366:72–75.
118. Petersen, J. S., Karlsen, A., Markholst, H., Worsaae, A., Dyrberg, T., and Michelsen, B., 1994, Neonatal tolerization with GAD but not with BSA delays the onset of diabetes in NOD mice, *Diabetes* 43:1478–1484.
119. Schwimmbeck, P. L., Dyrberg, T., Drachman, D. B., and Oldstone, M. B. A., 1989, Molecular mimicry and myasthenia gravis: An autoantigenic site of the acetylcholine receptor a-subunit that has biologic activity and reacts immunochemically with herpes simplex virus, *J. Clin. Invest.* 84:1174–1180.

120. Gautam, A. M., Lock, C. B., Smilek, D. E., Pearson, C. I., Steinman, L., and McDevitt, H. O., 1994, Minimum structural requirements for peptide presentation by major histocompatibility complex class II molecules, *Proc. Natl. Acad. Sci. U.S.A.* **91**:767–771.

121. Pak, C. Y., McArthur, R. G., Eun, H.-M., and Yoon, J.-W., 1988, Association of cytomegalovirus infection with autoimmune type 1 diabetes, *Lancet* **ii**:1–4.

122. Kaufmann, D. L., Erlander, M. G., Clare-Salzler, M., Atkinson, M. A., Maclaren, N. K., and Tobin, A. J., 1992, Autoimmunity to two forms of glutamate decarboxylase in insulin-dependent diabetes mellitus, *J. Clin. Invest.* **89**:283–292.

123. Richter, W., Mertens, T., Schoel, B., Muir, P., Ritzkowsky, A., Scherbaum, W. A., and Boehm, B. O., 1994, Sequence homology of the diabetes-associated autoantigen glutamate decarboxylase with Coxsackie B4-2C protein and heat shock protein 60 mediates no molecular mimicry of autoantibodies, *J. Exp. Med.* **180**:721–726.

124. Oldstone, M. B. A., Nerenberg, M., Southern, P., Price, J., and Lewicki, H., 1991, Virus infection triggers insulin-dependent diabetes mellitus in a transgenic model: Role of anti-self (virus) immune response, *Cell* **65**:319–331.

125. Ohashi, P. S., Oehen, S., Buerki, K., Pircher, H., Ohashi, C. T., Odermatt, B., Malissen, B., Zinkernagel, R. M., and Hengartner, H., 1991, Ablation of "tolerance" and induction of diabetes by virus infection in viral antigen transgenic mice, *Cell* **65**:305–317.

126. Oehen, S., Ohashi, P. S., Aichele, P., Buerki, K., Hengartner, H., and Zinkernagel, R. M., 1992, Vaccination or tolerance to prevent diabetes, *Eur. J. Immunol.* **22**:3149–3153.

127. Lo, D., Freedman, J., Hesse, S., Palmiter, R., Brinster, R., and Sherman, L. A., 1992, Peripheral tolerance to an islet cell–specific hemagglutinin transgene affects both CD4+ and CD8+ T cells, *Eur. J. Immunol.* **22**:1013–1022.

128. Adams, T. E., Alpert, S., and Hanahan, D., 1987, Non-tolerance and autoantibodies to a transgenic self antigen expressed in pancreatic beta-cells, *Nature* **325**:223–228.

129. Herrath, M., Dockter, J., and Oldstone, M. B. A., 1994, How virus induces a rapid or slow onset insulin-dependent diabetes mellitus in a transgenic model, *Immunity* **1**:231–242.

130. Ohashi, P. S., Oehen, S., Aichele, P., Pircher, H., Odermatt, B., Herrera, P., Higuchi, Y., Buerki, K., Hengartner, H., and Zinkernagel, R., 1993, Induction of diabetes is influenced by the infectious virus and local expression of MHC class I and tumor necrosis factor-α, *J. Immunol.* **150**:5185–5194.

131. Sutherland, D. E. R., 1987, Pancreas transplantation, an update, *Diabetes Annu.* **3**:159–188.

6

The Role of Coxsackie B Viruses in the Pathogenesis of Type I Diabetes

JI-WON YOON and HELEN KOMINEK

1. INTRODUCTION

Insulin-dependent diabetes mellitus (IDDM) is believed to result from the destruction of insulin-producing pancreatic beta cells over a lengthy asymptomatic period.[1-3] It is thought that a variety of etiological factors, including both genetic and nongenetic environmental factors, are associated with this destruction.[4,5] Viruses have been considered for many years as one possible environmental factor associated with the development of IDDM. It is thought that in some cases, viruses may act as injurious agents to beta cells and that in other cases viruses may trigger or somehow contribute to beta cell–specific autoimmunity (reviewed by Yoon).[6] Encephalomyocarditis (EMC) virus and mengovirus are viruses which can cause diabetes in genetically susceptible mice by cytolytic infection of beta cells, while retroviruses and Kilham's rat virus have been implicated in the initiation of beta cell–specific autoimmunity in non-obese diabetic (NOD) mice and diabetes-resistant BioBreeding (DRBB) rats, respec-

JI-WON YOON and HELEN KOMINEK • Laboratory of Viral and Immunopathogenesis of Diabetes, Julia McFarlane Diabetes Research Centre, Department of Microbiology and Infectious Diseases, Faculty of Medicine, University of Calgary, Calgary, Alberta T2N 4N1, Canada.

Microorganisms and Autoimmune Diseases, edited by Herman Friedman *et al.* Plenum Press, New York, 1996.

tively. In humans, mumps virus, rubella virus, reovirus, cytomegalovirus, and Coxsackie B viruses appear to be associated with the development of IDDM.

Indications that Coxsackie B viruses may be involved in the etiology of IDDM have come from epidemiological studies describing high frequencies of anti–Coxsackie B virus IgM antibody in newly diagnosed diabetic children,[7-23] as well as from anecdotal case reports associating recent-onset IDDM with a preceding or coinciding Coxsackie B viral infection. The presence of Coxsackie B virus-specific antigens in the islets,[24] and beta-cell damage and destruction in children who have died from severe Coxsackie B viral infections[25,26] further suggest the involvement of Coxsackie B viruses. Isolation of Coxsackie B4 and B5 viruses from the pancreata of patients with acute-onset IDDM, followed by the development of diabetes in susceptible mice infected with the isolates[25,27] provides further support for a role for the virus in the pathogenesis of the disease.

In this chapter, we will discuss studies on the involvement of Coxsackie B viruses in the induction of diabetes, the difficulties in studying the relationship between Coxsackie B viral infection and diabetes, the possible mechanisms for Coxsackie B viral involvement, as well as future research directions needed in this area of study.

2. EPIDEMIOLOGICAL STUDIES

2.1. Seroepidemiological Studies

Twenty-five years ago, Gamble *et al.*[7] conducted one of the first epidemiological studies to compare anti–Coxsackie B antibody levels in patients with recent-onset diabetes and those in nondiabetic controls. They reported that patients with recent-onset IDDM (less than 3 months duration) had higher titers of antibody to Coxsackie B virus, especially the B4 serotype, than the nondiabetic subjects. Since that time, researchers have continued to compare levels of anti–Coxsackie B virus neutralizing antibodies in sera from diabetic and nondiabetic subjects. Most studies have examined levels of Coxsackie B1–6 virus-specific IgM in patients with recently diagnosed IDDM, and many of these studies have found a positive correlation between a recent Coxsackie B viral infection and the onset of diabetes.[8-23] There have been several studies, however, that have found no difference in anti–Coxsackie B virus antibody titer levels between IDDM patients and nondiabetic controls,[28-33] while in other studies, some researchers have found that antibody titers to Coxsackie B viruses are actually lower in diabetic patients than in control subjects.[34,35] The results from the studies that find a negative correlation may indicate that some individuals with less previous exposure may have a more serious response and illness when finally exposed. The results from seroepidemiological studies are summarized in Table I.

2.2. Genetic Epidemiology

Other epidemiological studies have reported a definite correlation between HLA haplotype and the presence of IgM antibodies against Coxsackie B viruses. As well as assaying sera from 36 recent-onset IDDM patients for IgM antibodies against Coxsackie virus serotypes B1–5, Fohlman et al.[20] assayed the sera for restriction fragment length polymorphism (RFLP) patterns associated with HLA-DR3 or HLA-DR4, or HLA-DQ III or HLA-DQ IV beta (patterns thought to be associated with genetic predisposition to IDDM).[36,37] Eleven out of the 18 Coxsackie B–positive patients had HLA-DQ III and HLA-DR3 (61%) versus only 28% of the Coxsackie B–negative patients. All of the patients positive for antibodies against Coxsackie B2, B3, and B5 had HLA-DR4 and HLA-DR IV patterns. There was a statistically significant difference between these patients and all five of the Coxsackie B4–positive patients, who in contrast all had HLA-DR3 or HLA-DQ III patterns. These findings appear to support the view that susceptibility to Coxsackie B4 infection is associated with a different host genetic constitution from susceptibility to other serotypes of Coxsackie B virus.

The Pittsburgh IDDM study[38,39] examined 172 newly diagnosed patients with IDDM. It found that HLA-DR3 haplotypes were found more frequently among male subjects, while HLA-DR4 was found more frequently among female subjects. The patients with an HLA-DR4 haplotype more frequently reported recent viral infections and had a greater frequency of anti–Coxsackie B virus antibodies.

Field et al.[40] measured IgG antibody titers to Coxsackie B1–6 virus in sera from children with IDDM of several years duration (mean = 4.6 years) and 87 of their nondiabetic siblings. Compared with their nondiabetic siblings, the children with IDDM had a significantly increased frequency of high response (titers > 1:320) to Coxsackie B2 (8% for IDDM patients versus 1% for nondiabetic siblings), Coxsackie B4 (15% versus 1%), and Coxsackie B viruses in general (25% versus 5%). However, Field et al. found that the frequencies of HLA-DR and immunoglobulin (GM, KM) antigens did not differ between IDDM patients with or without high responses to Coxsackie B2, B4, or B viruses in general.

The 1989 Genetic Analysis Workshop 5 (GAW5)[41,42] assayed serum

TABLE I
Coxsackie B Virus Association with IDDM

Serotype	Findings	Reference
B2, B3	IDDM patients had higher titers to Coxsackie B virus than non-diabetic controls; titers were inversely proportional to the duration of the disease.	7
B4	Of 162 patients with recent-onset IDDM, 70% had Coxsackie B4 neutralizing antibody (titer > 1:4) compared to 58% of age-matched controls.	8
B1–6	IgM responses to Coxsackie B1–6 viruses were detected by ELISA in 39% of children with IDDM, while Coxsackie B virus–specific IgM responses were present in only 5.5% of age-matched controls.	9
B1–6	80% of 166 children with recent-onset IDDM and a clinical history of recent infectious illness had antibodies against at least one Coxsackie B serotype. In those with an antibody titer of greater than 1:256, 44% had specific neutralizing IgM antibodies against Coxsackie B viruses.	10
B4	2/24 recent-onset IDDM patients had IgM against Coxsackie B4 virus.	11
B3, B4, B5	A longitudinal study of viral antibodies against Coxsackie B3, B4, and B5 viruses in 17 IDDM patients showed that Coxsackie B4 antibody titers fell from the levels at diagnosis and 5 months after diagnosis to a lower level 2 years after diagnosis. Over the same period, titers of antibodies against Coxsackie B3 rose.	12
B[a]	Coxsackie B virus–specific IgM responses were detected in 67% of diabetic patients on the day of diagnosis, while none of the age-matched nondiabetic control subjects had these antibodies detected in their sera.	13
B1–5	30% of the diabetic patients had neutralizing antibodies present in their sera compared with only 6% of the nondiabetic controls.	14
B1–5	16/24 (67%) of recent-onset IDDM patients showed Coxsackie B virus–specific IgM responses, suggestive of a recent Coxsackie B viral infection.	15
B4	32% of new IDDM patients had Coxsackie B4 neutralizing antibody titers greater than 1:16, compared to only 10% of the control subjects.	16
B1–5	50% of recent-onset IDDM patients had IgM against Coxsackie B1–5.	17
B2, B4, B5	IgM antibodies against Coxsackie B2, B4, or B5 were found in 57% of children with IDDM versus 6% of healthy controls.	18, 19
B2	75.5% of diabetic children had antibodies against Coxsackie B2 virus versus 46.4% of controls.	20
B4	42% of children with recent-onset IDDM had significantly higher titers of neutralizing antibodies to Coxsackie B4 virus (1:64–1:256), compared to 14% of the control group.	21

TABLE I
(*Continued*)

Serotype	Findings	Reference
B1–5	Half of 35 newly diagnosed diabetic children and 93% of their siblings were Coxsackie B1–5 virus IgM antibody–positive, indicating an intrafamilial spread of the virus. The most frequent serotype against which the IgM was directed was Coxsackie B4.	22
No correlation		
B	No difference in titers between patients with diabetes and control subjects.	29
B	No difference in titers between patients with diabetes and control subjects.	30
B	No relationship between Coxsackie B viral antibodies and the onset of IDDM.	31
B1–6	No difference in levels of neutralizing antibody titer between groups.	32
B	No positive serotypes for Coxsackie B viruses in 91 newly diagnosed children with IDDM.	33
Reverse correlation		
B3, B4, B5	Antibody titer levels to Coxsackie B5 virus were comparable in diabetic and nondiabetic subjects, but levels against Coxsackie B3 and B4 virus were lower in diabetic patients than controls, and the lower antibody titer was associated with a significantly increased risk of IDDM.	34
B	15%–3% of children with IDDM had Coxsackie B virus–specific IgM compared to 19%–10% of healthy controls.	35

*a*B indicates serotype was not specified.

samples from North American IDDM families and found that Coxsackie B4 level was higher than that of any other B serotype and that the level of Coxsackie B4 antibodies in nondiabetic siblings appeared to differ based on the degree of haplotype sharing with the diabetic sibling.

In a three-year study, D'Alessio[43] also compared Coxsackie B IgM neutralizing antibody titers and HLA-DR typing in 194 newly diagnosed diabetic children and age- and sex-matched controls. The data from this study demonstrated an association between Coxsackie B viral infection and the onset of IDDM only in HLA-DR3-positive persons age 10 years or older and supports the theory that diabetogenic Coxsackie B strains circulate only periodically.

A further study of Coxsackie B virus–specific IgM responses and HLA-DR antigens in diabetic patients was done by Schernthaner *et al.*[44] It was found that 81% of the patients studied had HLA-DR3 or HLA-DR4, or both, and that 96% of patients with Coxsackie B virus–specific IgM responses had at least one of these two HLA haplotypes compared with 76% of patients without Coxsackie B virus–specific IgM responses. In addition, C-peptide secretion was significantly lower in patients with Coxsackie B virus–specific IgM antibodies.

Bruserud et al.[45,46] studied the relationships between responses to viral antigens and the HLA-DR3 and HLA-DR4 association with IDDM. They found that there was an increased frequency of T lymphocytes responding to Coxsackie B4 antigens together with DR4, compared with other DR determinants. Since similar results were seen in nondiabetic control subjects, it was suggested that elements on the DR4 molecule may control T-lymphocyte responses to Coxsackie B4 virus. Although some reports describe both genetic susceptibility to IDDM and immune responses to Coxsackie B4 infection as being linked to HLA haplotypes, no definitive pattern has emerged, though HLA-DR3 and HLA-DR4 haplotypes appear to be associated somehow with both of these phenomena. More studies on the relationship between specific HLA haplotypes, the development of Coxsackie B virus–associated IDDM, and immune responses to Coxsackie B viral infections are needed.

2.3. Other Epidemiological Studies

It was first noted almost 70 years ago[47] that there appeared to be a seasonal incidence in the onset of acute IDDM, with a peak in the autumn. Since diseases with seasonal incidences are often caused by viral infection, Gamble and Taylor[48] examined the incidence of the onset of IDDM in patients younger than 30 years of age and found a peak in the autumn. The pattern of incidence showed a positive correlation with the annual prevalence of Coxsackie B4 infection, but not for infection by other viruses. A later study by Gamble[49] found that peaks have also been seen in the onset of diabetes at school entry around age 5, at ages 8–9 when children enter elementary school, and at the age of high school entry. This also correlates with increased exposure to viruses known to take place at these times.

Wagenknecht et al.[50] studied the temporal relationship between the incidence of IDDM and an outbreak of Coxsackie B5 viral infections. They found that there was a significant increase in the incidence of IDDM corresponding to a worldwide Coxsackie B5 viral epidemic in the early 1980s, providing supporting evidence for a role of Coxsackie B5 virus in the etiology of IDDM. There was a worldwide increase in the incidence of IDDM in the same years, as evidenced by reports such as the one by

Rewers *et al.*,[51] and the correspondence of this increased incidence of IDDM with Coxsackie B viral epidemics provides some circumstantial evidence of the virus' involvement in the pathogenesis of IDDM. The correlation of outbreaks of infection by specific serotypes of Coxsackie B virus and the development of IDDM is interesting, but further studies are needed.

3. DIFFICULTIES ENCOUNTERED IN EPIDEMIOLOGICAL STUDIES ON THE RELATIONSHIP BETWEEN COXSACKIE B VIRAL INFECTIONS AND IDDM

Because results from epidemiological studies on the correlation between Coxsackie B virus–specific IgM responses and recent-onset IDDM are not consistent, there is some controversy as to the role of Coxsackie B viruses in the pathogenesis of IDDM. Many of the studies have found a fairly high correlation between recent infection with Coxsackie B viruses and development of IDDM, as opposed to the lack of correlation seen in the control population. Other studies have not found such a correlation. Reports describing both types of findings appear to be methodologically acceptable. The controversy may arise not from the science, but rather from the nature of the virus and genetically determined host factors. There are different variants of the virus found within each serotype. For example, Prabhakar *et al.*[52] isolated 13 variants of Coxsackie B4 virus. In another study, four variants of Coxsackie B4 virus were tested and one was found to be diabetogenic, while the remaining three were not.[53] This is an indication of the possible rarity of diabetogenic variants of Coxsackie B4 virus. Also, we are unable to distinguish serologically between diabetogenic and nondiabetogenic variants using routine neutralizing antibody or ELISA testing, since the variants are cross-reactive. Therefore, if a person is exposed to a more common nondiabetogenic variant of Coxsackie B4 virus prior to exposure to a more rare diabetogenic variant of the same serotype, the person will have already developed antibodies against the nondiabetogenic variant, which will neutralize the diabetogenic variant during the subsequent infection; thus, the person will not become diabetic, even if he or she is genetically predisposed to the disease. If this person is a subject in an epidemiological study, the results will not be meaningful, as the lack of diabetes seen will not be a result of lack of exposure to a diabetogenic Coxsackie B virus, and no correlation between Coxsackie B viral infection and the incidence of diabetes would be found. In contrast, in certain areas, outbreaks of diabetogenic virus prior to outbreaks of nondiabetogenic virus would result in a high correlation between Coxsackie B viral infection and the development of diabetes. In animal models, it has already been proven that prior infection with nondiabetogenic (EMC-B) virus results in no development of diabetes after subsequent infection by diabetogenic (EMC-D) virus.[54] In addition, there are genetically determined differences in susceptibility to virus-induced diabetes, as has been shown in experiments using different strains of mice infected with Coxsackie B4 virus.[55] It is believed that humans, as well, will not become diabetic when infected by diabetogenic Coxsackie B virus unless they are genetically predisposed to developing the disease. Thus the correlation between Coxsackie B virus infection and development of diabetes seen in some studies and the lack of correlation found in other studies may be dependent on the genetic makeup of the virus and genetic backgrounds of the patients.

4.   CASE REPORTS AND *IN VITRO* STUDIES ON HUMAN ISLET CELLS

4.1.   Anecdotal Reports

There have been a large number of anecdotal reports linking a Coxsackie B viral infection with the onset of IDDM. Wilson *et al.*[56] reported the case of an 18-month-old boy with acute-onset diabetes and high titers of Coxsackie B2 neutralizing antibody and Coxsackie B IgM, suggesting that the diabetes was associated with a recent Coxsackie B viral infection.

In studies of 11 children who died from picornavirus infections, Jenson *et al.*[26] found that seven of these infections had been caused by Coxsackie B viruses, and of the seven Coxsackie B–infected children, four showed lymphocytic infiltration of their pancreatic islets and destruction of beta cells. Coxsackie B1 virus was isolated from one child, Coxsackie B4 virus from the second, and untyped Coxsackie B viruses from the remaining two children. Ahmad and Abraham[57] reported the case of an infant who died of a Coxsackie B5 viral infection. The pancreas of the child exhibited islet cell damage in which the inflammatory response consisted of mononuclear cells. In this case, the exocrine acinar tissue was completely uninvolved.

Asplin *et al.*[58] reported the case of a family where serial measurements of islet cell antibodies and titers of antibodies against Coxsackie B3,

B4, and B5 had been determined for 3 years before one of the children developed IDDM. Viral titers to Coxsackie B3 and B5 were negative, while the child who subsequently became diabetic, as well as her sister and mother, all had raised titers of Coxsackie B4 antibodies ranging between 1:128 to 1:512 for the entire three-year period monitored. Orchard et al.[59] studied the development of IDDM in two siblings of diabetic children. One case involved a 13-year-old brother of a diabetic girl. The boy did not share his sister's HLA haplotype, had normal glucose tolerance 80 days before diagnosis, and showed serological evidence of Coxsackie B4 infection at the time of diagnosis. In the other case, the patient had had impaired glucose tolerance for over a year and a half, was HLA-identical to his diabetic brother, and did not present serological evidence of viral infection. Orchard et al. concluded that these two opposing cases showed the contrasting interaction of the main pathogenetic factors associated with IDDM.

Niklasson et al.[60] reported an outbreak of gastroenteritis in 22 United Nations soldiers stationed in Egypt. The majority of these patients showed a significant titer rise for Coxsackie B virus, as indicated by plaque reduction neutralization, and the serotype was determined to be most likely Coxsackie B4. One of the 22 patients developed IDDM 10 weeks after this infection. A fatal Coxsackie B6 infection in a 6-year-old girl was reported by Nigro et al.[61] The child initially exhibited severe meningoencephalitis, which was followed by hyperglycemia and glycosuria with complement-fixing islet cell antibodies, suggestive of IDDM.

While these anecdotal reports indirectly support the theory that Coxsackie B viral infections may result in destruction of pancreatic islets, leading to the development of IDDM, they do not provide proof that Coxsackie B viruses can unequivocally cause IDDM or that Coxsackie B viruses are definitely capable of infecting human beta cells. Because it is impossible to conduct in vivo studies on the ability of Coxsackie B viruses to replicate in human beta cells, investigations have been undertaken using in vitro systems to determine whether or not Coxsackie B viruses have the ability to infect and destroy human beta cells in culture.

4.2. In Vitro Studies on Human Islet Cells

The capacity of Coxsackie B3 and B4 viruses to infect insulin-containing beta cells was studied using human pancreatic cell cultures. Using a double-label immunofluorescent technique (since the cells were not pure beta cells), it was clearly demonstrated that Coxsackie B3 and B4 viruses could infect human beta cells.[25,62] Radioimmunoassays showed

that intracellular insulin in infected beta cells decreased rapidly beginning 24 hr after infection and that the decrease in insulin roughly
paralleled the increase in viral titer. These studies demonstrated that
human beta cells, at least *in vitro*, are subject to infection by Coxsackie B
viruses (Fig. 1).

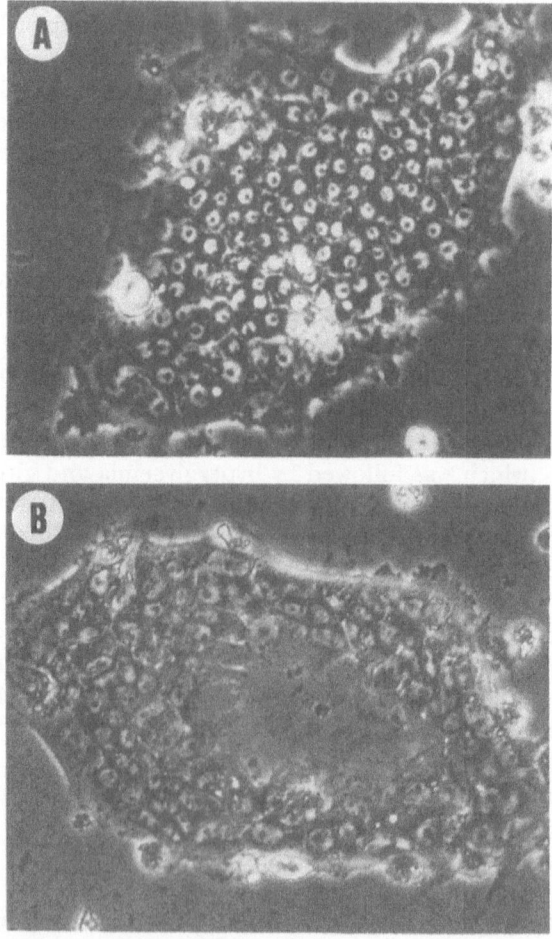

FIGURE 1. *In vitro* infection of human pancreatic beta cells. An uninfected monolayer
colony of pancreatic beta cells (A) and Coxsackie B4 virus–infected beta cell monolayer
(B). Severe cytopathic effects are noted in the center of the infected beta cell colony (×150).

The effect of Coxsackie B4 viral infection on human islet cells *in vitro* was studied by Szopa *et al.*[63] At 24 and 48 hr postinfection the islets were incubated in either a high or low concentration of glucose, and insulin release and insulin content were measured by radioimmunoassay. Coxsackie B4 infection increased basal insulin release at both 24 and 48 hr postinfection. Neither basal nor glucose-stimulated insulin biosynthesis was affected 24 hr postinfection, but 48 hr after infection, inhibition of this biosynthesis was observed at both glucose concentrations. Szopa *et al.* also found that there was a decrease in total protein synthesis 48 hr postinfection. It is interesting to note from this study's results that Coxsackie B4 viral infection can impair human islet cell metabolism *in vitro* without involvement of the immune system. It should be observed, however, that differences in beta-cell susceptibility to Coxsackie B viral infection *in vitro*, compared to *in vivo*, may exist, and the results from these studies do not provide conclusive evidence that Coxsackie B viruses can infect and destroy pancreatic beta cells *in vivo*, though they do suggest that beta cells are not inherently resistant to Coxsackie B viral infection.

4.3. Cases of Isolation of Coxsackie B Viruses from Pancreata

More direct supporting evidence of a role for Coxsackie B viral infection in the onset of IDDM has come from research reporting the isolation of Coxsackie B viruses from, or the presence of Coxsackie B viral antigens in, pancreata of recent-onset IDDM patients.

In 1976, Gladisch *et al.*[24] reported the case of a 5-year-old girl who developed myocarditis and diabetes 2 weeks after open heart surgery. At necropsy, her islets showed a lymphocytic infiltrate and beta-cell necrosis. Coxsackie B4 antigens were detected in the islets by immunofluorescence, and high levels of antibody against Coxsackie B4 virus were present in the child's serum. This study shows that Coxsackie B4 virus is able to infect human pancreatic islets.

More unequivocal support for the idea that Coxsackie B4 virus might trigger some cases of human IDDM came 3 years later when a variant of Coxsackie B4 virus isolated from the pancreas of a diabetic patient was found to induce diabetes in mice.[25] Less than 3 days after the onset of a flu-like illness, a previously healthy 10-year-old boy was admitted to the hospital with diabetic ketoacidosis. Seven days after admission the child died when all attempts to ameliorate his condition failed. At autopsy, lymphocytic infiltration of the islets and beta-cell necrosis was observed (Fig. 2). The condition of the child's islets were very similar to that of

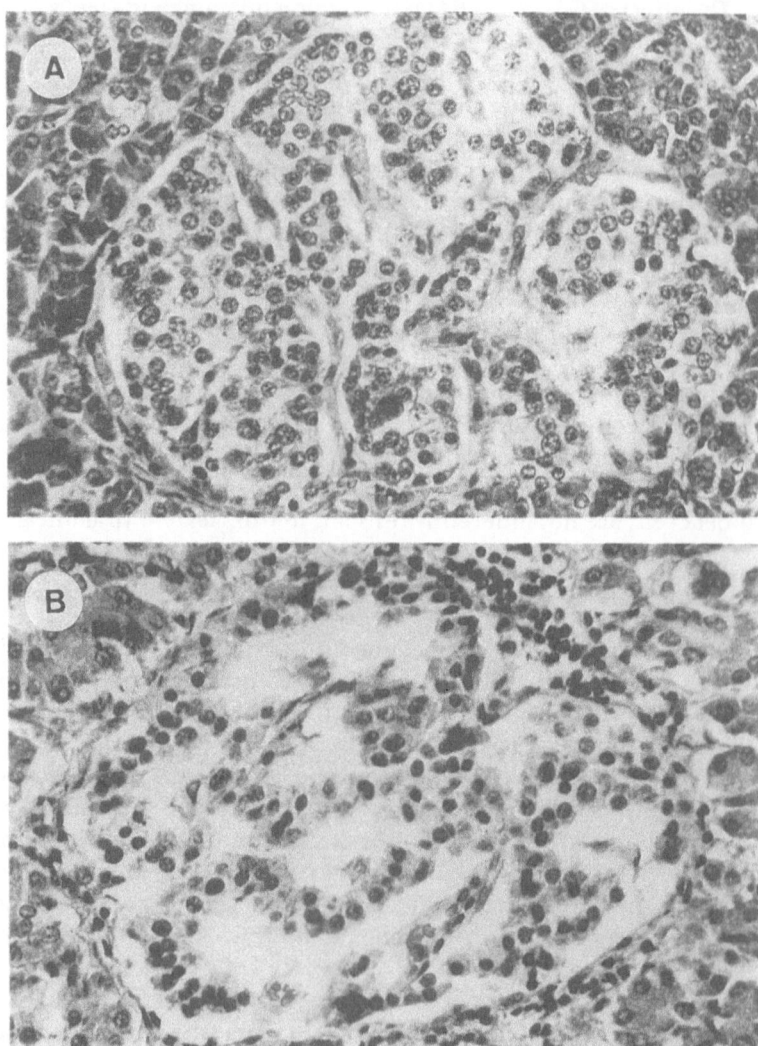

FIGURE 2. *In vivo* infection of human pancreatic beta cells. Section of a normal human pancreas (A), showing an islet of Langerhans surrounded by acinar cells (× 200); section of pancreas from a diabetic patient (B), showing moderate accumulation of inflammatory cells at the periphery of the islet (× 220); and a pancreatic section from a Coxsackie B virus–infected diabetic patient (C), showing extensive inflammatory infiltrate, loss of islet architecture, and severe islet-cell degeneration with little inflammation in surrounding acinar tissue (× 180).

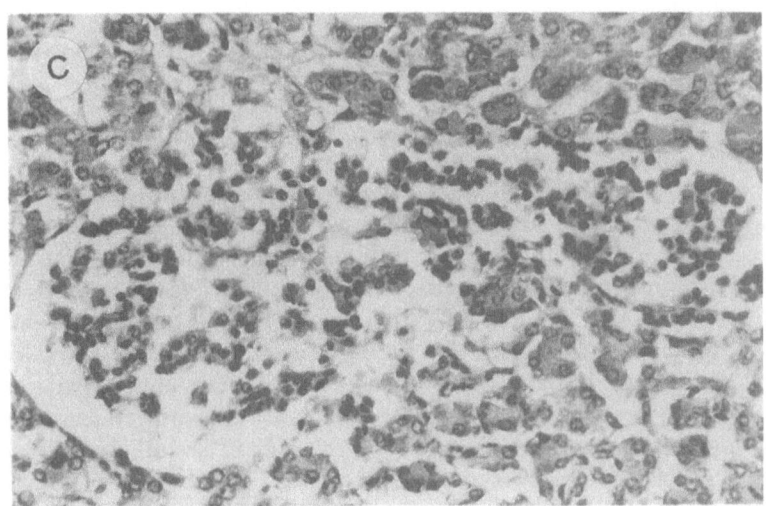

FIGURE 2. (*Continued*)

murine islets after Coxsackie B4 viral infection. When several inbred strains of mice were inoculated with the Coxsackie B4 variant isolated from the diabetic child, SJL/J male mice developed diabetes, while CBA/ J, C57BL/6J, and Balb/c mice did not. These studies fulfilled Koch's postulates, suggesting that Coxsackie B4 virus can induce IDDM in humans in certain cases.

An additional case report[27] has strengthened support for the notion that Coxsackie B viruses can trigger some cases of human IDDM. In this case study, Champsaur *et al.* reported that a 16-month-old girl with Coxsackie B5 infection developed diabetic symptoms for a 10-day period shortly after infection, went into remission for 2 months, then developed definite IDDM. In this case, the virus isolated from the girl's feces caused glucose intolerance in the same mouse strains used in the above study. Islet cell antibodies were found in the child a week before the onset of diabetes, and immunogenetic analysis revealed that the child had markers indicating a high risk for the development of IDDM.

On the basis of the above studies, which detected Coxsackie B viral antigens in the islets and which demonstrated that isolated viruses caused diabetes when injected into susceptible mice, it may be suggested that certain cases of IDDM can be caused by Coxsackie B viral infections.

5. ANIMAL STUDIES

5.1. Murine Studies

5.1.1. Diabetogenic Potential of Coxsackie B Viruses in Mice

Naturally acquired Coxsackie B viral infections of the pancreas normally produce a predominantly acinar cell pancreatitis and do not usually affect the endocrine pancreas.[64] Coleman et al.[65,66] infected CD1 mice, which are susceptible to diabetes, with a passaged Coxsackie B4 isolate, and 20–30% of the animals developed hyperglycemia. In other strains of mice, unadapted Coxsackie viruses did not infect pancreatic beta cells when inoculated; however, once repeatedly passaged in murine pancreatic beta cell–rich cultures, Coxsackie B4 virus was shown to be capable of producing hypoinsulinemia and hyperglycemia in several inbred strains of mice.[55] Histological examinations of pancreata from Coxsackie B4 virus–infected SJL/J mice revealed beta-cell necrosis and presence of viral antigens within the beta cells in some islets (Fig. 3). The ability of Coxsackie B4 virus to induce diabetes is influenced by the genetic background of the host, however, as only SJL/J, SWR/J, and NIH Swiss mice, but not other inbred strains of mouse such as C57BL/6J, CBA/J, AKR, Balb/c, C3H/J, DBA/1J, or DBA/2J, developed diabetes. There were also sex differences in the severity of hyperglycemia induced by Coxsackie B4 viral infection of mice and in the percentage of animals which became diabetic. A subsequent study was conducted by Toniolo et al.[67] in which all six B serotypes of Coxsackie virus were repeatedly passaged in beta cells. Once again, passaging of the viruses in beta-cell cultures changed their tropism for beta cells, and when SJL/J mice were infected with these passaged variants, they developed glucose intolerance, the severity and duration of which varied between different virus passages.

A Coxsackie B4 isolate of human origin, the Edwards isolate, was plaque-purified, and the resultant three "strains" were designated E1, E2, and E3.[68,69] It was noted after performing studies using these three strains of virus that E2 showed the most intense level of virus antigen accumulation in murine islet cells. This variant has been subsequently used in numerous experiments.

Cook et al.[70] found that the diabetogenic potential of the E2 variant of Coxsackie B4 did not correlate with the mortality it induced and that the inbred mouse strain SWR/J, with the most severe pancreatopathy and islet atrophy, suffered only intermediate susceptibility to virus-induced lethality. In these mice, problems in glucose homeostasis were seen 7 to 21

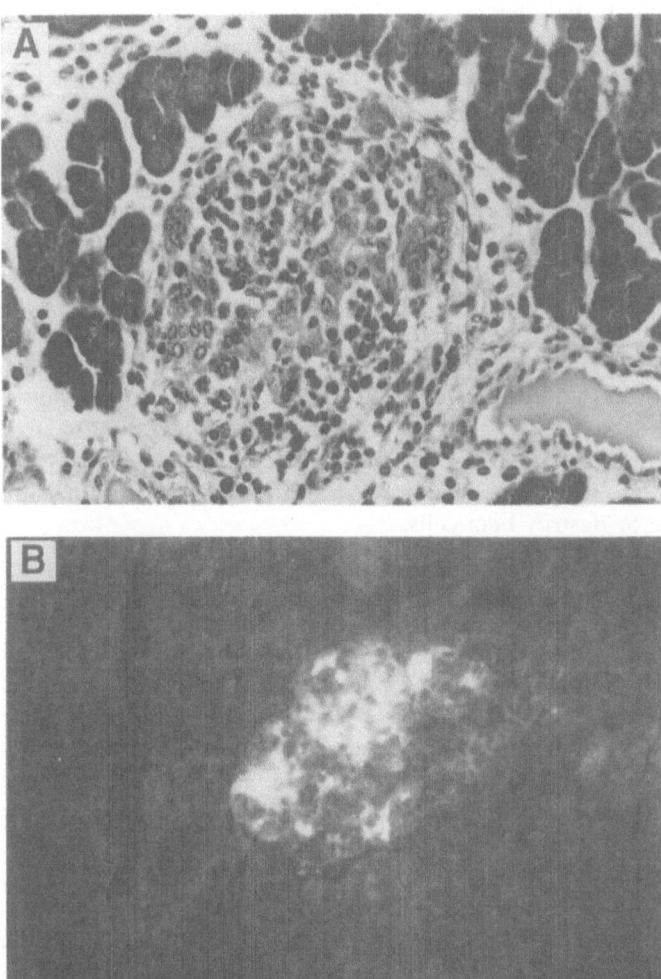

FIGURE 3. Coxsackie B4–infected murine beta cells. Pathological changes (A) in islets of Langerhans after infection with Coxsackie B4 virus. Section of a mouse pancreas obtained 5 days after infection shows extensive infiltration of the entire islet by mononuclear cells and the occasional polymorphonuclear leukocyte. Hematoxylin-eosin staining (× 350). Section of a mouse pancreas (B) obtained 3 days after infection with Coxsackie B4 virus and stained with FITC-labeled anti–Coxsackie B4 antibody. The majority of the cells in the islet contain viral antigens (× 200).

days after infection, while in C57BL/6 mice, in which virus infection was most lethal, diabetes was seen after 21 days postinfection. This study suggested that the genotype which determines glucose tolerance of a given mouse strain after Coxsackie B4 infection does not necessarily determine how lethal the virus is to that strain.

The diabetogenic potential of 37 clinical Coxsackie B1, B3, B4, and B5 isolates were studied by Jordan et al.[71] by their inoculation into SJL mice. Of the 37 isolates, 12 caused minor abnormalities of glucose metabolism. Only 25% of mice sequentially infected with Coxsackie B3, B4, and B5 virus exhibited abnormal glucose metabolism, suggesting that the diabetogenic potential of Coxsackie B strains in nature may be limited. Szopa et al.[72] have conducted similar studies using human isolates of Coxsackie B4 virus isolated from human throat swabs, feces, and cerebrospinal fluid. These isolates could directly infect mouse islets in vitro, leading to changes in islet-cell function. Szopa et al. concluded from these results that Coxsackie B4 variants with potential to damage beta cells occur quite frequently in nature and under certain circumstances may damage or destroy beta cells.

5.1.2.   Effect of Coxsackie B Viral Infections on Murine Beta-Cell Function

Studies on the influence of genetics on the response to Coxsackie B infections in mice have found that the "db" diabetic mutation on chromosome 4 had the most effect on susceptibility and host response to Coxsackie B4 virus[73,74] and was associated with an impaired humoral response to Coxsackie B4 infection, as these mice did not develop an adequate level of anti–Coxsackie B4 antibodies.[75] There was a general impairment in both total IgM and IgG production after Coxsackie B4 viral infection,[76] and the mice were shown to be unable to produce a virus-specific IgG response even though they did produce a high level of nonspecific antibody following Coxsackie B4 challenge, suggesting polyclonal activation. The animals were also found to be deficient in absolute and relative numbers of spleen lymphocyte subsets.[77] Similar findings that Coxsackie B4 virus infection alters thymic, splenic, and peripheral lymphocyte repertoire before the onset of hyperglycemia in mice have been reported.[78]

Chatterjee and Nejman[79] found that preproinsulin mRNA levels in islet fractions prepared from mice up to 8 weeks after Coxsackie B4 infection are reduced. Portwood and Taylor[80] found similar results in islets from DBA/2 mice infected in vitro, with preproinsulin mRNAs dropping to 9% of the noninfected control levels 96 hr after infection.

One investigation into the long-term effects of Coxsackie B4 infection on murine pancreatic islet function *in vivo* revealed that inoculation with a pancreas-adapted Coxsackie B4 virus caused a significant increase in insulin release in islets from inoculated mice 3 and 6 months after inoculation.[81] Histological examination of the islets revealed no changes and islet cell antibodies were not detected. The abnormal insulin release occurred with minimum changes in blood glucose concentration. Another study found that in Coxsackie B4 E2–infected mice, total protein and insulin synthesis decreased early in the infection, then increased at 8 weeks postinfection, but later dropped to levels lower than those seen in control mice.[82] There have been similar findings from *in vitro* studies on the effect of Coxsackie B4 infection in islet cultures.[83] It would appear from the results of both *in vivo* and *in vitro* studies that Coxsackie B4 infection may lead to lasting changes in islet metabolism with only very slight changes in blood glucose levels. Another interesting study[84] reported that chronic Coxsackie B viral infection of beta cells could result in the synthesis of interferon-α, which in turn induces major histocompatibility complex (MHC) class I hyperexpression on adjacent endocrine cells.

Gerling *et al.*[85] monitored the expression of the 64-kDa autoantigen glutamic acid decarboxylase (GAD)[86] in Coxsackie B4–infected SJL/J and CD1 mice. They found that the antigen's expression was increased two- to threefold before the onset of hyperglycemia, indicating that Coxsackie B4 virus infection may initiate or enhance an autoimmune reaction. This same group later found that 90% of Coxsackie B4 virus–infected mice developed antibodies to GAD by 4–6 weeks after infection.[87] Consistent with other reports, infectious virus was not detected after 72 hr postinfection. Since infection with Coxsackie B4 virus increased the expression of GAD, Hou *et al.*[88] analyzed immunoreactive GAD expression with a panel of antisera and polyclonal antisera against GAD and measured GAD activity in the brains, pancreata, and islets of the infected mice. Both GAD65 and GAD67 were detected in all these tissues in noninfected mice and also in the brains of infected mice; however, the pancreata from infected mice contained three times more GAD65 than the pancreata of the noninfected mice. There was virtually no detectable GAD67 in the pancreata of infected mice. Coxsackie B4 virus infection significantly reduced islet GAD activity, but not brain GAD activity. Also, investigation of GAD expression, anti-GAD antibody development, and pancreatic histopathology in Coxsackie B4 virus–infected, diabetes-resistant B10.BR/SgSnJ H-2kTIaa mice showed that these mice did not overexpress GAD at 72 hr postinfection, and only acinar tissue was de-

stroyed, not islets.[89] This may indicate that differences in genetic susceptibility to Coxsackie B4 infection may result in differences in response to
viral infection.

From these studies it appears that the interaction of pancreatic beta
cells with Coxsackie B viruses in infected mice may be dependent upon
the strain of virus, host genetics, immunological response, and other
factors such as age and sex.

5.2. Nonhuman Primate Studies

Coxsackie B4 virus was serially passaged in rhesus or cynomolgus
beta-cell cultures, then harvested and used to infect rhesus, cynomolgus,
cebus, and patas monkeys.[90] Glucose tolerance tests were performed
before and after infection, and an elevation of the glucose tolerance
curve and marked depression of the insulin secretion curve were seen
only in the Coxsackie B4 virus–infected patas monkey. Other species,
such as rhesus, cynomolgus, and cebus monkeys, did not show any
changes in insulin or blood glucose levels after Coxsackie B4 viral infection (Fig. 4). In a parallel experiment, patas monkeys were first treated
with a subdiabetogenic dose of streptozocin, followed first by Coxsackie
B4 viral infection and then by Coxsackie B3 viral infection. The glucose
tolerance curve was elevated after the first infection and continued to
increase after the second infection. These results suggest that cumulative
insults by different agents, including beta-cell toxins and different strains
of diabetogenic virus, lead to additional beta-cell damage. Immunoreactive insulin levels were also markedly depressed, as had been seen previously in patas monkeys infected with only Coxsackie B4 virus. These
results indicate that Coxsackie B4 virus alone is sufficient to produce
abnormalities in glucose tolerance tests and impaired insulin secretion in
only the patas monkey and that genetic factors are critical for glucose
homeostasis in monkeys infected with Coxsackie B4 virus.

5.3. *In Vitro* Studies in Other Animal Cells

To investigate an alternative pathogenic mechanism for Coxsackie B
virus, Montgomery *et al.*[91] studied the effect of infection by Coxsackie B4
virus isolated from a human pancreas on rat insulinoma (RINm5F) cells.
Following acute infection, virus was detectable in cells for 10 days; then,
virus could not be detected. However, viral antigens could be detected
using antibody and fluorescence activated cell sorter (FACS) within the
cells or at the cell surface. The Coxsackie B4–infected RINm5F (RIN
CB4) cells were readily passaged and grown for over 6 months. In the RIN

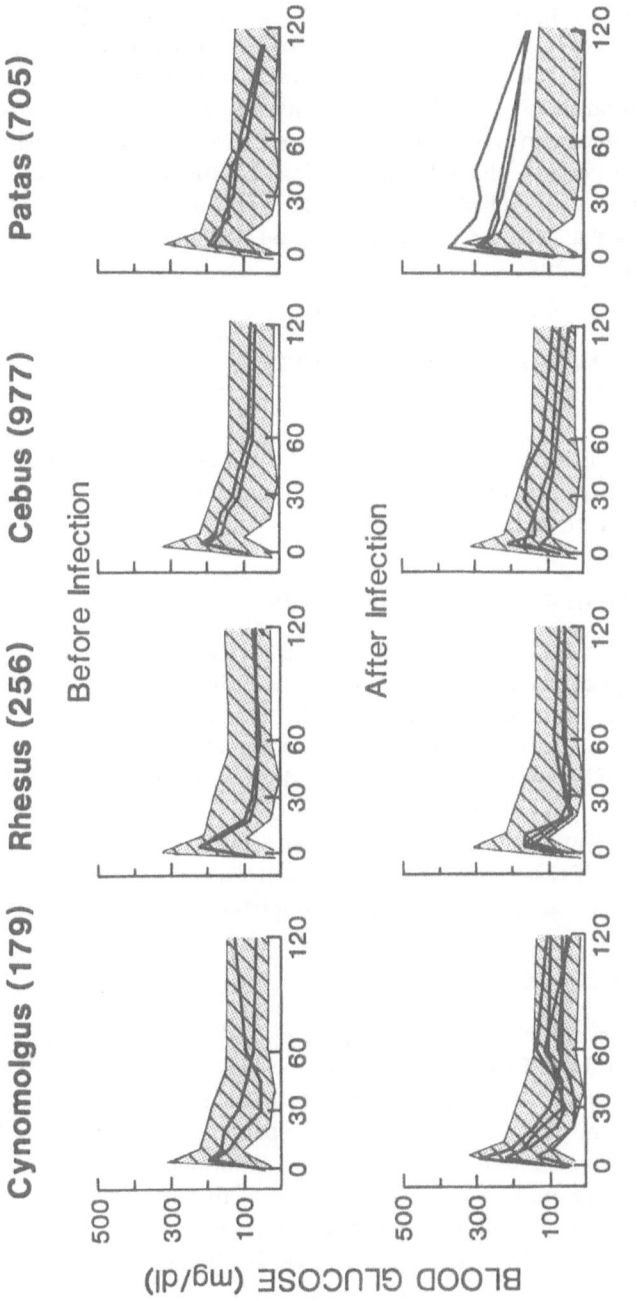

FIGURE 4. Glucose levels in blood after intravenous glucose tolerance tests in monkeys. Glucose tolerance tests before and after infection with Coxsackie B4 virus are shown for four different species of monkeys. It can be clearly seen that Coxsackie B4 infection affects glucose homeostasis in only the patas monkey.

CB4 cells, insulin secretion and intracellular insulin content were decreased 10–50% compared to control cells, and RIN CB4 cells had decreased numbers of insulin granules as seen by electron microscopy. It was also noted that MHC class I expression was increased by 50% on the RIN CB4 cells, compared to control cells. The investigators concluded that a normally lytic virus can persist in islet-derived tissue in a latent fashion and was associated with altered cell function. In this way, Coxsackie B4 virus infection may possibly lead to IDDM without direct cytotoxicity. Frank et al.[92] have also noted that Coxsackie B4 viral infection can persist in rat insulinoma cells.

6. MECHANISMS FOR COXSACKIE B VIRUS–INDUCED DIABETES

Regarding the induction of diabetes by Coxsackie B viruses, there are at least four possible mechanisms by which viruses can cause IDDM which will be discussed in turn.

6.1. Direct Cytolytic Infection of Beta Cells

One mechanism whereby Coxsackie B viruses may cause IDDM involves the direct infection of beta cells and viral replication within them. The replication of virus in beta cells results in cell lysis, and the infected animal or human subsequently becomes hypoinsulinemic and hyperglycemic. This mechanism has been shown to exist in several human cases, but appears to be uncommon. As discussed by Foulis et al.[93] in their report of a young man dying from synchronous, massive pancreatic islet destruction, if this was indeed a common mechanism, it would be overrepresented in autopsies of persons dying from acute-onset diabetes.

6.2. Molecular Mimicry

Over the last several years, it has been found that homology exists between GAD and the Coxsackie B4 virus noncapsid protein P2-C.[94] On the basis of this similarity, Kaufman and his co-workers initially proposed that molecular mimicry between the P2-C protein and GAD may be involved in viral induction of IDDM, whereby antibodies directed against the viral protein could cross-react with GAD on the beta cells. A more recent study by Kaufman and his colleagues[95] has shown, through peptide mapping of GAD for the fine specificity of T-cell responses, that the region of sequence similarity of GAD with Coxsackie B4 virus is not the region involved in the initial event in induction of autoimmune IDDM, but a region that subsequently reacts with T cells. It is therefore speculated that if Coxsackie B4 virus is able to trigger autoimmune IDDM, this ability may not be related to the homology between its P2-C protein and GAD, though this homology may play a role later in the disease process.

6.3. Generation of T Effector Cells

A third possible mechanism involves generation of viral antigen-specific T effector cells. It is speculated that Coxsackie B4 virus may generate viral antigen-specific cytotoxic T lymphocytes which then may recognize beta cell–specific autoantigens on the beta cells, if there is homology between them.[96,97] Under these circumstances, generation of viral antigen-specific T effector cells might be the initial step in T cell–mediated beta-cell destruction.

6.4. Persistent Infection

While the Coxsackie B viruses are normally considered to be highly lytic, there is some evidence that they may in certain cases establish a persistent infection.[91,92,98] Foulis et al.,[99] using an antiserum raised to the VP1 capsid protein of recombinant Coxsackie B3 virus, developed a technique to detect the presence of all Coxsackie B viruses in tissue. The technique was used to test autopsy specimens of heart and pancreas tissue from patients who had died from acute Coxsackie B myocarditis. Of 12 cases where the heart was available for study, seven had pancreata which had insulitis, and in all seven, islet endocrine cells contained Coxsackie B VP1 proteins. When pancreata from 88 patients who died from acute-onset diabetes were examined, none were found to contain Coxsackie B VP1 protein. Foulis et al. suggested that a persistent infection of beta cells by a defective enterovirus might result in their destruction by an auto-immune mechanism. This hypothesis is not incompatible with findings of continuing Coxsackie B viral infection in other diseases. In an earlier study, Foulis et al.[84] found that in three of four cases of infantile viral pancreatitis known to be caused by Coxsackie B viral infection, only the pancreata showed interferon-α related to hyperexpression of MHC class I antigens; this led them to suggest that Coxsackie B viral infection of the beta cells could result in the synthesis of interferon-α, which in turn could induce MHC class I hyperexpression on adjacent endocrine cells. This interesting hypothesis deserves further study.

6.5. Other Possible Mechanisms

One other way in which Coxsackie B viral infection may be involved in the pathogenesis of IDDM is that the virus may act as the final insult in an ongoing beta cell–specific autoimmune process. In this situation, Coxsackie B4 viral infection may destroy residual beta cells in individuals who have already undergone some beta-cell loss through an autoimmune process. The destruction of residual beta cells by Coxsackie B viral infection would result in the clinical onset of IDDM.

Whatever the mechanism, evidence from studies on mice, non-human primates, and humans indicates that Coxsackie B viruses can affect glucose homeostasis and that islet function may be impaired after Coxsackie B viral infection.

7. CONCLUDING REMARKS

Many epidemiological studies have shown a link between Coxsackie B viral infection, especially infection by the B4 serotype, and development of diabetes in humans; however, several studies have reported a negative correlation between Coxsackie B viral infection and diabetes. It is therefore difficult to draw any definite conclusions on the correlation between Coxsackie B viral infection and diabetes on the basis of epidemiological studies alone. The difficulties may arise from the fact that there are so many variants of each serotype of Coxsackie B virus and that only some of these variants may be diabetogenic. For example, research on Coxsackie B4 virus has demonstrated that antigenic changes at the epitope level occur at a frequency greater than 1/100.[52] This suggests that even within the same virus pool there may be many antigenic variants and that these variants have different tissue tropisms and different physiologic properties, accounting for the wide spectrum of clinical disease produced by Coxsackie B4 virus. Depending whether there was first an outbreak of a nondiabetogenic variant or a diabetogenic variant, the expectation of finding a correlation between Coxsackie B viral infection and diabetes would be different. If an outbreak of a diabetogenic variant occurred first, a link would be expected to be found, while if a nondiabetogenic variant first appeared no link would be expected, since the variants are cross-reactive and antibodies developed against the nondiabetogenic variant would neutralize the diabetogenic variant. The number of cases of diabetes seen following an initial outbreak of diabetogenic variant may also depend on how many genetically predisposed individuals were infected, as genetic susceptibility appears to play a critical role in the development of Coxsackie B virus–induced diabetes in animals.[55]

In addition to epidemiological studies, there have been several case reports which also support a link between Coxsackie B viral infection and the subsequent development of diabetes. These reports do not prove that this virus is actually involved in the destruction of beta cells and the consequent development of IDDM. However, *in vitro* studies clearly show that human pancreatic beta cells can be infected and destroyed by Coxsackie B virus, although there may be differences between *in vivo* and *in vitro* responses to viral infection. Therefore, *in vitro* studies do not unequivocally prove that Coxsackie B viruses can cause diabetes. The most convincing evidence that Coxsackie B viruses can cause some cases of IDDM comes from studies where the virus has been isolated from the pancreas of a patient with recent-onset IDDM, followed by the isolated virus causing diabetes when inoculated into susceptible animals. There have been two case reports where this has happened, indicating that Coxsackie B viruses can induce diabetes in humans under certain circumstances. The situation is more definitive in animals, where Coxsackie B viruses have induced a diabetes-like syndrome in genetically susceptible animals, including several strains of mice and patas monkeys.

It is unclear as yet how Coxsackie B viruses may induce diabetes. There have been several possible mechanisms proposed for how Coxsackie B viral infections interact with the host to effect pancreatic beta-cell destruction. The first mechanism is by direct cytolytic infection of the beta cells themselves, which leads to acutely developed IDDM. This mechanism appears to be supported by animal studies. The second mechanism by which Coxsackie B viruses may induce diabetes is molecular mimicry. If there is an antigenic determinant shared by the virus and a normally occurring beta-cell protein, a beta cell–specific immune response might possibly occur if virus-specific antibodies cross-react with the host determinant. The third possibility is that if the virus has amino acid homology with a normally occurring beta-cell protein, viral antigen-specific T effector cells may be generated which will recognize the proteins on the beta cells. A fourth mechanism is that persistent infection by Coxsackie B viruses may induce cytokines and interferons, thought to be associated with autoimmune IDDM. The fifth possibility is that a Coxsackie B viral infection may combine with other etiological factors to serve as the final insult in the development of IDDM. Whatever the mechanism proposed, it is clear that Coxsackie B viruses do appear to be associated with the development of some cases of IDDM, and the precise mechanisms by which the viruses act should be elucidated.

Recently the amino acid residue responsible for the virulence of Coxsackie B4 virus was discovered.[100,101] We have recently sequenced the entire genome of the diabetogenic E2 strain of Coxsackie B4 virus[102] and have compared it to the published sequence of the prototype non-diabetogenic JVB strain.[103] We noted 111 amino acid differences between the E2 and JVB strains. Titchener *et al.*[104] have also sequenced a beta-cell tropic, mouse pancreas–adapted variant of Coxsackie B4 JVB strain. Titchener's group found only seven amino acid differences between the passaged variant and the prototype strain. Since it is difficult to determine which of the amino acid changes are critical for diabetogenicity, further studies are needed, using nondiabetogenic variants isolated by additional passaging of the diabetogenic strain in tissue cultures.[105,106] It is expected that closely related nondiabetogenic variants would arise as there is an observed high mutation frequency of 10^{-4} per base[52] when Coxsackie B4 virus is passaged in cell cultures. Sequencing the genomes of these closely related nondiabetogenic variants and comparing these sequences to that of the diabetogenic strain may reveal the critical site (nucleotide or amino acid) responsible for diabetogenicity. The molecular identification of the specific diabetogenic Coxsackie B4 viral gene responsible for the selective cytolytic destruction of beta cells using recombinant Coxsackie B4 virus in animal models would be invaluable to gain a clear understanding of the mechanisms involved in the selective cytolytic destruction of beta cells by Coxsackie B4 virus in humans. Also, since Coxsackie B viruses have been suggested to be associated with autoimmune IDDM, studies are needed at the cellular and molecular levels to determine how Coxsackie B viruses may be involved in the development of beta cell–specific autoimmunity, leading to the disease. Once the pathogenic mechanisms for Coxsackie B virus–associated IDDM are more clearly understood, it may be possible to develop methods to prevent cytolytic destruction of beta cells by the virus or beta cell–specific autoimmune IDDM associated with Coxsackie B virus, if such an association is shown to actually exist.

REFERENCES

1. Gorsuch, A. N., Spencer, K. M., Lister, J., McNally, J. M., Dean, B. M., Bottazzo, G. F., and Cudworth, A. G., 1981, Evidence for a long prediabetic period in type 1 (insulin-dependent) diabetes mellitus, *Lancet* 2:1363–1365.
2. Rossini, A. A., Greiner, D., Friedman, H., and Mordes, J., 1993, Immunopathogenesis of diabetes mellitus, *Diabetes Rev.* 1:433–475.
3. Castano, L., and Eisenbarth, G. S., 1990, Type I diabetes: A chronic autoimmune disease of human, mouse and rat, *Ann. Rev. Immunol.* 8:647–679.

4. Green, A., 1990, Role of genetic factors in the development of insulin-dependent diabetes mellitus, in: *Human Diabetes: Genetic, Environmental and Autoimmune Etiology* (S. Baekkeskov and B. Hansen, eds.), Springer-Verlag, New York, pp. 3–16.

5. Yoon, J., 1990, Role of viruses and environmental factors in induction of diabetes, in: *Human Diabetes: Genetic, Environmental and Autoimmune Etiology* (S. Baekkeskov and B. Hansen, eds.), Springer-Verlag, New York, pp. 95–123.

6. Yoon, J. W., and Park, Y. H., 1993, Viruses as triggering agents of insulin-dependent diabetes mellitus, in: *The Causes of Diabetes* (R. D. G. Leslie, ed.), John Wiley and Sons, London, pp. 83–103.

7. Gamble, D. R., Kinsley, M. L., Fitzgerald, M. G., Bolton, R., and Taylor, K. W., 1969, Viral antibodies in diabetes mellitus, *Br. Med. J.* 3:627–630.

8. Gamble, D. R., Taylor, K. W., and Cumming, H., 1973, Coxsackie viruses and diabetes mellitus, *Br. Med. J.* 4:260–262.

9. King, M., Shaikh, A., Bidwell, D., Voller, A., and Banatvala, J., 1983, Coxsackie B virus specific IgM responses in children with insulin-dependent diabetes mellitus, *Lancet* 1:1397–1399.

10. Mertens, T., Gruneklee, D., and Eggers, H. J., 1983, Neutralizing antibodies against Coxsackie B viruses in patients with recent-onset of type I diabetes, *Eur. J. Pediatr.* 140:293–294.

11. Mirkovic, R. R., Varma, K. S., and Yoon, J. W., 1984, Incidence of coxsackievirus B type 4 (CB4) infection concomitant with onset of insulin-dependent diabetes mellitus, *J. Med. Virol.* 14:9–16.

12. Buschard, K., and Madsbad, S. A., 1984, Longitudinal study of virus antibodies in patients with newly diagnosed type 1 (insulin-dependent) diabetes mellitus, *J. Clin. Lab. Immunol.* 13:65–70.

13. Frisk, G., Fohlman, J., Kobbah, M., Ewald, U., Tuvemo, T., Diderholm, H., and Friman, G., 1985, High frequency of Coxsackie-B-virus-specific IgM in children developing type I diabetes during a period of high diabetes morbidity, *J. Med. Virol.* 17:219–227.

14. Banatvala, J., Schernthaner, G., Schober, E., DeSilva, L., Bryant, J., Borkenstein, M., Schober, E., Brown, D., DeSilva, L. M., Menser, M. A., and Silink, M., 1985, Coxsackie B., mumps, rubella and cytomegalovirus specific IgM responses in patients with juvenile-onset insulin-dependent diabetes mellitus in Britain, Austria and Australia, *Lancet* 1:1409–1412.

15. Friman, G., Fohlman, J., Frisk, G., Diderholm, H., Ewald, U., Kobbah, M., and Tuvemo, T., 1985, An incidence peak of juvenile diabetes. Relation to Coxsackie B virus immune response, *Acta Pediatr. Scand.* 320(Suppl.):14–19.

16. Alberti, A. M., Amato, C., Candela, A., Constantino, F., Grandolfo, M. E., Lombardi, F., Novello, F., Orsini, M., and Santoro, R., 1985, Serum antibodies against Coxsackie B1-6 viruses in type 1 diabetics, *Acta Diabetol. Lat.* 22:33–38.

17. Fohlman, J., Bohme, J., Rask, L., Frisk, G., Diderholm, H., Friman, G., and Tuvemo, T., 1987, Matching of host genotype and serotypes of Coxsackie B virus in the development of juvenile diabetes, *Scand. J. Immunol.* 26:105–110.

18. Michalkova, D., Petrovicova, A., Kolar, J., Jancova, E., and Silesova, J., 1989, The role of coxsackie virus infection in the development of type 1 diabetes mellitus and its effect on the postinitial course of the disease in childhood, *Cesk. Pediatr.* 44:257–262.

19. Michalkova, D., Kostal, M., Rajcani, J., Petrovicova, A., Bircak, J., Kolar, J., Barak, L., and Hrabcakova, D., 1989, Cytoplasmic islet cell antibodies in children with type I diabetes, *Bratisl. Lek. Listy* 90:159–167.

20. Verma, I. C., 1989, The challenge of childhood diabetes mellitus in India, *Indian J. Pediatr.* 56(Suppl. 1):S33–38.

21. Tuskiewicz-Misztal, E., 1991, Epidemiologic factors and serum antibody titer to Coxsackie B-4 virus in patients with type I diabetes, *Kinderaerztl. Prax.* **59:**88–91.

22. Frisk, G., Friman, G., Tuvemo, T., Fohlman, J., and Diderholm, H., 1992, Coxsackie B virus IgM in children at onset of type 1 (insulin-dependent) diabetes mellitus: Evidence for IgM induction by a recent or current infection, *Diabetologia* **35:**249–253.

23. Barrett-Connor, E., 1985, Is insulin-dependent diabetes mellitus caused by coxsackievirus B infection? A review of epidemiologic evidence, *Rev. Infect. Dis.* **7:**207–215.

24. Gladisch, R., Hoffmann, W., and Waldherr, R., 1976, Myocarditis and insulitis in Coxsackie virus infection, *Z. Kardiol.* **65:**873–881.

25. Yoon, J. W., Austin, M., Onodera, T., and Notkins, A. L., 1979, Virus-induced diabetes mellitus: Isolation of a virus from the pancreas of a child with diabetic ketoacidosis, *N. Engl. J. Med.* **300:**1173–1179.

26. Jenson, A., Rosenberg, H., and Notkins, A. L., 1980, Pancreatic islet cell damage in children with fatal viral infections, *Lancet* **2:**354–358.

27. Champsaur, H., Bottazzo, G., Bertrams, J., Assan, R., and Bach, C., 1982, Virologic, immunologic and genetic factors in insulin-dependent diabetes mellitus, *J. Pediatr.* **100:**15–20.

28. Palmer, J. P., Cooney, M. K., Crossley, J. R., Hollander, P. H., and Asplin, C. M., 1981, Antibodies to viruses and to pancreatic islets in nondiabetic and insulin-dependent diabetic patients, *Diabetes Care* **4:**525–528.

29. Pagano, G., Cavallo-Perin, P., Cavalot, F., Dall'Omo, A. M., Mascioloa, P., Surinani, R., Amoroso, A., Curtoni, S. E., Borelli, I., and Lenti, G., 1987, Genetic, immunologic, and environmental heterogeneity of IDDM. Incidence and 12 month follow-up of an Italian population, *Diabetes* **36:**859–863.

30. Ajuwon, Z. A., Olaleye, O. D., Omilabu, S. A., and Baba, S. S., 1992, Complement fixing antibodies against selected viruses in diabetic patients and non-diabetic control subjects in Ibadan Nigeria, *Rev. Roum. Virol.* **43:**3–5.

31. Pato, E., Cour, M. I., Gonzalez-Cuadrado, S., Gonzalez-Gomez, C., Munoz, J. J., and Figueredo, A., 1992, Coxsackie B4 and cytomegalovirus in patients with insulin-dependent diabetes, *Ann. Med. Intern.* **9:**30–32.

32. Emekdas, G., Rota, S., Kusitmur, S., and Kocabeyoglu, O., 1992, Antibody levels against coxsackie B viruses in patients with type 1 diabetes mellitus, *Mikrobiyol. Bul.* **26:**116–120.

33. Gunczler, P., Lanes, R., Layrisse, Z., Esparza, B., Salas, R., Hernandez, L., and Arnaiz-Villena, A., 1993, Epidemiology and immunogenetics in recently diagnosed Venezuelan children with insulin-dependent diabetes mellitus, *J. Pediatr. Endocrinol.* **6:**165–171.

34. Palmer, J. P., Cooney, M. K., Ward, R. H., Jansen, J. A., Brodsky, J. B., Ray, C. G., Crossley, J. R., Asplin, C. M., and Williams, R. H., 1982, Reduced Coxsackie antibody titres in type 1 (insulin-dependent) diabetic patients presenting during an outbreak of Coxsackie B3 and B4 infection, *Diabetologia* **22:**426–429.

35. Tuvemo, T., Dahlquist, G., Frisk, G., Blom, L., Friman, G., Landin-Olsson, M., and Diderholm, H., 1989, The Swedish childhood diabetes study III: IgM against coxsackie B viruses in newly diagnosed type 1 (insulin-dependent) diabetic children—No evidence of increased antibody frequency, *Diabetologia* **32:**745–747.

36. Winter, W. E., Obata, M., and Maclaren, N. K., 1992, Clinical and molecular aspects of autoimmune endocrine disease, in: *Clinical and Molecular Aspects of Autoimmune Diseases: Concepts in Immunopathology* (J. M. Cruse and R. E. Lewis Jr., eds.), Karger, Basel, Switzerland, pp. 189–221.

37. Nepom, G. T., 1993, Immunogenetics and IDDM, *Diabetes Rev.* **1:**93–103.

38. Eberhardt, M. S., Wagener, D. K., Orchard, T. J., LaPorte, R. E., Cavender, D. E., Rabin, B. S., Atchison, R. W., Kuller, L. H., Drash, A. L., and Becker, D. J., 1986, HLA heterogeneity of insulin-dependent diabetes mellitus at diagnosis, *Diabetes* **34**:1247–1252.

39. LaPorte, R. E., Fishbein, H. A., Drash, A. L., Kuller, L. H., Schneider, B. B., Orchard, T. J., and Wagener, D. K., 1981, The Pittsburgh insulin-dependent diabetes mellitus (IDDM) registry: The incidence of insulin-dependent diabetes mellitus in Allegheny County, Pennsylvania (1965–1976), *Diabetes* **30**:279–284.

40. Field, L. L., McArthur, R. G., Shin, S. Y., and Yoon, J. W., 1987, The relationship between Coxsackie-B-virus-specific IgG responses and genetic factors (HLA-DR, GM, KM) in insulin-dependent diabetes mellitus, *Diabetes Res.* **6**:169–173.

41. Yoon, J. W., Kim, K. W., Kim, H. M., Pak, C. Y., Kim, Y. W., and McArthur, R. G., 1989, Coxsackie B virus assays in IDDM families: The GAW5 data on antibody prevalence, *Genet. Epidemiol.* **6**:35–37.

42. Spielman, R. S., Baur, M. R., and Clerget-Darpoux, F., 1989, Genetic analysis of IDDM. Summary of GAW5 IDDM results, *Genet. Epidemiol.* **6**:43–58.

43. D'Alessio, D. J., 1992, A case-control study of group B Coxsackievirus immunoglobulin M antibody prevalence and HLA-DR antigens in newly diagnosed cases of insulin dependent diabetes mellitus, *Am. J. Epidemiol.* **135**:1331–1338.

44. Schernthaner, G., Banatvala, J. E., Scherbaum, W., Bryant, J., Borkenstein, M., Schober, E., and Mayer, W. R., 1985, Coxsackie-B-virus-specific IgM responses, complement-fixing islet-cell antibodies, HLA DR antigens, and C-peptide secretion in insulin-dependent diabetes mellitus, *Lancet* **2**:630–632.

45. Bruserud, O., Jervell, J., and Thorsby, E., 1985, HLA-DR3 and -DR4 control T-lymphocyte responses to mumps and Coxsackie B4 viruses: Studies on patients with type 1 (insulin-dependent) diabetes and healthy subjects, *Diabetologia* **28**:420–426.

46. Bruserud, O., and Thorsby, E., 1985, T lymphocyte responses to Coxsackie B4 and mumps virus. I. Influence of HLA-DR restriction elements, *Tissue Antigens* **26**:41–50.

47. Adams, S. F., 1926, The seasonal variation in the onset of acute diabetes, *Arch. Intern. Med.* **37**:861–862.

48. Gamble, D. R., and Taylor, K. W., 1969, Seasonal incidence of diabetes mellitus, *Br. Med. J.* **3**:631–633.

49. Gamble, D. R., 1975, Viral and epidemiological studies, *Proc. R. Soc. Med.* **68**:256.

50. Wagenknecht, L., Roseman, J., and Herman, W., 1991, Increased incidence of insulin-dependent diabetes mellitus following an epidemic of Coxsackie virus B5, *Am. J. Epidemiol.* **133**:1024–1031.

51. Rewers, M., LaPorte, R. E., Walczak, M., Dmochowski, K., and Bogaczynska, E., 1987, Apparent epidemic of insulin-dependent diabetes mellitus in midwestern Poland, *Diabetes* **36**:106–113.

52. Prabhakar, B. S., Haspel, M. V., McClintock, P. R., and Notkins, A. L., 1982, High frequency of antigenic variants among naturally occurring human Coxsackie B4 virus isolates identified by monoclonal antibodies, *Nature* **300**:374–376.

53. Yoon, J. W., Bachurski, C. J., and McArthur, R. G., 1986, Concept of virus as an etiological agent in the development of IDDM, *Diabetes Res. Clin. Pract.* **2**:365–366.

54. Notkins, A. L., and Yoon, J. W., 1983, Virus-induced diabetes in mice prevented by live attenuated vaccine, *N. Engl. J. Med.* **306**:486.

55. Yoon, J. W., Onodera, T., and Notkins, A. L., 1978, Virus-induced diabetes mellitus: Beta cell damage and insulin-dependent hyperglycaemia in mice infected with Coxsackie virus B4, *J. Exp. Med.* **148**:1068–1080.

56. Wilson, C., Connolly, J. M., and Thomson, D., 1977, Coxsackie B2 virus infection and acute onset diabetes in a child, *Br. Med. J.* 1:1008.
57. Ahmad, N., and Abraham, A. A., 1982, Pancreatic isleitis with coxsackie virus B5 infection, *Hum. Pathol.* 13:661–662.
58. Asplin, C. M., Cooney, M. K., Crossley, J. R., Dornan, T. L., Raghu, P., and Palmer, J. P., 1982, Coxsackie B4 infection and islet cell antibodies three years before overt diabetes, *J. Pediatr.* 101:398–400.
59. Orchard, T. J., Becker, A. J., Atchison, R. W., LaPorte, R. E., Wagener, D. K., Rabin, B. S., Kuller, L. H., and Drash, A. L., 1982, The development of type 1 (insulin-dependent) diabetes mellitus: Two contrasting presentations, *Diabetologia* 25:89–92.
60. Niklasson, B. S., Dobersen, M. J., Peters, C. J., Ennis, W. H., and Moller, E., 1985, An outbreak of coxsackievirus B infection followed by one case of diabetes mellitus, *Scand. J. Infect. Dis.* 17:15–18.
61. Nigro, G., Pacella, M. E., Patane, E., and Midulla, M., 1986, Multi-system coxsackievirus B-6 infection with findings suggestive of diabetes mellitus, *Eur. J. Paediatr.* 145:557–559.
62. Yoon, J. W., Onodera, T., and Notkins, A. L., 1978, Virus-induced diabetes mellitus. XI. Replication of Coxsackie virus B3 in human pancreatic beta cell cultures, *Diabetes* 27:778–782.
63. Szopa, T. M., Ward, T., and Taylor, K. W., 1986, Impaired metabolic functions in human pancreatic islets following infection with Coxsackie B4 virus in vitro, *Diabetologia* 30:587A.
64. Ross, M. E., Hayashi, K., and Notkins, A. L., 1974, Virus-induced pancreatic disease: Alterations in concentration of glucose and amylase in blood, *J. Infect. Dis.* 129:669–676.
65. Coleman, T. J., Gamble, D. R., and Taylor, K. W., 1973, Diabetes in mice after Coxsackie B4 virus infection, *Br. Med. J.* 3:25–27.
66. Coleman, T. J., Taylor, K. W., and Gamble, D. R., 1974, The development of diabetes following Coxsackie B virus infection in mice, *Diabetologia* 10:755–759.
67. Toniolo, A., Onodera, T., Jordan, G., Yoon, J. W., and Notkins, A. L., 1982, Virus-induced diabetes mellitus: Glucose abnormalities produced in mice by all six members of the Coxsackie B virus group, *Diabetes* 31:496–499.
68. Hartig, P. C., Madge, G. E., and Webb, S. R., 1983, Diversity within a human isolate of Coxsackie B4: Relationship to viral-induced diabetes, *J. Med. Virol.* 11:23–30.
69. Hartig, P. C., and Webb, S. R., 1983, Heterogeneity of a human isolate of Coxsackie B4: Biological differences, *J. Infect.* 6:43–48.
70. Cook, S. H., Loria, R. M., and Madge, G. E., 1982, Host factors in Coxsackievirus B4-induced pancreopathy, *Lab. Infect.* 46:377–382.
71. Jordan, G. W., Bolton, V., and Schmidt, N. J., 1985, Diabetogenic potential of coxsackie B viruses in nature, *Arch. Virol.* 86:213–221.
72. Szopa, T. M., Ward, T., Dronfield, D. M., Portwood, N. D., and Taylor, K. W., 1990, Coxsackie B4 viruses with the potential to damage beta cells of the islets are present in clinical isolates, *Diabetologia* 33:325–328.
73. Webb, S. R., Loria, R. M., Madge, G. F., and Kibrick, S., 1976, Susceptibility of mice to group B Coxsackievirus is influenced by the diabetic gene, *J. Exp. Med.* 143:1239–1248.
74. Loria, R. M., Montgomery, L. B., Corey, L. A., and Chinchilli, V. M., 1984, Influence of diabetes mellitus heredity on susceptibility to coxsackievirus B4, *Arch. Virol.* 81:251–262.
75. Loria, R. M., Montgomery, L. B., Tuttle-Fuller, N., and Gregg, H. M., 1986, Genetic predisposition to diabetes mellitus is associated with impaired humoral immunity to coxsackievirus B4, *Diabetes Res. Clin. Pract.* 2:91–96.

76. Montgomery, L. B., and Loria, R. M., 1986, Humoral immune response in hereditary and overt diabetes mellitus, *J. Med. Virol.* 19:255–268.
77. Montgomery, L. B., Loria, R. M., and Chinchilli, V. M., 1990, Immunodeficiency as primary phenotype of diabetes mutation db. Studies with coxsackievirus B4, *Diabetes* 39:675–682.
78. Catterjee, N. K., Hou, J., Dockstader, P., and Charbonnau, T., 1992, Coxsackievirus B4 infection alter thymic, splenic and peripheral repertoire preceding onset of hyperglycaemia in mice, *J. Med. Virol.* 38:124–131.
79. Chatterjee, N. K., and Nejman, C., 1988, Insulin mRNA content in pancreatic beta cells of coxsackievirus B4–induced diabetic mice, *Mol. Cell. Endocrinol.* 55:193–202.
80. Portwood, N. D., and Taylor, K. W., 1990, Coxsackie B4 virus–induced changes in mouse pancreatic-cell mRNAs, *Biochem. Soc. Trans.* 18:1264.
81. Szopa, T. M., Dronfield, D. M., Ward, T., and Taylor, K. W., 1989, In vivo infection of mice with Coxsackie B4 virus induces long-term functional changes in pancreatic islets with minimal alteration in blood glucose, *Diabetic Med.* 6:314–319.
82. Chatterjee, N. K., Haley, T. M., and Nejman, C., 1985, Functional alteration in pancreatic B cells as a factor in virus-induced hyperglycaemia in mice, *J. Biol. Chem.* 260:12786–12791.
83. Szopa, T. M., Gamble, D. R., and Taylor, K. W., 1985, Biochemical changes induced by Coxsackie B4 virus in short-term culture of mouse pancreatic islets, *Biosci. Rep.* 5: 63–69.
84. Foulis, A. K., Farquharson, M. A., and Meager, A., 1987, Immunoreactive α-interferon in insulin-producing β cells in type I diabetes mellitus, *Lancet* 2:1423–1427.
85. Gerling, I., Najman, C., and Chatterjee, N. K., 1988, Effect of coxsackievirus B4 infection in mice on expression of 64,000 Mr autoantigen and glucose sensitivity of islets before development of hyperglycaemia, *Diabetes* 37:1419–1425.
86. Baekkeskov, S., Anstoot, H.-J., Christgau, S., Reetz, A., Solimena, M., Cascalho, M., Folli, F., Richter-Olesen, H., and DeCamilli, P., 1990, Identification of the 64K autoantigen in insulin-dependent diabetes as the GABA-synthesizing enzyme glutamic acid decarboxylase, *Nature* 347:151–156.
87. Gerling, I., Chatterjee, N. K., and Nejman, C., 1991, Coxsackievirus B4-induced development of antibodies to 64,000 Mr islet autoantigen and hyperglycaemia in mice, *Autoimmunity* 10:49–56.
88. Hou, J., Sheikh, S., Martin, D. L., and Chatterjee, N. K., 1993, Coxsackievirus B4 alters pancreatic glutamate decarboxylase expression in mice soon after infection, *J. Autoimmun.* 6:529–542.
89. Gerling, I., and Chatterjee, N. K., 1993, Lack of 64,000 M(r) islet autoantigen overexpression and antibody development following coxsackievirus B4 infection in diabetes-resistant mice, *Autoimmunity* 14:197–203.
90. Yoon, J. W., London, W., Curfman, B., Brown, R., and Notkins, A. L., 1986, Coxsackie virus B$_4$ produces transient diabetes in nonhuman primates, *Diabetes* 35:712–716.
91. Montgomery, L., Gordon, D., George, K., and Maratos-Flier, E., 1991, Coxsackie infection of insulinoma cells leads to viral latency and altered class I MHC expression, *Diabetes* 40:150A.
92. Frank Jr., J. A., Schmidt, E. V., Smith, R. E., and Wifert, C. M., 1986, Persistent infections of rat insulinoma cells with Coxsackie B4 virus, *Arch. Virol.* 81:143–150.
93. Foulis, A. K., Frances, N. D., Farquharson, M. A., and Boylston, A., 1988, Massive synchronous B cell necrosis causing type 1 diabetes—A unique histopathological case report, *Diabetologia* 31:46–50.

94. Kaufman, D. L., Erlander, M. G., Clare-Salzler, M. J., Atkinson, M. A., Maclaren, N. K., and Tobin, A. J., 1992, Autoimmunity to two forms of glutamate decarboxylase in insulin-dependent diabetes mellitus, *J. Clin. Invest.* **89**:283–292.
95. Kaufman, D. L., Clare-Salzler, M. G., Tian, J., Forsthuber, T., Ting, G., Robinson, P., Atkinson, M. A., Sercarz, E. E., Tobin, A. J., and Lehmann, P. V., 1993, Spontaneous loss of T-cell tolerance to glutamic acid decarboxylase in murine insulin-dependent diabetes, *Nature* **366**:69–72.
96. Notkins, A. L., Onodera, T., and Prabhakar, B., 1984, Virus-induced autoimmunity, in: *Concepts in Viral Pathogenesis* (A. Notkins and M. B. A. Oldstone, eds.), Springer-Verlag, New York, pp. 211–217.
97. Oldstone, M. B. A., and Notkins, A. L., 1986, Molecular mimicry, in: *Concepts in Viral Pathogenesis II* (A. Notkins and M. B. A. Oldstone, eds.), Springer-Verlag, New York, pp. 195–201.
98. Matteucci, D., Paglianti, M., Giangregorio, A. M., Capobianchi, M. R., Dianzani, F., and Bendinelli, M., 1985, Group B Coxsackieviruses readily establish persistent infections in human lymphoid cell lines, *J. Virol.* **56**:651–654.
99. Foulis, A. K., Farquharson, M. A., Cameron, S. O., McGill, M., Schonke, H., and Kandolf, R., 1990, A search for the presence of the enteroviral capsid protein VP1 in pancreases of patients with type 1 (insulin-dependent) diabetes and pancreases and hearts of infants who died of coxsackieviral myocarditis, *Diabetologia* **33**:290–298.
100. Ramsingh, A., Araki, H., Bryant, S., and Hixson, A., 1992, Identification of candidate sequences that determine virulence in Coxsackievirus B4, *Virus Res.* **23**:281–292.
101. Caggana, M., Chan, P., and Ramsingh, A., 1993, Identification of a single amino acid residue in the capsid protein VP1 of Coxsackievirus B4 that determines the virulent phenotype, *J. Virol.* **67**:4797–4803.
102. Kang, Y., Chatterjee, N. K., Nodwell, M. J., and Yoon, J. W., 1994, Complete nucleotide sequence of a strain of Coxsackie B4 virus of human origin that induces diabetes in mice and its comparison with non-diabetogenic Coxsackie B4 JBV strain, *J. Med. Virol.* **44**:353–361.
103. Jenkins, O., Booth, J. D., Minor, P. D., and Almond, J. W., 1987, The complete nucleotide sequence of Coxsackievirus B4 and its comparison to other members of picornaviridae, *J. Gen. Virol.* **68**:1835–1848.
104. Titchener, P. A., Jenkins, O., Szopa, T. M., Taylor, K. W., and Almond, J. W., 1994, Complete nucleotide sequence of a beta-cell tropic variant of Coxsackievirus B4, *J. Med. Virol.* **42**:369–373.
105. Bae, Y. S., Eun, H. M., Pon, R. T., Giron, D., and Yoon, J. W., 1990, Two amino acids, Phe-16 and Ala-776 on the polyprotein are most likely to be responsible for diabetogenicity of encephalomyocarditis virus, *J. Gen. Virol.* **71**:639–645.
106. Bae, Y. S., and Yoon, J. W., 1993, Determination of diabetogenicity attributable to a single amino acid Ala-776, on the polyprotein of encephalomyocarditis virus, *Diabetes* **42**:435–443.

7

Neuropathic Viruses and Autoimmunity

CYNTHIA T. WELSH and ROBERT S. FUJINAMI

1. NEUROTROPIC INFECTION

The spectrum of viral infections of the central nervous system (CNS) or peripheral nervous system (PNS) includes acute, chronic, latent, and "slow" viral infections. There are relatively few host responses possible in the nervous system in a viral infection, so the results tend to be stereo-typed. Acute or subacute viral infections usually cause within the nervous system such diseases as meningitis, acute infective encephalomyelitis, and encephalitis, as well as acute postinfectious encephalitis. "Slow virus" infections are slow due to the long incubation periods, but may not be viral in origin. They continue to be discussed with viral infections as diseases caused by infectious agents unable to propagate in host cells (Table I). The severity and characteristics of the infections with various viruses that affect that CNS are thought to reflect tropism, virulence, and the host immune response. The incubation periods and length of disease are highly variable from one virus to another and will be discussed individually.

During CNS infection the brain may appear grossly normal or only slightly swollen and softened. The meninges may show some opacity or

CYNTHIA T. WELSH • Department of Pathology, University of Utah, Salt Lake City, Utah 84132.   ROBERT S. FUJINAMI • Department of Neurology, University of Utah, Salt Lake City, Utah 84132.

Microorganisms and Autoimmune Diseases, edited by Herman Friedman *et al.* Plenum Press, New York, 1996.

TABLE I
Virus Infections of the Central Nervous System

DNA Viruses
  Herpesviridae
    Herpes simplex virus type I, I (HSV); Varicella zoster virus; cytomegalovirus (CMV);
      Epstein-Barr virus (EBV)
  Papovaviridae
    JC virus; BK virus; adenoviridae

RNA Viruses
  Togaviridae
    Eastern equine encephalitis (EEE); Western equine encephalitis; rubella virus
  Picornaviridae
    Polioviruses; enteroviruses; Coxsackieviruses
  Flaviviridae ECHO viruses
    St. Louis encephalitis (SLE); Japanese encephalitis virus
  Paramyxoviridae
    Rubeola; mumps
  Bunyaviridae
    California encephalitis; LaCrosse virus
  Retroviridae
    Human immunodeficiency virus (HIV)
  Rhabdoviridae
    Rabies virus
  Orthomyxoviridae
    Influenza A virus

Unconventional agents
  Prions

congestion. Infection of the CNS by herpes simplex virus (HSV) or poliovirus infections are among the few infections which may provide specific gross clues as to the inciting agent. Mononuclear cells, which vary in distribution and intensity with the type of viral infection, comprise the inflammatory response. Arboviruses induce widespread encephalitis, in contrast to HSV, which tends to infect the limbic system and temporal lobes of the brain. Even in infections with selective features, such as HSV, more widespread inflammation is common. Encephalitis begins with a perivascular cuff, which is often a single layer of chronic inflammatory cells such as lymphocytes, plasma cells, and macrophages, around small venules and generally thicker cuffs around larger vessels (Fig. 1). These cells also appear in parenchyma, in areas which are beginning to show necrosis. The B cells and CD4+ T helper cells tend to remain in these perivascular Virchow-Robin spaces fairly close to each other for easier cell-to-cell interactions. Effector cells such as macrophages, NK cells,

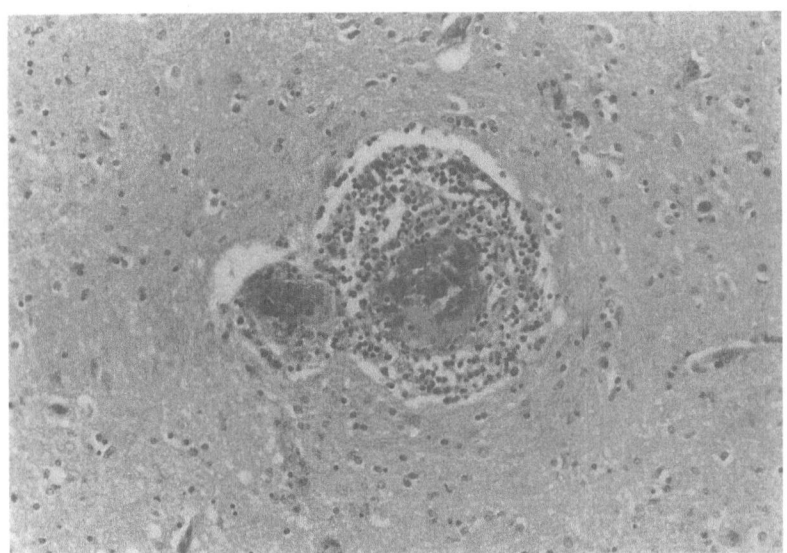

FIGURE 1. Perivascular cuffing (× 200).

and CD8+ T cells are also present but are usually further afield. After several days, rod cells derived from nervous system microglia appear, particularly in gray matter (whether cortical or deep gray).

Viruses can infect neurons and/or glial cells. This may result in direct viral cytolysis or cytopathology. For example, HSV infects all cell populations, in contrast to progressive multifocal leukoencephalopathy (PML), in which neurons are largely spared. Infection of neurons may incite nonspecific changes such as loss of Nissl substance and swelling or shrinkage of cells. These changes may alternatively be secondary to hypoxia or increased intracranial pressure. Inclusion bodies appear in some viral infections and may be intranuclear and/or intracytoplasmic. Cytomegalovirus (CMV) causes both (Fig. 2), whereas HSV typically causes intranuclear inclusions. The Negri body of rabies virus infections is cytoplasmic. The absence of inclusions in tissue sections does not exclude a viral etiology. Even in infections well known to cause inclusions, sampling and timing errors may occur. The best way to identify the specific type of viral inclusions is by electron microscopy. In poliomyelitis, neuronophagia (inflammatory cells surrounding and destroying neurons) occurs (Fig. 3). While the viral antigen or nucleic acid may not be detected, many viral infections leave glial nodules (collections of microglial cells) as evidence of former infection. Similarly, gliosis composed of reactive

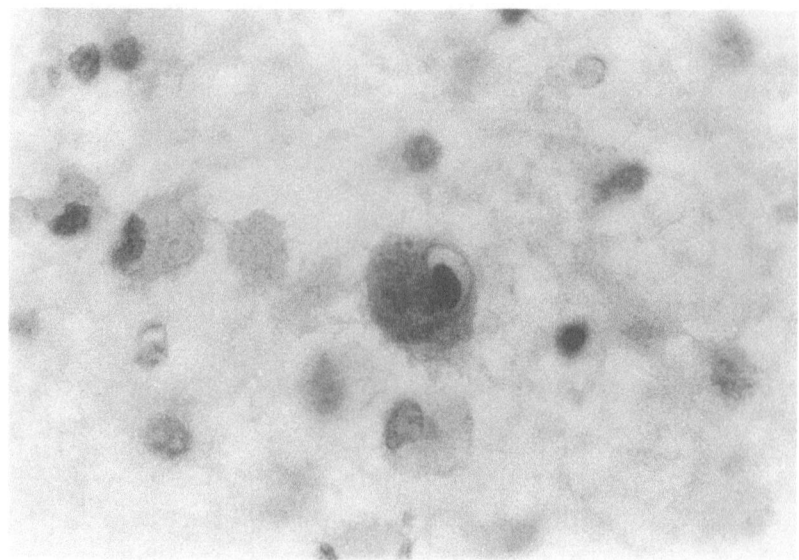

FIGURE 2. CMV. Cytoplasmic and nuclear inclusions (× 500).

astrocytes may appear focally or diffusely (Fig. 4). Gliosis can assume quite amazing proportions at times and can incite astoundingly large hypertrophied glial forms in diseases such as subacute sclerosing panencephalitis (SSPE) and PML. Microglial nodules and neuronophagia, however, can be seen in hypoxic brain damage, trauma, and neuronal degenerative diseases.

Demyelination can occur in viral infections and can be focal as in PML or diffuse as in human immunodeficiency virus (HIV) infection of the CNS. The demyelination seen in postinfectious encephalitis tends to be perivenular. Spongiform change (vacuolization in gray matter within neuronal cell processes) is seen in the slow viral infections, which do not demonstrate inflammation and therefore are encephalopathies, not encephalitides *per se*. Other circumstances that may result in little or no inflammation include infections in immunosuppressed patients.

2. VIRUS ENTRY AND INFECTION

Entry of virus can be through respiratory or gastrointestinal mucosa or the skin. The specific entry point depends on the virus and the vector.[1]

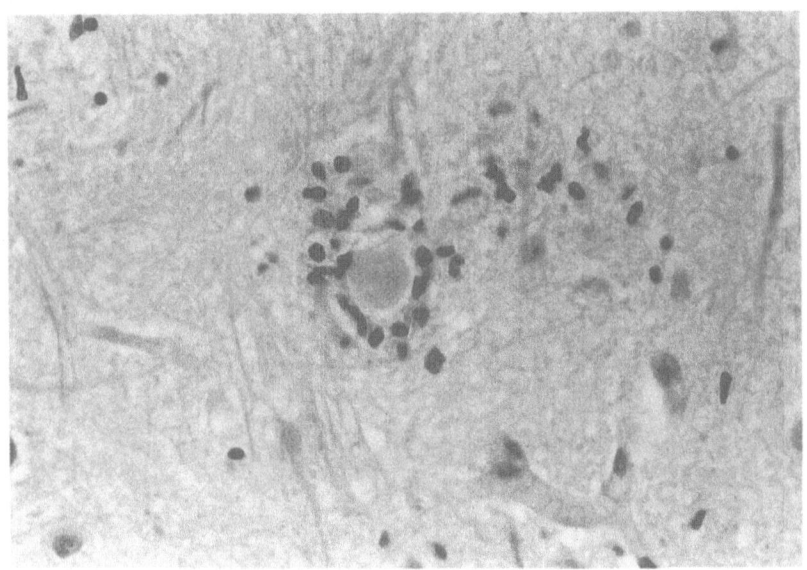

FIGURE 3. Neuronophagia (× 500).

Usually the entry point is quite distant from the CNS. Rabies and the arboviruses usually enter through skin. Enteroviruses generally enter through the gastric mucosa. Receptors on host target cells are thought to play a large role in susceptibility to different viral infections. Receptors may be cell membrane glycoproteins, thought to be part of the major histocompatibility complex (MHC),[2] and sometimes cell adhesion molecules[3,4] or hematopoietic membrane receptors.[5,6] Both cell-mediated and humoral immunity are important in protecting the nervous system from viral infections. Impaired cell-mediated immunity leads to persistent infections such as PML, SSPE, CMV, and varicella zoster virus (VZV). Persistent poliovirus and echovirus infections may result from agammaglobulinemia.

Viral invasion of the nervous system is usually through the blood system and rarely through direct invasion of a nerve. Viremia has to be of sufficient duration and magnitude to result in CNS infection. Virus in the blood is cleared by the reticuloendothelial system but may sometimes be protected from clearance. An example is HIV, which is sequestered in CD4+ T-cell macrophages. Some viruses may be inoculated directly into the bloodstream (e.g., arboviruses) by vectors such as insect bites, but others are released into the bloodstream from their primary replication

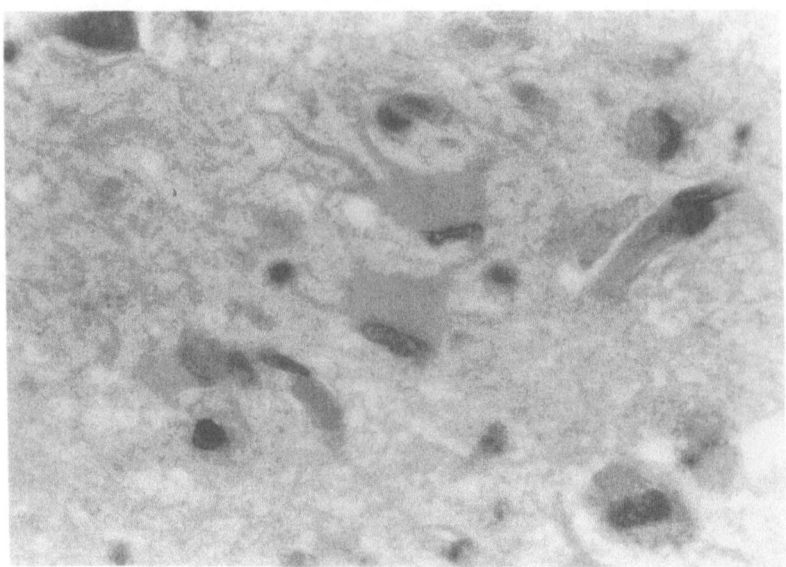

FIGURE 4. Gliosis (× 500).

site (e.g., poliovirus). For example, poliovirus infects and grows within the lymphoid areas of the human intestinal tract. Entry to the CNS is through endothelial cells into parenchyma or the choroid plexus. Rabies virus initially incubates in skeletal muscle and then enters the nerve at the neuromuscular junction and is transported to the dorsal root ganglia.[7] HSV and VZV also spread to the CNS through nerves.

3. ACUTE INFECTION

Most of the viruses that infect human beings can also invade the CNS. However, CNS involvement is rare[8] for all the systemic viral infections. Availability of a target cell for entry, the efficiency of the virus spread, and the immune response strength determines whether a virus results in CNS infection. Acute CNS syndromes include meningitis, encephalitis, poliomyelitis, transverse myelitis, encephalomyelitis, and myeloradiculitis as well as postinfectious encephalitis. Most acute syndromes occur as meningitis and/or encephalitis, but most patients with encephalitis also show evidence of meningeal involvement.

Aseptic viral meningitis is most often benign and self-limiting.[9] Typical symptoms include fever, headache, and meningismus (stiff neck). The symptoms tend to last about one week and disappear without sequelae. Enteroviruses (picornavirus, Coxsackievirus, and echovirus) are the most common causes of aseptic meningitis. Other causes of aseptic meningitis include mumps, lymphocytic choriomeningitis virus (LCMV), Epstein-Barr virus (EBV), and arboviruses.

Viral encephalitis is more severe than meningitis.[10] A variety of viruses cause encephalitis; arboviruses are the most common epidemic cause and HSV is the most common sporadic cause. Enteroviruses, mumps, HSV, rabies, arenaviruses (e.g., LCMV), adenoviruses, and EBV are other causes. With most viral encephalitides the clinical picture is not specific. The symptoms include seizures, delirium, focal neurological signs, and coma. Acute disease generally develops over days to weeks.

Arboviruses (*arthropod-borne*) are a group of several hundred enveloped positive-sense RNA viruses transmitted by arthropods such as mosquitoes, sandflies, and ticks.[11] They cause such diseases as hemorrhagic fevers, tropical fevers, and encephalitides. Arboviruses are the most common causes of epidemic encephalitis. The Japanese encephalitis virus is the most common cause of epidemic encephalitis worldwide. It commonly has a fatality rate of about 50% and serious sequelae.[12] In the United States, most cases are due to St. Louis encephalitis (SLE) virus, eastern equine encephalitis (EEE) virus, western equine encephalitis (WEE) virus, and California encephalitis virus. Pathologic changes include a mononuclear infiltrate in the leptomeninges and in parenchyma manifested chiefly as perivascular cuffing and microglial nodules. Severe cases may show neuronophagia. Typical viral inclusions do not generally appear.

The SLE virus is a Flavivirus and is transmitted by mosquitoes.[13] Cases in the midwestern United States tend to be rural and endemic, while cases in the East are usually urban and cause focal outbreaks which sometimes are widespread. These differences appear to be related to insect vectors. Epidemics tend to occur in summer and early fall. The severity of illness and level of clinically apparent illness tend to increase with age, and fatalities are most common in the elderly (20% of patients over 60). Approximately 75% of symptomatic patients develop encephalitis; the rest generally have aseptic meningitis or headache with fever. Approximately half of patients with neurologic symptoms develop a prodrome of headache, fever, gastrointestinal symptoms, and myalgias. The rest, particularly the elderly, have abrupt onset of neurologic dysfunction which includes stiff neck, confusion, tremors, seizures, myoclonic jerks, and cranial nerve palsies. Cerebrospinal fluid (CSF) examination usually shows an increase in white blood cells, particularly lymphocytes which are typical of viral encephalitis. Self-limited sequelae such as behavioral changes, tremor, lack of coordination, and weakness are not unusual.

The WEE virus is an alphavirus in the Togaviridae and is mainly transmitted by mosquitoes.[14] Cool wet spring weather favors the spread of the vector. Rural areas show the greatest risk for infection. Infants and the elderly are most at risk. The onset of systemic symptoms is generally abrupt and nonspecific, followed by neck stiffness, photophobia, and somnolence, but neurologic involvement is usually mild and may be asymptomatic. Paresis, abnormal reflexes, tremors, and seizures may occur. The CSF, as with SLE, shows a pleocytosis. Infants generally have the greatest risk for sequelae and suffer serious consequences in many infections. These include paralysis, spasticity, convulsions, and motor retardation. Adult complications include apnea, Parkinson's disease, and inappropriate antidiuretic hormone secretion.

The EEE virus, like WEE virus, is an alphavirus in the family Togaviridae and has a positive-sense, single-stranded RNA genome. The viruses are antigenically different in North and South America and have different transmission cycles. Mosquitoes transmit the virus; the areas affected on the Atlantic and Gulf coasts remain stable from year to year. The incidence of infection is high in young children. Contact with infected tissues is a risk factor in infection. These viruses cause the most severe form of arthropod-borne encephalitides.[15] The disease is more common in late summer and early fall. The onset of illness is often abrupt with severe systemic viral symptoms, with neurologic dysfunction occurring rapidly or days later. The longer prodrome appears to lead to a milder course and better prognosis. Nearly all infections result in encephalitis. Stupor, coma, and seizures are common. The very young and very old are at greater risk of serious illness and worst outcome. Infants may develop spasticity, seizures, paresis, or developmental delays. Neutrophils predominate in the marked pleocytosis in CSF at the beginning more so than in other viral encephalitides, followed by a shift to mononuclear cells. Computerized tomography (CT) scans may at best show diffuse edema, while electroencephalograms (EEG) can show a variety of patterns. The diagnosis can often be confirmed serologically at presentation since most of the population is nonimmune.

California encephalitis is a name given to disease caused by the California serogroup of the Bunyaviruses which are enveloped segmented RNA viruses.[16] There are at least 15 viruses within the serogroup, five of which cause human illness within the United States. The viruses are transmitted by mosquitoes and seem to be widely distributed across Canada and the United States. Nearly all California encephalitides are caused by LaCrosse virus. Cases of LaCrosse viral infections occur almost exclusively in children less than 15 years old, with a 2:1 male-to-female ratio, during the summer. The risk of illness is directly associated with proximity of residence to forests, making the illness most common outside the urban areas. In endemic areas, the annual incidence approaches that of bacterial meningitis. Greater than 99% of infections with LaCrosse virus are asymptomatic, and symptomatic infections range from a mild febrile illness to meningitis or encephalitis. Generally preceded by a systemic viral prodrome, the neurologic signs vary and can include weakness, seizures, and aphasia. Mortality is low but neurologic or psychiatric sequelae occur in up to 15% of patients, including inappropriate secretion of antidiuretic hormone and seizures. Imaging studies and EEG may show focal abnormalities, leading to misdiagnosis. The CSF shows a modest pleocytosis and sometimes elevated protein.

The herpes virus family includes HSV types I and II, VZV, EBV, and CMV. In the United States HSV encephalitis is the most common sporadic encephalitis. After infancy, the principal cause of HSV encephalitis is type I HSV, which is primarily a reactivation of latent herpes virus; this is discussed in Section 4.

Infants are primarily afflicted by HSV type II, typically acquired during birth.[17] Neonatal infections are a common problem in the United States and increasing in incidence. Babies with encephalitis have a mortality rate of only 15%, but 50% have neurologic sequelae compared to a mortality rate of 50% and complication rate of 15% in those with disseminated disease. These statistics were much grimmer prior to antiviral therapy. Many of the infected neonates are premature. The symptoms begin 1 to 3 weeks after birth usually with disseminated infection, but conditions may be limited to encephalitis or infection of only skin, eye, or mouth. Encephalitis secondary to dissemination is probably due to bloodborne seeding resulting in multiple areas of hemorrhagic cortical necrosis. In contrast, retrograde axonal transmission is probably responsible for cases with only encephalitis. Neonatal HSV encephalitis is diffuse and necrotizing. Cultures of CSF are commonly positive in affected newborns. Prevention is the best treatment.

The VZV causes chicken pox and zoster (shingles). The virus very rarely involves the CNS except as zoster.[18] Acute cerebellar disease is the most common form of varicella encephalitis. Nystagmus, dysarthria, and ataxia appear after a rash. The cerebellitis runs a benign course. Varicella can also cause an encephalitis with fever, stiff neck, neuromuscular dysfunction, behavioral changes, coma, and seizures. The mortality rate has been reported as high as 35%. The histology shows perivascular mononuclear infiltrates with demyelination and axonal degeneration. Transverse myelitis, meningitis, and polyradiculopathy may occur. Varicella zoster infections can show inflammation and hemorrhagic necrosis of dorsal root ganglia, degeneration of motor and sensory roots, and intranuclear inclusions in dorsal root ganglia neurons. Virus has been recovered by polymerase chain reaction (PCR) methods, supporting the theory of direct invasion as opposed to an immunological mechanism.

The EBV rarely causes symptomatic nervous system disease, but death is most frequently due to neurological complications.[19] Guillain-Barré, Bell's palsy, meningitis, diffuse or focal encephalitis, cerebellitis, peripheral neuropathy, and transverse myelitis have all been reported as neurological manifestations of EBV. Usually the neurologic symptoms appear as the systemic infection peaks, but sometimes the picture is not classic for infectious mononucleosis. Epstein-Barr virus encephalitis can be the first or only manifestation and can be focal or diffuse. It is usually self-limited and benign but is occasionally fatal and has been reported to cause mild deficits. The virus has been detected in brain tissue and CSF by PCR techniques, suggesting that the disease is due to direct invasion.

Most cases of CMV encephalitis occur in immunosuppressed patients, premature infants, or fetuses.[20] The manifestations of the disease in infants depend on whether the virus was transmitted transplacentally, perinatally, or by blood transfusion. Also important is whether the maternal infection is primary (which is most likely to cause overt disease) or a reactivation. Many anomalies are sequelae of congenital CMV infections, including microcephaly, hydrocephalus, and microphthalmia. The meningoencephalitis, which is prominent in subependymal areas, is characterized by microglial nodules and mononuclear inflammatory infiltrates as well as Cowdry type A inclusions in neuronal, glial, and ependymal cells. Tissue destruction and dystrophic calcification are common residuals. In adults, encephalitis can be acquired through transplantation or blood transfusion, but it is more often a reactivation of a latent infection. Intranuclear inclusions are less common in adult cases. Cytomegalovirus encephalitis appears to be a direct viral effect, as opposed to some of the other disease processes caused by CMV which have extensive host-driven immunological damage.

Enteroviruses, which are picornaviruses, cause encephalitis infrequently; aseptic meningitis is the most common presentation.[21] Transverse myelitis and cranial nerve palsies are also seen. Typically, nonpolio myelitis is characterized by a milder paralysis with no residual effects.

Echoviruses and Coxsackie viruses are the major causes in the United States, since vaccination has decreased the cases of poliomyelitis.[22] As with poliomyelitis, the other enteroviral diseases typically occur during the late summer and early fall, most often in children. Symptoms include fever, seizures, focal neurological deficits, and obtundation. The encephalitides are usually mild and benign except in neonates with Coxsackie B viral infections or patients with agammaglobulinemia.[23] Coxsackie and echoviruses can be recovered from CSF early in the disease, unlike poliovirus. Poliovirus remains a serious health problem in many parts of the world, but cases of paralytic poliomyelitis in the United States are few and usually vaccination-related. Out of five suspected cases of poliomyelitis in 1992, four were confirmed and were vaccine-associated.[8] Anterior horn cells in the spinal cord are selectively vulnerable, due to the large number of surface viral receptors. Other neurons in the brain stem and reticular activating system, as well as Betz cells in the motor cortex, can also be infected. Viral inclusions are not present, but neuronophagia is prominent in poliomyelitis. Perivascular infiltrates and microglial nodules are also found.

4. PERSISTENT VIRAL INFECTIONS

Persistent viral infections can be categorized as chronic, latent, or slow.[24] Chronic infections are here defined as infections with continuous production of virus and latent infections as those where virus is undetectable for a period of time. Latent infections reappear from time to time due to renewed virus replication and spread from the latent site. Persistent infections include PML, SSPE, progressive rubella panencephalitis (PRP), Creutzfelt-Jacob, HTLV-1-associated myelopathy/tropical spastic paraparesis (HAM/TSP), HIV, adenoviruses, and herpes viruses. Both host and viral factors contribute to viral persistence. Viruses may be able to develop more than one pattern of disease based on differing host factors. The virus must be able to evade the immune system in order to persist.

Not all viruses kill the cells they infect. It is actually in their best interest not to kill their hosts.[25] Cells may have slowed growth or be unable to function well, which, for example, can cause serious problems in a fetus with a rubella infection. Viruses that remain latent are undetectable by the host immune system until reactivated. This allows them to be persistent. Prions (infectious proteinaceous material) may persist because they are nonimmunogenic. Viruses may persist by infecting hematopoietic cells, as HIV does, or by infecting nonpermissive cells, as perhaps occurs in SSPE. Antigenic drift may make it possible for the virus to escape detection. Viruses produce proteins which can inhibit immune responses.

Progressive multifocal leukoencephalopathy is a CNS disease caused by an opportunitistic papovavirus infection of a patient with an underlying immune dysfunction such as malignancy, immunosuppressive therapy, or HIV infection.[26] Two papovaviruses, JC virus and SV40 have been recovered from PML brain. The brain shows softening, discoloration, and a granular change in white matter. The disease causes demyelination, enlarged oligodendroglial cell nuclei with amphophilic intranuclear inclusions that push the chromatin to the nuclear membrane, and large bizarre astrocytes. Many oligodendroglial cells die, and the survivors with inclusions are frequently around the edges of the lesions. Papovavirus particles are frequently found in oligodendroglial cells and rarely in astrocytes on electron microscopy. This shows that the bizarre astrocytes are infected and not just reacting to the process, though the lower rate of infectivity brings up the question of whether astrocytes are nonpermissive to infection. Patients usually show signs such as paralysis, personality changes progressing to dementia, aphasia, visual loss, ataxia, and sensory abnormalities. The disease progresses to death of the patient usually within one year. Imaging studies show widespread involvement of white matter, most often in the occipital lobes. Multifocal, nonenhancing lesions without mass effect appear on imaging.

Herpes simplex virus may cause aseptic meningitis, encephalitis, ascending myelitis, and keratitis. A reactivation of HSV type I is the major cause of HSV encephalitis after infancy and is the most common sporadic fatal encephalitis.[27] The incidence is distributed throughout the year. The virus has been found in the trigeminal ganglia in a large percentage of adults. The basal frontal lobes and temporal lobes are the areas most commonly affected, possibly due to tracking up from the trigeminal ganglia to the basilar dura of the anterior and middle fossae. Alternatively, the virus may follow the olfactory bulbs to the brain. HSV encephalitis can occur in healthy people of all ages but has a predilection for immunosuppressed patients. The usual signs and symptoms may develop rapidly or insidiously and include fever, headache, coma, focal neurological deficits, and nuchal rigidity. The CSF usually shows an increase in white blood cells (mainly lymphocytes), increased protein, normal glucose, and red blood cells. Cultures are seldom positive in adults. Imaging studies and EEG are of great diagnostic value in showing localization and valuable for long-term follow-up. Age and the Glasgow

Coma Score are the best predictors of outcome. Antiviral therapy has drastically improved survival and complication rate, but there are long-term cognitive effects in many of the survivors which may be subtle. The typical pathology includes multiple small sites of hemorrhagic necrosis involving the cortical ventral frontal lobes and mesial temporal lobes. Typical Cowdry type A inclusions (eosinophilic intranuclear inclusions) appear only in the first few days in neurons and glial cells. An abundant inflammatory infiltrate is present, consisting chiefly of lymphocytic cuffing. Destruction of tissue may become widespread.

Measles virus is a paramyxovirus. Measles virus rarely causes encephalitis acutely.[28] It does cause SSPE, a chronic encephalitis of childhood. The usual age of onset of symptoms is between 4 and 14 years. The original measles infection is usually before the age of 2. Measles virus vaccination programs have reduced the incidence of SSPE. There is an insidious onset of dementia, behavioral changes, and myotonic seizures, progressing to obtundation and death. The clinical course may range from one month to several years. Widespread lesions in both gray and white matter of the hemispheres show a mixture of demyelination, necrosis, and gliosis. The inflammatory reaction is chronic, consisting of lymphocytes and plasma cells mostly, much of it perivascular. The perivascular edema, inflammation, and demyelination closely resemble the pathology seen in postinfectious encephalomyelitis. Eosinophilic intranuclear inclusions appear within nuclei and cytoplasm of neurons and oligodendrocytes. Because the virus obviously persists for years, some mechanism for immune system evasion must be postulated. Antibody titers to measles virus are elevated in both CSF and blood.[29] This vigorous immune response, coupled with the pathology, has led many to suspect that SSPE is an immune-mediated disease.

Rubella virus is also a paramyxovirus which rarely causes acute encephalitis. Rubella also rarely causes congenital rubella syndrome since the institution of immunization.[30] This syndrome consists neurologically of growth retardation, ocular abnormalities, sensorineural hearing loss, and signs of meningoencephalitis. Clinically similar to SSPE is PRP, usually following an acquired infection of childhood. The disease is very rare and demonstrates progressive neurological symptoms such as dementia, ataxia, spasticity, and death within a few years. No inclusions are present in brain tissue sections. Histology shows perivascular cuffs of lymphocytes and microglial nodules within parenchyma. There is greater white matter involvement than gray, and the cerebellum is greatly affected.[31] The virus has been recovered from brain and from peripheral blood lymphocytes.

Slow virus infections (spongiform encephalopathies) are thought to be caused by agents simpler than viruses, known as prions.[32] It has been proposed that an abnormal form of a normal cellular protein can cause the normal protein (PrP) to adopt an aberrant conformation. The abnormal form may be a mutation as in the familial forms or introduced as a pathogen. The normal function of the prion protein is not yet known. Prion diseases in humans include Creutzfeldt-Jacob disease (CJD) and Gerstmann-Sträussler-Scheinker syndrome (GSS). Discovery of mutations in the PrP genes in GSS and familial CJD shows a genetic as well as infectious basis for the disease. Mice lacking the normal PrP gene show normal development and behavior and are resistant to infection with prions.[33] Creutzfeldt-Jacob disease is a transmissible, relentlessly progressive dementia of fatal outcome most often seen between the ages of 50 and 65 years. Most patients die within 6 months after the onset of symptoms. A minority of cases are familial. There is often a prodrome of malaise, altered personality, emotional liability, and sleep disturbance followed by progressive dementia, ataxia, and myoclonus. The CSF is usually normal and scans may show some atrophy, particularly in patients with longer survival. The EEG is characteristic in about two thirds of cases. The gray matter shows vacuolization, concurrent with and followed by neuronal loss and gliosis.[34] The changes are most prominent in the cerebral and cerebellar cortices, putamen, and thalamus. Amyloid plaques are seen in a small number of patients. There is no inflammation but the disease is transmissible and has been classically categorized as an encephalitis. The process may be quite focal, requiring many histologic sections.

Human immunodeficiency virus is a retrovirus affecting the CNS and PNS directly and through secondary infections with opportunistic agents. Neurologic dysfunction is the presenting symptom in 10% of patients. Over time, 30–40% will have clinical evidence of nervous system involvement, but at autopsy more than 75–90% will have nervous system pathology. Secondary viral infections include PML, CMV, HSV, and VZV. The virus itself can cause a plethora of CNS and PNS diseases such as acute meningitis, myelitis, vacuolar myelopathy, mononeuritis multiplex, demyelinating polyneuropathies, ganglioneuritis, inflammatory myopathy, and AIDS-dementia complex.[35] AIDS-dementia complex is the most common neurological disorder and is characterized by impaired memory, psychomotor slowing, and behavioral changes. Motor deficits and progressive global dysfunction of the CNS appear later. The brain shows perivascular and parenchymal infiltrates of mononuclear cells in white matter of cerebrum, basal ganglia, and brain stem. Macrophages and multinucleated giant cells can be found. Children may show, in addition, calcific vasculopathy. Demyelination and white matter necrosis may be seen in areas of inflammation. Vacuolar myelopathy leads to varying degrees of spasticity, paraparesis, and ataxia. Vacuolar degeneration of the spinal cord, particularly of the lateral and posterior columns, is present, most frequently at the cervical and thoracic levels.[36] The HTLV-1 is associated with HAM/TSP. This disorder is characterized by weakness, unsteadiness of gait, and paresthesia of the lower limbs. Deafness and vision loss are also common. Spasticity may occur.[37] Perivascular inflammatory cell cuffing in the cord and around nerve roots and ganglia is seen. Fibroblasts proliferate around vessels and in the meninges. Loss of myelin is found in the dorsal columns, spinocerebellar tracts and lateral pyramidal tracts. Neuronal degeneration is usually seen. Axonal loss and gliosis eventually develop. Neuronoaxonal spheroids have been reported as prominent in some cases.[38] The degree of inflammation and fibrosis varies. Chronicity, in particular, lends itself to a shift from inflammation to sometimes prominent fibrosis. The cerebellum, brain stem, and cerebrum may show less severe changes. The HTLV-1 has been amplified by PCR from spinal cord tissue.[39]

5. VIRAL COMPLICATIONS OF IMMUNOSUPPRESSION

Opportunistic viral infections include CMV, HSV, VZV, EBV, measles virus, papovaviruses, enteroviruses, and adenoviruses. Syndromes include encephalitis, meningitis, myelitis, PML, and radiculitis. Immunosuppression may be due to chemotherapy for cancer or posttransplant immunosuppression, congenital defects in the immune system, or infection by an immunosuppressing virus such as HIV. Infections can be acute or indolent. Common clinical neurologic abnormalities after transplantation include encephalopathy, seizures, obtundation, and coma. A low-grade encephalitis with microglial nodules, but without inclusions, has been reported in 22% of liver transplant patients.[40] The most common recognized viral infection in the first 30 days after bone-marrow transplant is caused by HSV. Most serious infections between successful engraftment and day 120 after transplants are viral; after 120 days many of the infections are viral.[41]

6. PARAINFECTIOUS SYNDROMES

Parainfectious syndromes are characterized by emergence of neurological signs late in the course or after apparent recovery from an infection. In 1993, 151 cases of postinfectious encephalitis were reported to the Centers for Disease Control and Prevention.[8] Syndromes include classic acute disseminated encephalomyelitis (ADEM), hyperacute acute hemorrhagic leukoencephalitis (AHLE), Guillain-Barré syndrome (GBS), and Reye's syndrome. In the past the childhood diseases of measles and varicella were the major culprits causing ADEM. Worldwide, they continue to be important causes, but the use of vaccines has decreased the incidence of measles as a cause in this country. The virus has not been isolated from the CNS. Although frequently associated with antecedent viral illness, postinfectious encephalomyelitis can follow immunization. The incidence of complications from immunization has decreased considerably since the improvement of rabies vaccines and the discontinuation of vaccination for smallpox. The organisms that cause GBS include CMV, EBV, HZV, mycoplasma, and *Campylobacter jejuni*. The syndrome may also follow influenza immunization.

Guillain-Barré syndrome is an acute or subacute illness that causes predominantly motor problems. Relapse has been reported.[42] Classically, the patient has suffered a mild upper respiratory or gastrointestinal illness, then within a few weeks begins to experience an ascending paralysis. Tendon reflexes are lost, autonomic abnormalities may develop, and assisted ventilation may become necessary. In the CSF, protein is elevated, nerve conduction velocities are reduced, and denervation is demonstrable. The full-blown syndrome may develop over days to weeks.[43]

The histologic features of GBS vary, suggesting that the syndrome may not be homogeneous. Widespread demyelination with variable axonal degeneration is most common, but axonal loss with little demyelination has been reported. Some cases show prominent lymphocytic infiltration, suggesting that the demyelination may be related to a cell-mediated process. In others, the predominant cell is the macrophage, raising the possibility of an antibody-mediated process.[41] Studies have shown circulating lymphocytes sensitized to peripheral nerve antigens and circulating antibodies.

The onset of symptoms in ADEM is often sudden following within a week or two of a viral infection or vaccination. The CSF may be normal or nonspecific. Death occurs in approximately 20% of cases. Of those who recover, some have permanent neurological damage, and persistent MRI abnormalities can be seen even in patients without clinical deficits.[44]

The brain and spinal cord may appear normal or only show congestion or edema. Histologically, there is a diffuse inflammatory process involving perivenular areas. The white matter is more severely involved than the gray. With time, the major inflammatory cell present shifts from neutrophil to lymphocyte. The most characteristic finding is perivascular demyelination, with total myelin loss, microglial proliferation, and phagocytosis. Older lesions show gliosis.

AHLE is a fulminate hemorrhagic form of postinfectious encephalitis, with feverish patients quickly becoming comatose.[45] Mortality is high in the first few days. The CSF may show a high protein level. The brain will be edematous and soft. Multiple small hemorrhages which are usually symmetrical appear on gross sections. Because of the speed of the process the demyelination of ADEM may not be seen. The process remains perivascular with necrosis of vessel walls, thrombosed vessels, perivascular edema, PMNs, and hemorrhages.

Demyelination is a major pathological finding in postinfectious encephalomyelitis, as well as SSPE and PML. Viruses can cause demyelination by various mechanisms. Direct infection of oligodendrocytes may cause demyelination through cell damage or death. Myelin may be destroyed by virus or viral products. Viruses may also initiate autoimmune antibody or cell-mediated responses which can damage tissue. The autoimmune theory for parainfectious encephalitis is supported by the lack of culturable virus, the timing of disease as the systemic infection has subsided, and the perivascular chronic inflammatory cells and demyelination, which can be replicated experimentally in autoimmune disease animal models. Autoantibodies in low titer can occur after viral infections. This may be due to antiidiotype antibodies, polyclonal B-cell proliferation, immune dysfunction due to lymphocyte infection, or alteration or exposure of host proteins due to infection of host cells. The symptoms of parainfectious encephalitis cannot be separated from those of viral encephalitides themselves but usually develop as the patient is recovering from a systemic infection. The mortality rates and complications in survivors vary with the antecedent infection. Reye's syndrome is also a parainfectious problem with high morbidity and many sequelae in survivors.

Experimental allergic encephalomyelitis (EAE) is an animal model for postinfectious encephalitis which can be induced by the injection of myelin components. Acute and chronic forms of the disease may be developed by variation in protocols. Passive transfer of lymphocytes sensitized by myelin basic protein (MBP) or other spinal cord components causes chronic relapsing EAE (crEAE). This in turn can cause a relapsing, remitting clinical course similar to the most common type of multiple sclerosis (MS).[46] The lesions resemble those of MS pathologically. Perivascular infiltrates, demyelination, and gliosis are encountered in both crEAE and MS. The types of infiltrating inflammatory cells are comparable.

Animal models of virally induced demyelination are numerous and include mouse hepatitis virus, Theiler's murine encephalomyelitis virus,[47] sheep visna virus, simian virus 40, and canine distemper virus. A viral etiology for MS has been repeatedly postulated. Without question, viruses can have a long incubation time and cause demyelinating disease. Examples discussed in this chapter have included PML, postinfectious encephalomyelitis, and HIV. In Theiler's virus infections, T cells initiate the immune-mediated demyelination. The cause of multiple sclerosis has long been attributed to infectious agents, primarily viruses. The increase of risk of MS in someone spending their youth in particular areas leads to the assumption of exposure during childhood to an infectious agent. Infections seem to exacerbate preexisting MS.[48]

7. DISCUSSION AND CONCLUSIONS

The pathogenesis of many viral infections is well known. There is no definitive evidence that viruses cause autoimmune diseases, although this has been postulated many times over the past few decades. It is well known that many diseases of man considered to be autoimmune in nature, especially multiple sclerosis exacerbations, are related to recent infection with a microorganism, such as the influenza virus or even the common cold virus. However, some virologists and immunologists believe the temporal relationship between infection and autoimmune disease may be due to stress, which results in release of cytokines and catecholamines that activate immune cells, especially macrophages and T cells thought to be important in autoimmune reactions. For example, it is now widely believed that infections, especially by viruses, seem to exacerbate existing MS. One of the early concepts concerning autoimmune diseases, based upon experimental animal studies with inbred mice, was that autoimmune diseases are multifactorial and have a genetic component, an immune component, and a viral infectious disease component. It was also widely speculated that malignancies, especially of the lymphoid system, such as lymphomas and leukemias, were associated with genetic factors controlling the "strength" of the immune response. For example, experimental animals infected with a leukemogenic virus often develop an autoimmune reaction, and the immune competence of the animal is strong enough to inhibit the virus and prevent leukemia. The immune response to possible cross-reactive antigens of normal tissue would result in autoimmunity. For example, neurological tissues and myelin basic protein antigens may contain cross-reactive antigen to a virus which first infects the CNS. In contrast, individuals with a less robust immune response due to genetic factors would not be able to destroy an infecting virus, such as a tumor-inducing retrovirus, and would develop leukemia or lymphoma rather than an autoimmune disease. Therefore, the best of two possibilities would be development of an immune response against "self" rather than development of a malignancy. This obviously is a very oversimplified concept, but, as with many other hypotheses in biology, there may be some correct aspects in this scenario.

It is important to note that no one has yet shown that a specific virus is associated with any specific autoimmune disease. Nevertheless, as indicated previously, demyelination of the CNS is a major pathological finding in many viral infections, especially encephalomyelitis. Although it is known that viruses may cause demyelination by various mechanisms, such as direct destruction by the viruses or viral products, such intracellular pathogens may also initiate an autoimmune response, especially cell-based responses on activated T cells and macrophages which can cause the tissue damage. As indicated above, the autoimmune theory for such viral infections appears to be supported by the absence of detection of culturable viruses, regardless of method or technology used, including newly developed highly sensitive techniques such as RT-PCR. It is now believed by some biomedical scientists that certain diseases which are considered autoimmune in nature may be related to infection several weeks earlier with a virus such as measles or influenza. Thus, an exacerbation or initiation of a demyelinating disease is often associated with a prior systemic infection by a virus when the acute symptoms of infection have already subsided. Accumulation of inflammatory cells around the demyelinated lesions in the CNS also supports the presence of such virus. Experimental autoimmune diseases in animals, such as EAE, have shown that tissue antigens, including spinal or brain tissue, can induce disease and this is related to activation of specific cytokines which defined select subsets of T cells, especially Th1 cells. Nevertheless, no one has been able to show that even such experimental autoimmune diseases can be induced by a virus, although specific antiviral antibody titers have been found in the circulation and even in the spinal fluid of patients who have MS or other demyelinating autoimmune diseases. It is widely believed, however, that animal models for postviral autoimmunity, including demyelinating disease, are useful in determining a possible relationship and even a mechanism for the role of microbial infection and autoimmunity. For example, it has been shown that passive transfer of lymphocytes sensitized by encephalomyelitis-inducing spinal tissue or even purified myelin basic protein as well as other spinal cord components results in relapsing-remitting allergic encephalomyelitis with lesions resembling those of MS pathology in humans. However, it is not known whether such animal models are applicable to human demyelinating diseases such as MS. Furthermore, it is not known whether there is a direct relationship between the viral infection and such activation of reactive T cells, macrophages, or even humoral antibody to myelin protein. However, it appears there is now a wide consensus that postviral infections may be related to autoimmune disease, and furthermore, it seems likely that the immune response to a virus, especially cellular immunity, can result in typical pathogenesis of autoimmune-like infection in an otherwise normal individual. It is apparent that much work still must be done to demonstrate a relationship, either direct or indirect, between virus infection and autoimmune diseases such as MS. However, with the advent of modern neurobiologic and immunologic techniques, it seems likely that within the near future much new information will be obtained to permit a more definite understanding of the mechanisms of viral infection and autoimmunity, especially of the CNS.

REFERENCES

1. Haywood, A. M., 1994, Virus receptors: Binding, adhesion strengthening, and changes in viral structure, *J. Virol.* **68**:1–5.
2. Collins, A. R., 1993, HLA class I antigen serves as a receptor for human cornavirus OC43, *Immunol. Invest.* **22**:95–103.
3. Giranda, V. L., Chapman, M. S., and Rossmann, M. G., 1990, Modeling of the human intercellular adhesion molecule. 1. The human rhinovirus major group receptor, *Proteins: Struct. Funct. Genet.* **7**:227–233.
4. Bergelson, J. M., Chan, B. M., Finberg, R. W., and Hemler, M. E. I., 1993, The integrin VLA-2 binds echovirus 1 and extracellular matrix ligands by different mechanisms, *J. Clin. Invest.* **92**:232–239.
5. Maddon, P. J., Dalgleish, A. G., McDougal, J. S., Clapham, P. R., Weiss, R. S., and Axell, R., 1986, The T4 gene encodes the AIDS virus receptor and is expressed in the immune system and the brain, *Cell* **47**:333–348.
6. McClure, J. E., 1992, Cellular receptor for Epstein-Barr virus, Prog. Med. Virol. **39**: 116–138.
7. Mrak, R. E., and Young, L., 1994, Rabies encephalitis in humans: Pathology, pathogenesis and pathophysiology, *J. Neuropathol. Exp. Neurol.* **53**:1–10.
8. Centers for Disease Control and Prevention, 1994, Notifiable disease reports, MMWR **42**(51, 52):1002–1003.
9. Connolly, K. J., and Hammer, S. M., 1990, The acute aseptic meningitis syndrome, *Infect. Dis. Clin. North Am.* **4**:599–622.
10. Bale, J. F., 1993, Viral encephalitis, *Med. Clin. North Am.* **77**:25–42.
11. Tsai, T. F., 1991, Arboviral infections in the United States, *Infect. Dis. Clin. North Am.* **5**: 73–102.

12. Vaughn, D. W., and Hoke Jr., C. H., 1992, The epidemiology of Japanese encephalitis: Prospects for prevention, *Epidemiol. Rev.* 14:197–218.

13. Bleed, D. M., Marfin, A. A., Karabatsos, N., Moore, P., Tsai, T., Olin, A. C., Lofgren, J. P., Higdem, B., and Townsend, T. E., 1992, St. Louis encephalitis in Arkansas, *Sci. Am.* 89:127–130.

14. Peters, C., and Delrymple, J., 1990, Alphaviruses, in: *Fields Virology* (N. F. Barnard *et al.*, eds.), Raven Press, New York, pp. 742–746.

15. McHugh, C. P., 1994, Arthropods: Vectors of disease agents, *Lab. Med.* 25:429–437.

16. Gonzalez-Scarano, F., and Nathanson, N., 1990, Bunyaviruses, in: *Fields Virology* (B. N. Fields *et al.*, eds.), Raven Press, New York, pp. 1203–1208.

17. Whitley, R. J., 1993, Neonatal herpes simplex virus infections, *J. Med. Virol.* 1:13–21.

18. Krywanio, M. L., 1991, Varicella encephalitis, *J. Neurosci.* 23:363–368.

19. Imai, S., Usui, N., Sugiura, M., Osato, T., Tsutsumi, H., Tachi, N., Nakata, S., Yamanaka, T., Chiba, S., and Shimada, M., 1993, Epstein-Barr virus genomic sequences and specific antibodies in cerebrospinal fluid in children with neurologic complications of acute and reactivated EBV infections, *J. Med. Virol.* 40:278–284.

20. Zaia, J. A., 1990, Epidemiology and pathogenesis of cytomegalovirus disease, *Semin. Hematol.* 27:5–10.

21. Moore, M., 1982, Enteroviral disease in the United States, 1970–1979. *J. Infect. Dis.* 146:103–108.

22. Centers for Disease Control and Prevention, 1994, Progress toward global eradication of poliomyelitis 1988–1993, MMWR 43:499–503.

23. Hertel, N. T., Pedersen, F. K., and Heilmann, C., 1989, Coxsackie B3 virus encephalitis in a patient with agammaglobulinaemia, *Eur. J. Pediatr.* 148:642–643.

24. Garcia-Blanco, M. A., and Cullen, B. R., 1994, Molecular basis of latency in pathogenic human viruses, *Science* 254:815–820.

25. Marrack, P., and Kappler, J., 1994, Subversion of the immune system by pathogens, *Cell* 76:323–332.

26. Schlitt, M., Morawetz, R. B., Bonnin, J., Chandra-Sekar, B., Curtiss, J. J., Diethelm, A. G., Whelchel, J. D., and Whitley, R. J., 1986, Progressive multifocal leukoencephalopathy: Three patients diagnosed by brain biopsy, with prolonged survival in two, *Neurosurgery* 18:407–414.

27. Skoldenberg, B., 1991, Herpes simplex encephalitis, *Scand. J. Infect.* 78:40–46.

28. Gindler, J. S., Atkinson, W. L., and Markowitz, L. E., 1992, Update—The United States measles epidemic, 1989–1990, *Epidemiol. Rev.* 14:270–276.

29. Swoveland, P. T., 1991, Molecular events in measles virus infection of the central nervous system, *Int. Rev. Exp. Pathol.* 32:255–275.

30. Lindegren, M. L., Fehrs, L. J., Hadler, S. C., and Hinman, A. R., 1991, Update: Rubella and congenital rubella syndrome, 1980–1990, *Epidemiol. Rev.* 13:341–348.

31. Townsend, J. J., 1982, Neuropathology of progressive rubella panencephalitis after childhood rubella, *Neurology* 32:185–190.

32. Prusiner, S. B., 1993, Genetic and infectious prion diseases, *Arch. Neurol.* 50:1129–1153.

33. Bueler, H., Aguizzi, A., Sailer, A., Greiner, R.-A., Autenried, P., Aguet, M., and Weissmann, C., 1993, Mice devoid of PrP are resistent to scrapie, *Cell* 73:1339–1347.

34. Bell, J. E., and Ironside, J. W., 1993, Neuropathology of spongiform encephalopathies in humans, *Br. Med. Bull.* 49:738–777.

35. Berger, J. R., and Levy, R. M., 1993, The neurologic complications of human immunodeficiency virus infection, *Med. Clin. North Am.* 77:1–23.

36. Schmidbauer, M., Huemer, M., Christina, S., Trabattoni, G. R., and Budka, H., 1992,

Morphological spectrum, distribution and clinical correlation of white matter lesions in AIDS brains, *Neuropathol. Appl. Neurobiol.* **18**:489–501.

37. Kruickshank, J. K., Rudge, P., Dalgleish, A. G., Newton, M., McLean, B. N., Barnard, R. O., Kendall, B. E., and Miller, D. H., 1989, Tropical spastic paraparesis and human T cell lymphotropic virus type 1 in the United Kingdom, *Brain* **112**:1057–1090.

38. Wu, E., Dickson, D. W., Jacobson, S., and Raine, C. S., 1993, Neuroaxonal dystrophy in HTLV-1-associated myelopathy/tropical spastic paraparesis: Neuropathologic and neuroimmunologic correlations, *Acta Neuropathol.* **86**:224–235.

39. Power, C., Weinshenker, B. G., Dekaban, G. A., Kaufmann, J. C. E., Shandling, M., and Rice, G. P. A., 1991, Pathological and molecular biological features of a myelopathy associated with HTLV-1 infection, *Can. J. Neurol. Sci.* **18**:352–355.

40. Ferreiro, J. A., Robert, M. A., Townsend, J., and Vinters, H. V., 1992, Neuropathologic findings after liver transplantation, *Acta Neuropathol.* **84**:1–14.

41. Skinner, J., Finlay, J. L., Sondel, P. M., and Trigg, M. E., 1986, Infectious complication in pediatric patients undergoing transplantation with T lymphocyte-depleted bone marrow, *Pediatr. Infect. Dis.* **5**:319–324.

42. Al-Hakim, A., Cohen, M., and Daroff, R. B., 1993, Postmortem examination of relapsing acute Guillain-Barre syndrome, *Muscle Nerve* **16**:173–176.

43. Ropper, A. H., 1992, The Guillain-Barre syndrome, *N. Engl. J. Med.* **326**:1130–1136.

44. Kesselring, J., Miller, D. H., Robb, S. A., Kendall, B. E., Moseley, I. F., Kingsley, D., Du Boulay, E. P. G., and McDonald, I., 1990, Acute disseminated encephalomyelitis, *Brain* **113**:291–302.

45. Seales, D., and Greer, M., 1991, Acute hemorrhagic leukoencephalitis, *Arch. Neurol.* **48**:1086–1088.

46. Mokhtarian, F., McFarlin, D. E., and Raine, C. S., 1984, Adoptive transfer of myelin basic protein–sensitized T cells produces chronic relapsing demyelinating disease in mice, *Nature* **309**:356–358.

47. Welsh, C. J. R., Tonks, P., Barrow, P., and Nash, A. A., 1990, Theiler's virus: An experimental model of virus-induced demyelination, *Autoimmunity* **6**:105–112.

48. Sibley, W. A., Bamford, C. R., and Clark, K., 1985, Clinical viral infections and multiple sclerosis, *Lancet* **1**:1313–1315.

8

Postmeasles Encephalomyelitis

LISA M. ESOLEN and DIANE E. GRIFFIN

1. INTRODUCTION

Measles virus (MV) is a highly contagious human pathogen responsible for significant worldwide morbidity and mortality. The illness is spread via the respiratory route after an incubation period of 10–14 days. Most commonly, measles begins with a prodrome of fever, cough, and conjunctivitis. This is followed by a distinctive maculopapular skin rash which is coincident with the development of the immune response and clearance of virus. Most patients recover inconsequentially. Secondary bacterial infections are problematic, particularly in malnourished children of developing countries, and play a major role in the 1.5 million measles-related deaths per year.[1] Other complications of measles infection include giant cell pneumonia, subacute sclerosing panencephalitis, acute inclusion body encephalitis (in immunocompromised hosts), and postmeasles encephalomyelitis (PMEM). In addition, measles virus has been implicated in a wide variety of other illnesses including Paget's disease,[2] multiple sclerosis,[3] chronic hepatitis,[4] glomerulonephritis, and systemic

LISA M. ESOLEN • Department of Medicine, Division of Infectious Diseases, The Johns Hopkins University School of Medicine, Baltimore, Maryland 21205.   DIANE E. GRIFFIN • Department of Molecular Microbiology and Immunology, The Johns Hopkins School of Hygiene and Public Health, Baltimore, Maryland 21205.

Microorganisms and Autoimmune Diseases, edited by Herman Friedman *et al.* Plenum Press, New York, 1996.

lupus erythematosus,[5,6] although no definitive evidence exists for a role in these diseases.

2. POSTMEASLES ENCEPHALOMYELITIS

2.1. Clinical Description

Postmeasles encephalomyelitis is an acute demyelinating condition of the central nervous system (CNS) and spinal cord accounting for 95% of all neurologic complications of measles. It occurs after other infections as well, and postinfectious encephalomyelitis is the most common of the human demyelinating diseases. Postmeasles encephalomyelitis complicates 1 in 1000 cases of measles[7] and rarely occurs in children under the age of 2.[8] There is no male or female predominance, and occurrence of disease does not correlate with severity of acute measles infection.[9] Seventy-five percent of patients have a normal episode of measles and begin to recover 2–7 days after the onset of the rash when there is recrudescence of fever and diminished consciousness, with or without seizures.[10] On rare occasions neurological symptoms may precede the rash or occur as long as 3 weeks after.[11] Other focal neurologic symptoms often occur, such as involuntary movement (18%), acute hemiplegia (12%), cerebellar ataxia (10%), and paraparesis (< 10%).[8,9] Usually a diffuse cerebral disturbance, often leading to coma, is seen. Mortality estimates range from 10–20%.[7,12] Sequelae do not reliably correlate with severity of the encephalitis. Of those who survive, 50–60% suffer permanent neurological deficits, most commonly diminished intelligence, motor handicaps, paraparesis, or emotional lability.[9,12]

Laboratory examination shows no consistent diagnostically helpful abnormalities. Examination of the cerebrospinal fluid (CSF) may show a moderate mononuclear cell pleocytosis (< 200 cells/ml), mild elevation in protein (rarely > 100 mg/dl), and elevated glucose, especially in comatose patients.[12–15] Measles antibody in CSF is detected in approximately 6–15% of patients, but rising antibody titers in encephalitic patients (CSF or serum) have only rarely been described.[9,16–18] In one study, CSF was positive by radioimmunoassay for myelin basic protein (MBP) in six out of ten patients (Table I).[19,47] Overall, one third of all patients have completely normal CSF.[19]

Other diagnostic tests are of limited usefulness. Urine studies are not helpful. Electroencephalography shows nonspecific diffuse slowing. Enhanced computerized tomography (CT) scans are normal. Magnetic

TABLE 1
Cerebrospinal Fluid Abnormalities in Patients with Neurological Complications after Semple Rabies Vaccine and after Measles[a]

Complication	Cells/mm³	Protein (mg %)	MBP (+/total)
Semple rabies vaccine encephalitis	82 ± 30	73 ± 6	11/23
Measles encephalitis	71 ± 29	66 ± 15	6/10

[a]From Ref. 19 and 47.

resonance imaging, however, may show multiple diffuse foci of brain stem and cerebellar demyelination.

2.2. Pathological Findings

Acutely, the brain is edematous, and vessels are prominent in the white matter. Lymphocytic infiltration of small veins in both gray and white matter is seen even in early deaths. Classically, perivenular lymphocytic infiltration is seen with perivascular and predominantly perivenular demyelination resulting in nerve cell damage and occasional hemorrhage.[9,12] There may be little evidence of demyelination in children with a rapidly fatal course, suggesting that the demyelination may be a relatively late phenomenon (illness greater than 3 days).[9] The demyelinating lesions show a concordant loss of MBP and myelin-associated glycoprotein (MAG) as occurs in experimental autoimmune encephalomyelitis (EAE), rather than the discordant loss of these two myelin proteins as occurs in acute viral infection of myelin-maintaining oligodendrocytes.[20] Rarely have viral inclusion bodies been seen.[21] Immunocytochemical staining has failed to identify viral antigen, and *in situ* hybridization has also been negative for viral messenger RNA.[22–24]

2.3. Pathogenesis

The pathogenesis of PMEM is poorly understood. Of note are two fundamental issues. First, as previously mentioned, the virus has never been reliably demonstrated in affected CNS tissue. The majority of patients have antibody to measles virus in CSF, but demonstration of rising CSF antibody has been extremely rare, suggesting that the virus is not replicating locally.[25] Secondly, the pathological lesions of PMEM, its latency period, and its monophasic clinical presentation closely resemble the CNS complications of Semple rabies vaccine.

2.3.1. Semple Rabies Vaccine

Neuroparalytic accidents were documented in the late nineteenth century in some patients after receiving a rabies vaccine prepared from the spinal cords of infected rabbits.[26] With subsequent viral inactivation and preparation from the brains of infected sheep and goats, the vaccine became known as the Semple rabies vaccine. Symptoms begin 6–18 days after the first injection. Initially they are nonspecific, with fever, headache, and myalgia predominating.[26,27] Neurological deficits occur shortly thereafter and include symptoms and signs of clinical encephalitis with seizures, movement disorders, and depressed consciousness; rarely, the disease is primarily paralytic resembling the Guillain-Barré syndrome.[28] Initially, the cause was thought to be persistent live virus, but this was later disproved.

In 1932, Hurst hypothesized that these complications were due to an immune reaction induced by nervous tissue present in the vaccine.[29] Subsequently, a similar disease was induced in monkeys when brain tissue was injected systemically.[30,31] It was later demonstrated that the encephalitogenic component of brain was MBP. The injection of MBP with adjuvant into certain species of animals reproducibly causes the perivenular demyelinating CNS disease EAE.[32,33] Lymphoproliferative responses to myelin have been demonstrated in three out of six patients with neurological complications from rabies vaccine, and high levels of antibodies to MBP correlate with disease severity.[34] Antibody to MBP is synthesized intrathecally and is present at the onset of symptoms, suggesting that the immune response to myelin is not a late phenomenon, as would be likely to occur with direct viral destruction of myelin-maintaining oligodendrocytes.[34]

2.3.2. Experimental Autoimmune Encephalomyelitis

Experimental autoimmune encephalomyelitis is the prototypical autoimmune disease that occurs by injection of MBP or proteolytic apoproteins, in combination with Freund's adjuvant, systemically into animals. Neurologic deficits include hind limb weakness and paralysis. Perivascular cuffs of inflammatory cells predominate in the white matter of the spinal cord. Lymphocytic cells infiltrate the CNS, followed by monocytes and macrophages.[35] An intact cellular immune system is necessary for development of the disease.[36–38] Animals deprived of T-cell function by thymectomy do not develop EAE after the appropriate challenge, and the disease may be passively transferred by CD4+ lymphocytes specific for

MBP.[36] Immunoglobulin production is also essential as evidenced by the fact that IgM/G-deficient mice do not develop EAE when immunized.[39,40]

In some species, injection of MBP is not enough to cause EAE; it must be combined with lipids from the CNS such as galactocerebroside, sulfatide, or ethanolamine phosphoglyceride.[37,38,41] Immunization with galactocerebroside or proteolipid protein alone can also produce a CNS demyelinating disease, implicating these antigens as particularly important.[42–45] Notably, immune responses to galactocerebrosides may also augment neurologic disease following Semple rabies vaccine, since patients with neurologic disease were likely to have antibody to these lipids as well as to MBP.[34] Therefore, in many experimental situations, demyelination requires antibody, macrophages, CD4+ T lymphocytes, and perhaps accessory molecules for full manifestation of disease.[39,40,46]

2.3.3. Comparison of EAE and PMEM

The predominant hypothesis—that PMEM results from an autoimmune response to myelin—is based on the clinical and pathological similarities with EAE and postvaccinal rabies encephalitis. Both EAE and postrabies neuroparalytic complications require exposure of peripheral CD4+ T lymphocytes to the previously sequestered MBP.[34] In PMEM, evidence exists for abnormal immune responses to MBP. Lymphocytes cultured from 47% of patients with PMEM showed a proliferative response to MBP as compared with only 15% of uncomplicated measles cases.[47,48] Intrathecal presence of MBP, indicative of myelin destruction, was found in patients who experienced encephalomyelitis and polyneuritis.[25]

Evidence of immune activation or dysregulation exists. Lower levels of plasma-soluble interleukin (IL)-2 receptor are present compared with uncomplicated measles.[49] Elevations of serum IgE levels, often subject to careful immunoregulation, are prolonged in PMEM, possibly indicating perturbations in T- and B-cell responses.[50] In addition, C-reactive protein (CRP), the prototypical acute phase reactant synthesized in response to IL-1, is elevated in measles and is elevated longer when neurologic complications exist.[51] Data are suggestive of a second peak of CRP when demyelinating encephalomyelitis occurs. These data indicate a reactive if not hyperreactive immunologic response, probably to MBP and possibly to other myelin constituents.

Experimentally, a subacute measles encephalomyelitis can be induced by injecting a neurotropic strain of measles virus intracerebrally into rats.[52] Four to 8 weeks after inoculation, CNS changes similar to those of EAE are seen. Also, MBP-specific class II restricted T-cell clones can be isolated from these animals and used to induce EAE in naive recipients following adoptive transfer.[53] The specificity was compared with that of clones isolated from MBP-challenged rats that developed EAE: both the MV-infected and the MBP-challenged rats developed MBP-specific T cells which responded to *in vitro* stimulation with encephalitogenic peptides of MBP but not to nonencephalitogenic peptides.[52,53] The T-cell clones failed to proliferate in response to MV antigens, and isolated MV-specific clones only responded to MV antigens and not to MBP.[53]

3. VIRAL INDUCTION OF AUTOIMMUNE DISEASE

There are basically three mechanisms by which viruses may induce autoimmunity: activation of previously suppressed T-cell clones, molecular mimicry, or modification and/or exposure of host antigens.

3.1. Activation of Autoreactive T-Cell Clones

It is possible that a virus may interact with lymphocytes in such a way that previously suppressed "autoreactive" T-cell clones become activated. It is thought that MV predominantly infects monocytes *in vivo*,[54] and in the rat infected with neurotropic strains of MV, T and B cells are not infected.[53] It is possible, however, that subsequent to infection, a cascade of cytokines or other regulatory proteins are released which secondarily activate, or remove the suppression of, an autoreactive T-cell clone capable of responding to MBP. It is known in natural measles, despite lack of evidence for direct lymphocyte infection, that lymphocyte functions are profoundly affected. Delayed-type hypersensitivity skin test responses are inhibited, and relapses of latent tuberculosis have been described.[55] There is spontaneous proliferation of lymphocytes, but mitogenic lymphoproliferative responses are suppressed.[56,57] Examination of the release of cytokines into the circulation during measles has shown an early elevation of IL-2 and interferon (IFN)-γ and a late increase in IL-4.[58] Studies of cytokines in PMEM are limited but generally show the same trends. The CSF in PMEM does show increased levels of soluble CD8, neopterin, and β2 microglobulin, indicative of local T-cell and macrophage activation.[59,60]

Recently, IL-2 has been shown to reverse anergy of human and mouse T-cell clones, resulting in autoimmune manifestations such as DNA autoantibodies and rheumatoid factor.[61,62] Autoreactive T cells that ignore self antigens may cause autoimmune diabetes when provided with exogenous help from IL-2, for example.[63] Also, therapeutic administration of IL-2 has been associated with the induction of an acute demyelinating encephalitis.[64] Systemic production and, perhaps more importantly, local production of IL-2 in PMEM has not been examined. Systemic levels of IL-2 receptor (thought to be an indicator of IL-2 utilization) are elevated but are lower in PMEM than in uncomplicated measles.[49]

3.2. Molecular Mimicry

Molecular mimicry is often considered a potential mechanism of autoimmunity. This hypothesis is based on the proposal that a viral antigen may have enough homology with a self antigen that an immune response to the virus initiates an immune response against the homologous self antigen. This "self-sensitization" may then continue even after the virus has been cleared. Several viruses have sequences homologous to MBP, including adenovirus, influenza, canine distemper, respiratory syncytial virus, Epstein-Barr virus, hepatitis B, and measles virus.[65,66] The biological significance of these sequence similarities is unclear. Fujinami and Oldstone showed in 1985 that a peptide of the hepatitis B virus polymerase, which resembles an encephalitogenic region of MBP, was able to induce an EAE-like disease in rabbits.[66] However, when the known encephalitogenic sequences of MBP were analyzed for similarity to MV, no significant homologies were found.[67,68] Moreover, MV-specific T-cell clones do not proliferate in response to MBP, and MBP-specific T-cell clones do not proliferate in response to MV antigen.[53]

3.3. Viral Modification or Exposure of Host Antigens

Finally, when viruses replicate in host cells, they may incorporate host cell antigens into the viral envelope and in so doing modify or expose these antigens such that they are recognized by the host as foreign. With regard to PMEM, this would require virus infection of myelin-containing or myelin-producing cells. While evidence for this is lacking, it is possible that the virus infects these cells early and is then cleared by a cellular immune response with the hypersensitivity occurring subsequently.

4. AUTOIMMUNE DESTRUCTION OF MYELIN

In considering the above possibilities as potentially important to the pathogenesis of postinfectious encephalomyelitis and the generation of

MBP-specific CD4+ T-cell clones, the role of infection, antibody production, cytokine production, regulation of adhesion molecules and MHC
antigens, and possibly breach of the blood-brain barrier must all be
considered.

4.1. Integrity of Blood-Brain Barrier during Infection

The blood-brain barrier is formed by endothelium of the CNS capil-
laries surrounded by astrocytic processes which help to form the tight
junctions. These junctions serve as an important barrier to the passage of
most molecules except the smallest. Infection of the endothelium and/or
astrocytic processes may physically disrupt the integrity of the barrier. As
mentioned previously, endothelial cell infection can occur in acute fatal
measles.[24] At the time of death in PMEM, no infection can be demonstrated, though this does not rule out an earlier infection of these cells.

4.2. Expression of Adhesion Molecules

Brain specimens from patients with multiple sclerosis (considered
an autoimmune and possibly measles-related disease) show upregulation
of intercellular adhesion molecule (ICAM)-1 on both endothelial cells
and astrocytes.[69] Expression of ICAM-1 is also upregulated in EAE, but this
does not correlate with disease severity or inflammation, and anti-ICAM-1
antibodies do not protect against passive transfer of EAE.[70] Conversely,
vascular cell adhesion molecule (VCAM)-1 is also upregulated in EAE,
and antibodies to VCAM-1 as well as VLA-2 (its ligand) do inhibit EAE.[71,72]

4.3. Role of Antibody Production

Antibody has been implicated in the development of PMEM since
antibody-deficient children get uncomplicated measles disease with normal recovery, and immunoglobulin-deficient rats fail to develop EAE.[39]
Conversely, intravenous administration of human immunoglobulin inhibits the induction of EAE in Lewis rats[73] and reportedly has been useful
in the treatment of some forms of multiple sclerosis.[74] The exact role of
antibody is unclear. Some hypothesize that exogenous administration of
antibody regulates the production of the autoantibody, thus having a
beneficial effect,[75] while others report that the immunoglobulin decreases tumor necrosis factor (TNF)-α levels.[73] Endogenous production,
as opposed to high-dose exogenous administration, of antibody may be
deleterious in EAE if deposition of antibody on infected endothelial cells

causes activation of the complement cascade, possibly damaging the integrity of the blood-brain barrier; this, however, is unknown.

4.4. Role of Cytokines

TNF-α has been implicated in the pathogenesis of EAE. Injection of TNF-α or IFN-γ (generated by T helper type 1 cells) into lumbosacral rat spinal cords produces meningitis with mononuclear inflammation similar to EAE.[76] Also, TNF-α augments EAE both clinically and histologically in Lewis rats.[77] The factor induces a reactive astrocytosis *in vitro*,[78] and *in vivo* reactive astrocytes are the predominant cell type in demyelinating plaques.[79] MBP-specific T-cell clones spontaneously produce high levels of TNF-α and TNF-β (lymphotoxin).[80] Most importantly, antibodies to TNF-α (and lymphotoxin) prevent passive transfer of EAE by T-cell clones, and in diseased rats, administration of the anti-TNF antibody decreases the average length of disease.[78,80] Pentoxyfylline, a phosphodiesterase inhibitor that inhibits TNF-α production and effects, decreases the clinical severity as well as the histological severity of EAE in rats.[81] Conversely, IL-10 (generated from T helper 2 cells) prevents EAE in rats and totally abrogates TNF-α-induced generation of encephalitogenic T lymphocytes.[82] Of note, during recovery of EAE, there is a dramatic rise in IL-10 levels.[83] Finally, transforming growth factor (TGF)-β protects against EAE, and anti-TGF-β antibody accelerates the disease in mice.[84,85] Expression of TNF-α, TGF-β, IFN-γ, and IL-10 systemically and in the CNS in PMEM is not known but would be an interesting area for future investigation.

4.5. Expression of Major Histocompatibility Antigens

T cells can only recognize processed antigen presented on the surface with an MHC antigen. CD4+ T cells recognize antigen presented with MHC class II, which is normally not expressed in the CNS. While the level of MHC class II expression is not known in PMEM, class II is upregulated on endothelial cells in multiple sclerosis,[86,87] and EAE.[86] MHC class II expression is known to increase prior to the inflammatory cell infiltration in EAE.[87] Interferon-γ and TNF-α act synergistically to increase class II expression on astrocytes and vascular endothelial cells *in vitro* which are then capable of presenting MBP to CD4+ T cells.[88,89] Finally, mice are protected from EAE if anti-Ia monoclonal antibodies are administered before or after sensitization with myelin.[90]

Since PMEM is an infrequent complication of measles, it is possible that genetic determinants play an important role in susceptibility. Lebon *et al.* addressed the issue of genetic susceptibility in a small study characterizing the prevalence of HLA antigens.[91] There was no correlation between encephalomyelitis and HLA-A, HLA-B, and HLA-C (class I molecules). While there was a suggestion of an increase in DR4 and DR5 (class II molecules), it was not statistically significant.

4.6. Unifying Hypothesis

One unifying hypothesis centers on early viral infection of endothelium. Endothelial cell infection and subsequent antibody deposition may injure the blood-brain barrier by stimulating the complement cascade and antibody-dependent cell-mediated cytotoxicity, making it more permissive for lymphocytic passage. Infection of endothelium and/or release of cytokines may upregulate the expression of key adhesion molecules which would also facilitate passage of lymphocytes into the CNS. Finally, viral infection may cause upregulation of IFN-γ and TNF-α levels, leading to MHC class II antigen expression on both endothelial and glial cells. The T cell, now capable of physically accessing the CNS and with the help of class II antigen expression, recognizes the previously sequestered myelin as antigenic, and demyelination begins.

5. ADDITIONAL MECHANISMS OF VIRAL DEMYELINATION

Autoimmunity is just one of the mechanisms by which a virus may cause demyelination. It is worth briefly mentioning two others.

A virus may infect oligodendroglial cells, the myelin-producing cells in the CNS. Demyelination may ensue because of cell lysis or altered cell metabolism. The JC virus, which causes progressive multifocal leukoencephalopathy in immunocompromised patients, is an example of this type of demyelination. In addition, viruses may destroy the myelin membrane directly. MBP is an excellent substrate for the protein kinase of vaccinia virus, for example. Neither of these mechanisms seems to apply to PMEM, since they would require ongoing CNS infection, rather than a strictly systemic infection. This seems highly unlikely since immunocytochemical and *in situ* hybridization techniques have failed to identify MV antigen in formalin-fixed paraffin-embedded brain tissue of seven patients who died with PMEM.[23,24]

Finally, remote viral infection may stimulate cytokines which are directly or indirectly damaging to oligodendroglial cells. For example,

IL-2 dimers are cytotoxic to oligodendroglial cells, and IL-2 therapy also has caused a case of acute fatal leukoencephalopathy. As mentioned previously, soluble IL-2 receptor is lower in PMEM than in uncomplicated measles.[49]

6. CONCLUSION

Postmeasles encephalomyelitis is a rare complication of acute measles infection that is thought to be a secondary autoimmune response to a component of myelin, namely, myelin basic protein. Clinically and pathologically it resembles experimental autoimmune encephalomyelitis, a prototypical autoimmune disease. To clearly define the pathogenesis, definitive studies are needed to characterize genetic susceptibility, possible early presence of virus, regulation of adhesion molecules, cytokine profiles, and the expression of MHC class II molecules in the brain.

REFERENCES

1. Beckford, A. P., Kaschula, R. O. C., and Stephen, C., 1985, Factors associated with fatal cases of measles: A retrospective autopsy study, S. Afr. Med. J. 68:858–863.
2. Basle, M. F., Fournier, J. G., Rozenblatt, S., Rebel, A., and Bouteille, M., 1986, Measles virus RNA detected in Paget's disease bone tissue by in situ hybridization, J. Gen. Virol. 67:907–913.
3. Dubois-Dalcq, M., 1979, Pathology of measles virus infection of the nervous system: Comparison with multiple sclerosis, Int. Rev. Exp. Pathol. 19:101–135.
4. Robertson, D. A. F., Guy, E. C., Zhang, S. L., and Wright, R., 1987, Persistent measles virus genome in autoimmune chronic active hepatitis, Lancet 2:9–11.
5. Andjaparidze, O. G., Chaplygina, N. M., Bogomolova, N. N., Koptyaeva, I. B., Nevryaeva, E. G., Filimova, R. G., and Tareeva, I. E., 1989, Detection of measles virus genome in blood leukocytes of patients with certain autoimmune diseases, Arch. Virol. 105:287–291.
6. ter Meulen, V., and Liebert, U. G., 1993, Measles virus–induced autoimmune reaction against brain antigen, Intervirology 35:86–94.
7. La Boccetta, A. C., and Tornay, A. S., 1964, Measles encephalitis. Report of 61 cases, Am. J. Dis. Child 107:247–255.
8. Reik, L., 1982, Immune-mediated central nervous system disorders in childhood viral infection, Semin. Neurol. 2:106–114.
9. Johnson, R. T., Griffin, D. E., Hirsch, R. L., Wolinsky, J. S., Roedenbeck, S., Lindo de Soriano, I., and Vaisberg, A., 1984, Measles encephalomyelitis—clinical and immunological studies, N. Engl. J. Med. 310:137–141.
10. Litvak, A. M., Sands, I. J., and Gibel, H., 1943, Encephalitis complicating measles: Report of fifty-six cases with followup studies in thirty-two, Am. J. Dis. Child. 65:265–295.
11. Holliday, P. B., 1950, Pre-eruptive neurological complications of the common contagious diseases—Rubella, rubeola, roseola, and varicella, J. Pediatr. 36:185–198.

12. Miller, D. L., 1964, Frequency of complications of measles, *Br. Med. J.* **2**:75–78.
13. Tyler, H. R., 1957, Neurologic complications of rubeola (measles), *Medicine* **36**:147–167.
14. Ford, F. R., 1928, The nervous complications of measles, *Bull. Johns Hopkins Hosp.* **43**:141–185.
15. Ojala, A., 1947, On changes in the cerebrospinal fluid during measles, *Ann. Intern. Med. Finl.* **36**:321–331.
16. Hanninen, P., Arshila, P., Lang, H., Salmi, A., and Panelius, M., 1980, Involvement of the central nervous system in acute, uncomplicated measles virus infection, *J. Clin. Microbiol.* **11**:610–613.
17. Kennedy, C. R., and Webster, A. D. B., 1984, Measles encephalitis, *N. Engl. J. Med.* **311**:330.
18. McClean, D. M., Best, J. M., Smith, P. A., Larke, R. P. B., and McNaughton, G. A., 1966, Viral infections of Toronto children during 1965: Measles encephalitis and other complications, *Can. Med. Assoc. J.* **94**:905–910.
19. Johnson, R. T., and Griffin, D. E., 1987, Postinfectious encephalomyelitis, in: *Infections of the Nervous System* (R. T. Johnson and P. G. E. Kennedy, eds.), Butterworths, London, pp. 209–226.
20. Itoyama, Y., Webster, H., Sternberger, N. H., Richardson, E. P., Walker, D. L., Quarles, R. H., and Padgett, B. L., 1982, Distribution of papovavirus, myelin-associated glycoprotein and myelin basic protein in progressive multifocal leukoencephalopathy lesions, *Ann. Neurol.* **11**:396–407.
21. Poser, C. M., 1969, Disseminated vasculomyelinopathy, *Acta Neurol. Scand.* **45**:7–44.
22. Gendelman, H. E., Wolinsky, J. S., Johnson, R. T., Pressman, J. N. J., Pezeshkpour, G. J., and Boisset, G. F., 1984, Measles encephalomyelitis: Lack of evidence of viral invasion of the central nervous system and quantitative study of the nature of demyelination, *Ann. Neurol.* **15**:353–360.
23. Moench, T. R., Griffin, D. E., Obriecht, C. R., Vaisberg, A. J., and Johnson, R. T., 1988, Acute measles in patients with and without neurological involvement: Distribution of measles virus antigen and RNA, *J. Infect. Dis.* **158**:433–442.
24. Esolen, L. M., Takahashi, K., Johnson, R. T., Vaisberg, A., Moench, T. R., Wesselingh, S. L., and Griffin, D. E., 1995, Brain endothelial cell infection in children with acute fatal measles, *J. Clin. Invest.* **96**:2478–2481.
25. Johnson, R. T., Hirsch, R. L., Griffin, D. E., Wolinsky, J. S., Roedenbeck, S., Lindo de Soriano, I., and Vaisberg, A., 1981, Clinical and immunological studies of measles encephalitis, *Trans. Am. Neurol. Assoc.* **106**:1–4.
26. Bareggi, C., 1889, Su cinque casi di rabbia paralitica (di laboratorio) nell'uomo, *Gaz. Med. Lomb.* **48**:217–219.
27. Swamy, H. S., Shankar, S. K., Chandra, P. S., Aroor, S. R., Krishna, A. S., and Perumal, V. G. K., 1984, Neurological complications due to B-propiolactone (BPL)-inactivated antirabies vaccination, *J. Neurol. Sci.* **63**:111–128.
28. Appelbaum, E., Greenberg, M., and Nelson, J., 1953, Neurological complications following antirabies vaccination, *J. Am. Med. Assoc.* **151**:188–191.
29. Hurst, E. W., 1932, The effect of the injection of normal brain emulsion into rabbits with special reference to the aetiology of the paralytic accidents of antirabies treatment, *J. Hyg.* **32**:3–44.
30. Kabat, E. A., Wolf, A., and Bezer, A. E., 1947, The rapid production of acute disseminated encephalomyelitis in rhesus monkey by injection of heterologous and homologous brain tissue with adjuvants, *J. Exp. Med.* **85**:117–130.
31. Rivers, T. M., and Schwentker, F. F., 1935, Encephalomyelitis accompanied by myelin destruction experimentally produced in monkeys, *J. Exp. Med.* **61**:689–702.

32. Einstein, E. R., Robertson, D. M., DiCaprio, J., and Moore, W., 1962, The isolation from bovine spinal cord of homogenous protein with encephalitogenic activity, *J. Neurochem.* 9:353–361.
33. Eyler, E. H., Salk, J., Beveridge, G. C., and Brown, L. V., 1969, Experimental allergic encephalomyelitis, *Arch. Biochem. Biophys.* 132:34–48.
34. Hemachudha, T., Griffin, D. E., Giffels, J., Johnson, R. T., Moser, A. B., and Phanuphak, P., 1987, Myelin basic protein as an encephalitogen in encephalomyelitis and poly-neuritis following rabies vaccination, *N. Engl. J. Med.* 316:369–374.
35. Lampert, P. W., 1969, Mechanism of demyelination in experimental allergic neuritis, *Lab. Invest.* 20:127–139.
36. Paterson, P. Y., 1960, Transfer of allergic encephalomyelitis in rats by means of lymph node cells, *J. Exp. Med.* 111:119–135.
37. Raine, C. S., Traugott, U., Farooq, M. S., Bornstein, M. B., and Norton, W. T., 1981, Augmentation of immune-mediated demyelination by lipid haptens, *Lab. Invest.* 45: 174–182.
38. Moore, G. R. W., Traugott, U., Farooq, M., Bornstein, M. B., Norton, W. T., and Raine, C. S., 1984, Experimental autoimmune encephalitis, *Lab. Invest.* 51:416–424.
39. Willenborg, D. O., and Prowse, S. J., 1983, Immunoglobulin deficient rats fail to develop experimental allergic encephalomyelitis, *J. Neuroimmunol.* 5:99–109.
40. Willenborg, D. O., Sjollema, P., and Danta, G., 1986, Immunoglobulin deficient rats as donors and recipients of effector cells of allergic encephalomyelitis, *J. Neuroimmunol.* 11:93–103.
41. Hoseim, A. A., Gilbert, J. J., and Strejan, G. H., 1986, The role of myelin lipids in experimental allergic encephalomyelitis, *J. Neuroimmunol.* 10:219–233.
42. Williams, R. M., Less, M. B., Cambi, F., and Macklin, W. B., 1982, Chronic experimental allergic encephalomyelitis induced in rabbits with bovine white matter proteolipid apoprotein, *J. Neuropathol. Exp. Neurol.* 41:508–521.
43. Yamamura, T., Namikawa, T., Endoh, M., Kunishila, T., and Tabira, T., 1986, Experimen-tal allergic encephalomyelitis induced by proteolipid apoprotein in Lewis rats, *J. Neuro-immunol.* 12:143–153.
44. Goban, Y., Saida, T., Nishitani, H., and Kameyama, M., 1986, Ultrastructural study of central nervous system demyelination in galactocerebroside sensitized rabbits, *Lab. Invest.* 55:86–90.
45. Van der Veen, R. C., Sobel, R. A., and Lees, M. B., 1986, Chronic experimental allergic encephalomyelitis and antibody responses in rabbits immunized with bovine pro-teolipid apoprotein, *J. Neuroimmunol.* 11:321–333.
46. Brosnan, C. F., Bornstein, M. B., and Bloom, B. R., 1981, The effects of macrophage depletion on the clinical and pathological expression of experimental allergic encepha-lomyelitis, *J. Immunol.* 126:614–620.
47. Griffin, D. E., 1988, Post-infectious and post-vaccinal disorders of the central nervous system, *Immunol. Allerg. Clin. North Am.* 8:239–249.
48. Behan, P. O., Geschwind, N., Lamarche, J. B., Lisak, R. P., and Kies, M. W., 1968, Delayed hypersensitivity to encephalitogenic protein in disseminated encephalomyelitis, *Lancet* 2:1009–1012.
49. Griffin, D. E., Ward, B. J., Jauregui, E., Johnson, R. T., and Vaisberg, A., 1989, Immune activation in measles, *N. Engl. J. Med.* 320:1667–1672.
50. Griffin, D. E., Cooper, S. J., Hirsch, R. L., Johnson, R. T., Lindo de Soriano, I., Roedenbeck, S., and Vaisberg, A., 1984, Changes in plasma IgE levels during compli-cated and uncomplicated measles virus infections, *J. Allerg. Immunol.* 76:206–213.

51. Griffin, D. E., Hirsch, R. L., Johnson, R. T., Lindo de Soriano, I., Roedenbeck, S., and Vaisberg, A., 1983, Changes in serum C-reactive protein during complicated and uncomplicated measles virus infections, *Infect. Immun.* **41**:861–864.

52. Liebert, U. G., Linington, C., and ter Meulen, V., 1988, Induction of autoimmune reactions to myelin basic protein in measles virus encephalitis in Lewis rats, *J. Neuroimmunol.* **17**:103–118.

53. ter Meulen, V., and Liebert, U. G., 1993, Measles virus-induced autoimmune reactions against brain antigen, *Intervirology* **35**:86–94.

54. Esolen, L. M., Ward, B. J., Moench, T. R., and Griffin, D. E., 1993, Infection of monocytes during measles, *J. Infect. Dis.* **168**:47–52.

55. Von Pirquet, C., 1908, Tuberkulin-reaktion wahrend der Masern, *Dtsch. Med. Wochenschr.* **34**:1297–1300.

56. Ward, B. J., Johnson, R. T., Vaisberg, A., Jauregui, E., and Griffin, D. E., 1991, Cytokine production in vitro and the lymphoproliferative defect of natural measles virus infection, *Clin. Immunol. Immunopathol.* **61**:236–248.

57. Munyer, T. P., Mangi, R. J., Dolan, R., and Kantor, F. S., 1975, Depressed lymphocyte function after measles-mumps-rubella vaccination, *J. Infect. Dis.* **132**:75–78.

58. Griffin, D. E., and Ward, B. J., 1993, Differential CD4 T cell activation in measles, *J. Infect. Dis.* **168**:275–281.

59. Griffin, D. E., Ward, B. J., Jauregui, E., Johnson, R. T., and Vaisberg, A., 1992, Immune activation during measles: β_2-microglobulin in plasma and cerebrospinal fluid in complicated and uncomplicated disease, *J. Infect. Dis.* **166**:1170–1173.

60. Griffin, D. E., Ward, B. J., Jauregui, E., Johnson, R. T., and Vaisberg, A., 1990, Immune activation during measles: Interferon-γ and neopterin in plasma and cerebrospinal fluid in complicated and uncomplicated disease, *J. Infect. Dis.* **161**:449–453.

61. Gutierrez-Ramos, J., Morena de Alboran, I., and Martinez, A. C., 1992, In vivo administration of interleukin-2 turns on anergic self-reactive T cells and leads to autoimmune disease, *Eur. J. Immunol.* **22**:2867–2872.

62. Krömer, G., and Wick, G., 1989 The role of interleukin 2 in autoimmunity, *Immunol. Today* **7**:199–200.

63. Heath, W. R., Allison, J., Hoffmann, M. W., Schonrich, G., Hammerling, G., Arnold, B., and Miller, J. F. A. P., 1992, Autoimmune diabetes as a consequence of locally produced interleukin-2, *Nature* **359**:547–550.

64. Vecht, C. J., Keohane, C., Menon, R. S., Henzen-Logmans, S. C., Punt, C. J. A., and Stoter, G., 1990, Acute fatal leukoencephalopathy after interleukin-2 therapy, *N. Engl. J. Med.* **323**:1146–1147.

65. Jahnke, U., Fischer, E. H., and Alvord, E. C., 1985, Sequence homology between certain viral proteins and proteins related to encephalomyelitis and neuritis, *Science* **229**:282–284.

66. Fujinami, R. S., and Oldstone, M. B. A., 1985, Amino acid homology between the encephalitogenic site of myelin basic protein and virus: Mechanism for autoimmunity, *Science* **230**:1043–1045.

67. Rubio, N., and Cuesta, A., 1989, Lack of cross-reaction between myelin basic proteins and putative demyelinating virus, *Mol. Immunol.* **26**:663–668.

68. Richert, J., Robinson, E. D., Reuben-Burnside, C. A., Johnson, A. J., McFarland, J. F., McFarlin, D. E., and Hartzman, R. J., 1988, Measles virus–specific human T cell clones: Studies of alloreactivity, *J. Neuroimmunol.* **19**:59–68.

69. Sobel, R. A., Mitchell, M. E., and Fondren, G., 1990, Intercellular adhesion molecule-1

(ICAM-1) in cellular immune reactions in the human central nervous system, *Am. J. Pathol.* **136**:1309–1316.

70. Willenborg, D. O., Simmons, R. D., Tamatani, T., and Miyasaka, M., 1993, ICAM-1-dependent pathway is not critically involved in the inflammatory process of autoimmune encephalomyelitis or in cytokine-induced inflammation of the central nervous system, *J. Neuroimmunol.* **45**:147–154.

71. Barten, L. D. M., and Ruddle, N. H., 1994, Vascular cell adhesion molecule-1 modulation by tumor necrosis factor in experimental allergic encephalomyelitis, *J. Neuroimmunol.* **51**:123–133.

72. Yednick, T. A., Cannon, C., Fritz, L. C., Sanchez-Madrid, F., Steinman, L., and Karin, N., 1992, Prevention of experimental autoimmune encephalomyelitis by antibodies against a4β1 integrin, *Nature* **356**:63–66.

73. Achiron, A., Margalit, R., Hershkoviz, R., Markovits, D., Reshef, T., Melamed, E., Cohen, I. R., and Lider, O., 1994, Intravenous immunoglobulin treatment of experimental T cell–mediated autoimmune disease, *J. Clin. Invest.* **93**:600–605.

74. Achiron, A., Pras, E., Gilad, R., Mandel, M., Gordon, C. R., Noy, S., Sarova-Pinhas, I., and Melamed, E., 1992, Open controlled therapeutic trial of intravenous immune globulin in relapsing-remitting multiple sclerosis, *Arch. Neurol.* **49**:1233–1236.

75. Basta, M., Kirshbom, P., Frank, M. M., and Fries, L. F., 1989, Mechanism of therapeutic effect of high-dose intravenous immunoglobulin, *J. Clin. Invest.* **84**:1974–1981.

76. Simmons, R. D., and Willenborg, D. O., 1990, Direct injection of cytokines into the spinal cord causes autoimmune encephalomyelitis-like inflammation, *J. Neurol. Sci.* **100**:37–42.

77. Kuroda, Y., and Shimamoto, Y., 1991, Human tumor necrosis factor-α augments experimental allergic encephalomyelitis in rats, *J. Neuroimmunol.* **34**:159–164.

78. Selmaj, K. W., Farooq, M., Norton, W. T., Raine, C. S., and Brosnan, C. F., 1990, Proliferation of astrocytes in vitro in response to cytokines, *J. Immunol.* **144**:129–135.

79. Raine, C. S., 1984, Biology of disease: Analysis of autoimmune demyelination, its impact upon multiple sclerosis, *Lab. Invest.* **50**:608–635.

80. Ruddle, N. H., Bergman, C. M., McGrath, K. M., Lingenheld, E. G., Grunnet, M. L., Padula, S. J., and Clark, R. B., 1990, An antibody to lymphotoxin and tumor necrosis factor prevents transfer of experimental allergic encephalomyelitis, *J. Exp. Med.* **172**:1193–1200.

81. Nataf, S., Louboutin, J. P., Chabannes, D., Feve, J. R., and Muller, J. Y., 1993, Pentoxifylline inhibits experimental allergic encephalomyelitis, *Acta Neurol. Scand.* **88**:97–99.

82. Rott, O., Fleischer, B., and Cash, E., 1994, Interleukin-10 prevents experimental allergic encephalomyelitis in rats, *Eur. J. Immunol.* **24**:1434–1440.

83. Kennedy, M. K., Torrance D. S., Picha, K. S., and Mohler, K. M., 1992, Analysis of cytokine mRNA expression in the central nervous system of mice with experimental autoimmune encephalomyelitis reveals that IL-10 mRNA expression correlates with recovery, *J. Immunol.* **149**:2496–2505.

84. Santambrogio, L., Hochwald, G. M., Saxena, B., Leu, C., Martz, J. E., Carlino, J. A., Ruddle, N. H., Palladino, M. A., Gold, L. I., and Thorbecke, G. J., 1993, Studies on the mechanisms by which transforming growth factor-β (TGF-β) protects against allergic encephalomyelitis, *J. Immunol.* **151**:1116–1127.

85. Kuruvilla, A. P., Shah, R., Hochwald, G. M., Liggitt, H. D., Palladino, M. A., and Thorbecke, G. J., 1991, Protective effect of transforming growth factor-β1 on experimental autoimmune diseases in mice. *Proc. Natl. Acad. Sci. U.S.A.* **88**:2918–2921.

86. Traugott, U., and Lebon, P., 1988, Interferon-γ and Ia antigen are present on astrocytes in active chronic multiple sclerosis lesions, *J. Neurol. Sci.* **84**:257–264.

87. Traugott, U., Scheinberg, L. C., and Raine, C. S., 1985, On the presence of Ia-positive endothelial cells and astrocytes in multiple sclerosis lesions and its relevance to antigen presentation, *J. Neuroimmunol.* **8**:1–14.

88. Massa, P. T., Schimpl, A., Wecker, E., and ter Meulen, V., 1987, Tumor necrosis factor amplifies measles virus–mediated Ia induction on astrocytes, *Proc. Natl. Acad. Sci. U.S.A.* **84**:7242–7245.

89. Leeuwenber, J. F. M., Van Damme, J., Meager, T., Jeunhomme, T. M. A. A., and Buurman, W. A., 1988, Effects of tumor necrosis factor on the interferon-γ induced major histocompatibility complex class II antigen expression by human endothelial cells, *Eur. J. Immunol.* **18**:1469–1472.

90. Steinman, L., Rosenbaum, J. T., Sriram, S., and McDevitt, H. O., 1981, In vivo effects of antibodies to immune response gene products: Prevention of experimental allergic encephalitis, *Proc. Natl. Acad. Sci. U.S.A.* **78**:7111–7114.

91. Lebon, P., Ponsot, G., Gony, J., and Hors, J., 1986, HLA antigens in acute measles encephalitis, *Tissue Antigens* **27**:75–77.

9

Epstein-Barr Virus and Autoimmunity

CARLO GARZELLI

1. INTRODUCTION

Epstein-Barr virus (EBV) is a ubiquitous member of the human herpesvirus group. The virus is the agent of infectious mononucleosis (IM) and shows the peculiar capability of immortalizing human B lymphocytes *in vivo* and *in vitro*; EBV is also associated with the development of African Burkitt's lymphoma (BL), undifferentiated nasopharyngeal carcinoma (NPC), and lymphoproliferative disorders and lymphomas in individuals with severe congenital or acquired immunodeficiencies.

This virus seems also to be involved in the etiopathogenesis of some diseases with autoimmune components, such as rheumatoid arthritis (RA), systemic lupus erythematosus (SLE), Sjögren's syndrome (SS), and, as recently suggested, in degenerative diseases of the nervous system, such as multiple sclerosis (MS). Infectious mononucleosis itself is characterized by the transient appearance of autoantibodies and, therefore, represents an appropriate model for the study of the mechanism(s) by which EBV induces autoimmune responses and, possibly, for investigating the relationships of EBV to autoimmune diseases.

This chapter will review our present understanding of the relationships between EBV and autoimmune diseases and discuss the mechanism(s) by which the virus initiates autoimmunity.

CARLO GARZELLI • Department of Biomedicine, University of Pisa, I-56127 Pisa, Italy.

Microorganisms and Autoimmune Diseases, edited by Herman Friedman *et al.* Plenum Press, New York, 1996.

2. THE EPSTEIN-BARR VIRUS

2.1. The Virus, Viral Genome, and Virus-Encoded Antigens

The mature virion, approximately 120 nm in diameter, has four components: the *nucleoid*, a ring-shaped protein core that is wrapped with the viral DNA; the *capsid*, icosahedron of 100 nm diameter formed by 162 capsomers; the *tegument*, made up of globular protein material that surrounds the nucleocapsid; and the *envelope*, with external glycoproteins and spikes.

The viral genome consists of linear double-stranded DNA of 172×10^3 base pairs (bp), approximately 10^5 kDa molecular weight. The terminal ends of the molecule contain homologous repeated sequences (TR, terminal repeats), which likely facilitate the circularization of DNA needed for replication. The genome also has four nonhomologous repeated internal sequences (IR, internal repeats), known as IR1, IR2, IR3, and IR4. Sequences IR1 and IR2 are separated by a short unrepeated sequence (U, unique DNA) named U2, which seems to play an important role in EBV-induced growth transformation.[1] Nucleotide sequence analysis between different EBV isolates shows very little differences, with the major exception of the U2 region encoding the nuclear antigen *EBNA-2*. There are in fact two genetic variants of U2 that define two distinct viral types known as type 1 (or A), of Caucasian origin, and type 2 (or B), of African origin.[1]

The viral genome encodes about 200 proteins, only a few of which have been identified. Infected cells express several virus-encoded antigens. Five groups of EBV-encoded antigens, detectable by immunological methods, are known: 1) *EBNA* (*EBV nuclear antigen*), a complex of six DNA-binding nuclear proteins named *EBNA-1, 2, 3A, 3B, 3C*, and *LP* (*leader protein*); 2) *LMP* (*latent membrane protein*), which likely consists of the antigen formerly named *LYDMA* (*lymphocyte-detected membrane antigen*), detectable by means of EBV-specific cytotoxic T lymphocytes (CTL). Two molecular forms of *LMP*, designated *LMP-1* and *LMP 2A/B* and encoded by the U5 region of the viral genome, are known: *LMP-1*, studied in more detail, is present on the cell membrane, is associated with the cell cytoskeleton, and probably constitutes or is part of a growth-factor receptor; 3) *MA* (*membrane antigen*), consisting of a complex of proteins located both in the viral envelope and in the membranes of infected cells and involved in virus binding to the cell receptor, membrane fusion, and viral penetration; 4) *EA* (*early antigen*), a complex of early antigens expressed solely in cells that have entered the lytic cycle and detectable only in the cytoplasm (*EA-R, restricted*) or in the cytoplasm and the nucleus (*EA-D, diffuse*); and 5) *VCA* (*viral capsid antigen*), located both in the cytoplasm and the nucleus and including the structural components of viral capsid synthesized at the end of the lytic cycle.[1]

2.2. Cell Infection

Epstein-Barr virus naturally infects humans and certain subhuman primates. In humans, EBV shows a peculiar tropism for B lymphocytes,[2] but it is also capable of infecting epithelial cells of the oropharynx.[3] The virus receptor is the same as the receptor for the complement C3d component, a 140-kDa protein known as CR2 or CD21,[4] and is expressed on the surface of all mature B lymphocytes and, to a lesser extent, in a subpopulation of oropharyngeal epithelial cells.

Epstein-Barr virus absorption to the receptor is followed by entrance into the cell by fusion of the viral envelope with the cell membrane. Viral uncoating then occurs and viral DNA is transferred to the nucleus where the viral genome is transcripted. Infection can evolve differently depending on the cell type infected. In nonpermissive cells, EBV establishes a latent nonproductive infection followed by cell immortalization; the nonpermissive cycle occurs in the majority of human B lymphocytes infected *in vitro* or *in vivo*, as well as in neoplastic cells such as those of BL and NPC. In permissive cells, EBV causes a productive infection leading to cell death and release of mature infectious virions. During natural infection, the productive cycle occurs in the oropharyngeal epithelium and in a small proportion of B lymphocytes. *In vitro*, some cells of latently infected lymphoblastoid cell lines can spontaneously undergo a productive cycle.[1]

During cell infection, EBV-encoded antigens are expressed in an ordered sequence: *EBNAs* and *LMP* are synthesized first, followed by early and late antigens. The antigenic complex *EA*, corresponding to enzymes necessary for viral replication, and certain *MAs* are encoded by early genes, while *VCA* and the other *MAs* are encoded by late genes.

3. EBV-RELATED DISEASES

3.1. Infectious Mononucleosis

Epstein-Barr virus infection is distributed worldwide and affects more than 90% of individuals by adulthood. Especially in early childhood, primary EBV infection occurs in subclinical form. In adolescents, especially in developed countries, and rarely in children less than 2 years old, infections occur in approximately 50% of cases as IM. Infectious mononucleosis is a benign disease that only rarely is accompanied by severe or fatal complications due to extensive lymphoproliferation, spleen rupture, or, possibly, autoimmune phenomena, including hemolytic anemia, neurologic complications, or impairment of liver functions (see below). Severe or fatal consequences often occur in individuals with genetic disorders of immune functions or in immunosuppressed hosts.[5]

Oropharyngeal epithelial cells and parotid duct cells seem to be the main source of infectious virus, which is shed in the saliva. In the susceptible host EBV infects oropharyngeal epithelial cells, where it establishes a lytic cycle characterized by expression of virus-encoded antigens and production of mature viral particles. The virus then infects B lymphocytes in the Waldeyer ring, which are largely nonpermissive. A proportion of B lymphocytes is growth-transformed, yielding typical immunoglobulin-secreting lymphoblastoid cells. As a consequence, an extensive lymphoproliferation occurs, leading to the spread of EBV-infected B lymphocytes to lymphoid organs and the circulatory system. These growth-transformed B lymphocytes represent a minor proportion of the atypical lymphocytes seen in peripheral blood of IM patients. Most are CTL triggered by the extensive proliferation of EBV-infected B lymphocytes. This cell-mediated immune response, which is essential for recovery, is also supported by NK cells and antibodies to virus-encoded antigens.[5]

After clinical recovery, the virus persists for up to 18 months in the oropharyngeal secretions; in about 10–20% of individuals it is subsequently intermittently shed in the saliva. After primary infection, the virus persists in a latent form in a proportion of B lymphocytes for the individual's life span and can spontaneously reactivate.[6]

Very rarely, symptoms persist for years and the virus is shed at high titers in the saliva (chronic IM); death seldom occurs as a consequence of lymphoproliferative disease. Chronic IM is likely due to ineffective immune control of viral infection, as suggested by the lack of circulating EBV-specific T lymphocytes.

3.2. EBV-Related Malignancies

Epstein-Barr virus is associated with the development of African BL and undifferentiated NPC, although the precise role the virus plays in the etiology and pathogenesis of these tumors is still controversial.[5]

Burkitt's lymphoma is a monoclonal neoplasia and one of the most prevalent tumors in certain parts of Africa and Papua New Guinea. Concomitantly with malaria infection, which induces a strong polyclonal lymphoid activation, EBV-infected B lymphocytes constitute a polyclonal population of immature cells among which a single cell, for reasons not yet known, undergoes malignant transformation. A necessary event for oncogenesis seems to be one of three specific chromosome translocations involving a sequence on the long arm of chromosome 8 and containing the oncogene *c-myc*; this sequence is translocated to chromosome 14, 2, or 22, next to transcriptionally active loci, such as the genes coding for the heavy or light (κ and λ) chains of immunoglobulins. This leads to activation of *c-myc*, whose product is a nuclear protein involved in cell activation and proliferation.[7]

Nasopharyngeal carcinoma is a neoplasia particularly common in South China, and its development seems to require genetic and environmental factors as well as EBV infection. The occurrence of the tumor appears to be influenced by dietary habits, such as ingestion of tumorigenic compounds (e.g., nitrosamines are abundant in smoked fish, a popular food among the poorest Chinese people). The association between EBV and NPC was initially suggested by the demonstration of high titers of antibodies precipitating EBV antigens such as *EA* and *VCA* in the sera of NPC patients. Mature EBV virions and genomes, either as episomes or integrated DNA, have also been observed in NPC cells.[5]

3.3. Lymphoproliferative Disorders

In individuals with primary or acquired immunodeficiencies, such as transplant patients undergoing immunosuppressive therapy or patients with acquired immunodeficiency syndrome (AIDS), primary EBV infection or reactivation can lead to extensive outgrowth of EBV-transformed lymphocytes, causing B-cell proliferative diseases or polyclonal or monoclonal lymphomas.[1,5] For example, in subjects with the X-linked lymphoproliferative disease Duncan's syndrome, which occurs in males with a congenital defect of B and T lymphocytes, EBV causes a fatal form of IM or the development of lymphomas.[8] Furthermore, B-cell lymphomas, largely EBV-positive and with the histopathological features of BL, occur in 1–5% of AIDS patients.[9] In these patients the so-called hairy cell leukoplakia, an EBV-positive precancerous lesion at the sides of the tongue, can also develop.[10] Finally, EBV DNA has been detected in Hodgkin's lymphoma cells, T-cell lymphomas, and malignant thymomas, although the role of EBV in these tumors remains controversial.

4. AUTOIMMUNE DISEASES POSSIBLY RELATED TO EBV

In addition to the clinical conditions listed above, a variety of diseases of unknown etiology, including autoimmune disorders such as RA, SS, SLE, and MS, have been associated with EBV on the basis of epidemiological, serological, and/or experimental evidence which is discussed below and summarized in Table I.

4.1. Rheumatoid Arthritis

Rheumatoid arthritis is a chronic inflammatory disease of the synovial membrane characterized by the production of rheumatoid factor (RF). Initial observations noted that RA patients suffered more severe infections with EBV than normal individuals. Antibodies to one antigen, the RA nuclear antigen (*RANA*), were initially reported only in RA,[11,12] but were subsequently shown to be identical to anti-*EBNA-1* antibody and present also in normal individuals.[13] However, RA patients have increased numbers of EBV-positive circulating B lymphocytes, probably due to a reduced ability of their T lymphocytes to control the outgrowth of EBV-infected cells;[14] these EBV-positive B cells are capable of producing auto-antibodies, including RF.[15,16] Moreover, sera from RA patients have higher antibody titers than age-matched controls to synthetic peptides derived from the glycine-alanine repeat regions of *EBNA-1*, as well as to *EBNA-3A* and *EBNA-3B*.[17] This is potentially important since a monoclonal antibody directed against the glycine-alanine repeat region of the *EBNA-1* molecule cross-reacts with a normal cellular protein of approximately 62 kDa[18] and has been found immunohistologically to react with the synovial lining cells of RA patients.[19] The chronic inflammation in the synovial membrane of RA patients also suggests that local viral antigens play an important role in pathogenesis, however, attempts to demonstrate EBV DNA or antigens directly in synovial biopsies have consistently failed.

More recently, a computer search of the EBV genome revealed of 6–amino acid region (EQKRAA) of sequence similarity between the protein gp110 of EBV and a region of the human class II major histocompatibility complex (MHC) antigens HLA-DR4 and HLA-DR1.[20] Importantly, the shared sequence is only present in a subset of HLA-DR molecules that confer a very high relative risk of developing RA;[21] moreover, synthetic peptides containing these amino acids were found to stimulate the T cells of RA patients. Since the synovial lining cells become intensely HLA-DR4-positive in RA patients,[22] it is possible that T lymphocytes sensitized against the viral antigen mistakenly direct their immune effector func-

TABLE I

Evidence Linking EBV to Autoimmune Diseases[a]

Rheumatoid arthritis	Sjögren's syndrome	Systemic lupus erythematosus	Multiple sclerosis
High antibody titers to EBV-encoded antigens in serum	Moderately elevated titers of anti-EBNAs antibodies in serum	High antibody titers to EBV-encoded antigens in serum	High antibody titers to EBV-encoded antigens in serum
Enhanced numbers of EBV-positive circulating B cells	Salivary glands are a site of EBV infection and reactivation	Enhanced numbers of EBV-positive circulating B cells	Epidemiological association with delayed symptomatic or complicated IM
EBV-infected B cells produce autoantibodies and rheumatoid factor	High levels of EBV-encoded antigens and DNA in saliva	Epitope homology between snRNP and EBNA-1 and EBNA-2	Anti-EBNA-1 antibody in cerebrospinal fluid
Epitope homology between EBNA-1 and synovial cell antigen	SSA and SSB autoantigens complex with EBV-encoded RNAs		Epitope homology between EBNA-1 and myelin basic protein
Sequence similarity between EBV-encoded antigens and HLA-DR of patients			

[a]For references, see text under Section 4.

tions against these cells by mechanisms of molecular mimicry. It has also been speculated that in individuals with a specific genetic predisposition (i.e., HLA-DR4), the chronic antigenic stimulation of T cells by EBV antigens might influence the repertoire of T cells that control differentiation of synovial cells, thus causing the synovial cell hypertrophy characteristic of RA.[23]

4.2. Sjögren's Syndrome

Sjögren's syndrome is an autoimmune disorder characterized by lymphocyte infiltration and destruction of the salivary and lacrimal glands. Special attention has been focused on this disease, as salivary glands not only are the site of primary EBV infection, but also serve as a reservoir for latent EBV infection and recurrent reactivation.

Data indicates increased EBV reactivation in SS patients. In fact, patients have recurrent parotid swelling associated with EBV, resembling the mumps-like swelling of IM;[24] moreover, their saliva contains increased levels of EBV-encoded antigens and DNA, as detected by DNA hybridization methods and polymerase chain reaction.[25,26] Furthermore, sera of the majority of SS patients contain specific and characteristic autoantibodies, termed anti-SSA (Ro) and anti-SSB (La), which bind ribonuclear proteins that complex with small EBV-encoded RNAs.[27] These patients also have moderately elevated titers of anti-*EBNAs* antibodies,[28] altered anti-EBV T-cell responses, and polyclonal B-cell activation.[23] However, the relevance of these abnormalities to SS pathogenesis remains unknown.

As SS occurs at a site where EBV normally has its latency and reactivation, it has been suggested that SS involves an intense T-cell response to viral or local tissue antigens associated with MHC class II molecules (especially HLA-DR3). The inability of the immune response to eradicate the EBV-containing cells would lead to the chronic immune stimulation and the salivary gland destruction that characterize SS. In other words, EBV might play a role in perpetuating a local immune response generated within the salivary gland of genetically predisposed individuals.[23]

4.3. Systemic Lupus Erythematosus

Systemic lupus erythematosus is a systemic autoimmune disease characterized by circulating autoantibodies that are reactive with several autoantigens, including small nuclear ribonucleoproteins (snRNP), DNA, and histones.

Like RA patients, SLE patients have high titers of anti-EBV antibodies and increased numbers of circulating B cells containing EBV.[29,30] Quite recently, it has been proposed that molecular mimicry between EBV-encoded antigens, particularly the *EBNAs*, and certain snRNP might play a role in the onset of SLE autoimmunity. In fact, the C-terminal region of one snRNP, namely, the D polypeptide (SmD) ($_{96}$Ala–Arg$_{119}$), which contains a ninefold Gly-Arg repeat, shows a strong sequence homology with *EBNA-1* ($_{35}$Gly–Gly$_{58}$). The C-terminal portion of SmD is indeed recognized by autoantibodies of SLE patients. In addition, antibodies raised against a synthetic peptide covering the SmD $_{96}$Ala–Arg$_{119}$ region react with the homologous *EBNA-1* peptide ($_{35}$Gly–Gly$_{58}$), and conversely, antibodies elicited by immunizing with *EBNA-1* peptide react with SmD.[31,32] Epitope homology has been also observed between *EBNA-2* and SmD. We have isolated a human monoclonal antibody reactive with both a 20–amino acid synthetic peptide derived from *EBNA-2* ($_{354}$Gly–Pro$_{373}$) and recombinant SmD; computer alignment analysis showed a high degree of homology between the *EBNA-2* synthetic peptide and SmD ($_{101}$Gly–Arg$_{119}$); moreover, we found that a significant proportion of SLE patients produce IgG antibody to the *EBNA-2* synthetic peptide.[33] Epitope homologies between *EBNAs* and SmD are shown in Fig. 1.

Further evidence for a link between EBV and SLE autoimmunity derives from the observation that a large number of SLE patients produce antibodies to histones.[34,35] Antibodies to histones have been reported only in a few other diseases, most of them related to EBV, including RA, IM, and BL. Normal individuals from tropical Africa, where there is a high incidence of EBV-positive BL, also possess these antibodies.[36] Since the EBV-encoded nuclear antigens (*EBNA-1, 2, 3A, 3B,* and *3C*) that are expressed in virus-transformed B cells have several short amino acid sequences similar to those of histones (personal observation), it has been

FIGURE 1. Epitope homology between *EBNAs* and SmD. (Identical amino acids are enclosed in boxes, conservative changes are indicated by asterisks.)

suggested that in autoimmune disease–prone individuals antihistone antibodies nonspecifically induced by EBV infection might be maintained by means of a specific antigen-driven mechanism based on molecular mimicry with EBNAs.[37,38]

4.4. Other Diseases

Autoimmunity induced by EBV has also been suggested in other diseases with autoimmune components,[39,40] including several neurologic diseases that evolve as complications of the acute phase of IM, such as peripheral nerve neuropathy, cranial neuropathy, acute cerebellar ataxia, and acute autonomic dysfunction.[41–45] In these disorders it is unclear whether the acute neurologic abnormalities reflect viral replication or are secondary to immunologically mediated events. Among the neurologic disorders that possibly stem from an EBV infection, multiple sclerosis, the most common progressive demyelinating disease in the economically privileged Western world, deserves special attention. Several lines of evidence suggest the existence of an association between this disease and EBV infection, delayed symptomatic EBV infection, or dual EBV and retrovirus infection.[46–48] Multiple sclerosis patients are 100% EBV-positive and have serum antibody titers to VCA and EBNA that are markedly higher than those of controls.[49–51] A significant proportion of patients with neurologically complicated primary EBV infection develop MS.[43] A cell line producing both EBV and retrovirus-like particles has been established from one patient with progressive myelopathy resembling MS.[52] Moreover, quite recently it has been shown that the cerebrospinal fluid (CSF) of most MS patients contains an antibody that reacts with a 70-kDa protein shown to be EBNA-1. Interestingly, EBNA-1 and myelin basic protein (MBP) share two pentapeptide identities, namely, QKRPS and PRHRD, which in individuals genetically susceptible to MS might be the target of T lymphocytes, causing MBP damage.[53]

5. AUTOIMMUNITY IN INFECTIOUS MONONUCLEOSIS

Perhaps, the main factor linking EBV to the onset of autoimmune disorders is that the virus has the potential to induce autoimmune phenomena, as shown by the appearance of numerous autoantibodies with varied specificity during the acute phase of IM. Only a few of the many autoantibodies described in IM patients, however, have been proposed as clinically significant.

5.1. Serum Autoantibodies

As shown in Table II, a high proportion of patients with EBV-induced IM show the transient occurrence of IgM class antibodies to several autoantigens. Approximately 70–80% of IM patients produce autoantibodies against smooth muscle, detectable by indirect immunofluorescence on frozen sections of human stomach.[54-56] These antibodies are usually present in the acute phase of the disease with titers up to 1:80, generally disappear by 7 weeks, but sometimes persist for 1–2 years after recovery; it has been suggested that they are directed, at least in part, at cytoskeleton tubulin subunits.[57,58]

Sera from IM patients may also exhibit autoantibodies against cytoskeleton components, such as intermediate filaments of epithelial cells;[59] they are usually assayed by immunofluorescence on human epithelioid cell lines, such as the Hep-2 cell line, or by immunoenzymatic and radioimmunologic assays using purified antigens. Anti–intermediate filaments autoantibodies have been found in 30–90% of IM cases, depending on the assay used; they are detectable only in the acute phase of the disease and appear to be directed against cytoskeletal filaments of prekeratin or vimentin type.[60]

In about half the IM patients autoantibodies to membrane antigens of T lymphocytes appear transiently.[55,61] IgM antibodies to cardiolipin, cross-reactive with cell-membrane antigens found in EBV-infected B cells, have been reported in 37% of IM patients.[62] Moreover, approximately

TABLE II
Serum Autoantibodies in Infectious Mononucleosis[a]

Autoantibody specificity	Isotype	Frequency (%)
Smooth muscle (tubulin subunits)	IgM	70–80
Cytoskeleton intermediate filaments (prekeratin, vimentin)	IgM	30–90
T lymphocyte (membrane antigens)	IgM	40–50
Cardiolipin	IgM	37
Histones	IgM	26
Rheumatoid factor	IgM	Occasional
Nucleus	IgM	Occasional
Organ-specific (thyroid, stomach)	IgM	Occasional
Erythrocyte (i-antigen)	IgM	8–70
p29 (triosephosphate isomerase)	IgM	100
p26 (manganese superoxide dismutase)	IgM	100

[a]For references, see text under Section 5.1.

one fourth of IM sera samples show increased IgM binding values to histones.[63] Occasionally, RF,[64] antinuclear,[64] and organ-specific autoantibodies, such as antithyroid and antistomach autoantibodies, are also found.[55]

Patients with IM may transiently produce cold agglutinins (CA),[65] antibodies that bind autologous red blood cells at low temperature; the specificity of these CA has been reported to be mainly anti-i erythrocyte antigen, and there is some suggestion that although IgM antibodies are prevalent, IgG antibodies with CA activity are also synthesized. The reported incidence of the anti-i CA in IM patients varies from 8% to 60–70%.[66,67] These antibodies seem especially important because they are apparently associated with a pathologic condition, that is, the hemolytic anemia which is one of the rare severe complications of IM.[68]

Moreover, quite recently, it has been reported that 100% of IM patients produce, transiently and concomitantly with the clinical symptoms, high titers of two IgM autoantibodies against cellular antigens, designated p26 and p29. The autoantigen p29 has been identified as triosephosphate isomerase (TPI),[69] and anti-TPI antibody has been proposed as clinically relevant in that it is capable of causing hemolysis.[70] The autoantigen p26 has been identified as the enzyme manganese superoxide dismutase (MnSOD).[71] It has been suggested that anti-MnSOD antibody might contribute to the immunopathogenesis of IM by inhibiting the removal of the increased oxygen radicals produced during IM and believed to be responsible for common physical signs of the disease, such as disturbed liver function, gelatinous appearance of the soft palate, supraorbital edema, and exanthema.[71]

5.2. Mechanisms by which EBV Induces Autoantibodies

Since EBV infects human B lymphocytes and causes lymphoblastoid cell transformation and polyclonal B-cell activation, it was initially thought that EBV-infected and growth-transformed B lymphocytes are triggered to express their antibody repertoire, including autoantibodies.[15,72] In fact, when infected *in vitro* with EBV, resting B lymphocytes from normal individuals release autoantibodies without any requirement for autoantigen.[73,74]

Many IgM monoclonal autoantibodies secreted by EBV-transformed B cells have been studied in detail, and most of them turned out to be polyreactive, that is, capable of reacting with multiple unrelated antigens;[73,75–77] based on studies performed with IM polyclonal sera, serum IgM autoantibodies also were found to be broadly cross-reactive.[78] It appeared unlikely, therefore, that polyclonal B-cell activation in its simplest form, because of its inherently random nature, might account for the appearance of antibodies with a relatively limited autoantigen specificity in most IM patients.[58–60,62,78] A virus-induced polyclonal activation of germ-line encoded antibodies possessing broad cross-reactivity was more likely. As these antibodies require few, if any, mutations, they should have high precursor frequencies and be very similar in all individuals. We actually demonstrated this in the case of a human IgM(κ) monoclonal antibody, designated HY 5488, that was isolated from EBV-transformed B lymphocytes of a patient with acute IM: similar to the patient's serum, the antibody reacts with smooth muscle, cytoskeleton filaments, and histones.[63,79,80] Sequence analysis of HY 5488 monoclonal antibody V_H and V_L genes (Tables III and IV) actually showed that V_H and V_κ nucleotide sequences closely matched the germ-line V-A77 gene, a member of the V_H3 family, and the germ-line kVAC gene, a member of Vκ IIIa family, respectively. The HY 5488-like antibodies seem to be involved in EBV-induced autoimmunity, since IgM bearing an idiotype of HY 5488 monoclonal antibody (Id 5488), defined by a murine antiidiotype monoclonal antibody, are produced by a large proportion of B lymphocytes infected *in vitro* with EBV,[38] are increased in the serum of IM patients, and in the sera correlate statistically with autoantibody activity to smooth muscle, cytoskeleton, and histones.[38,80] Thus, the EBV-induced Id 5488–positive IgM autoantibodies found in sera of IM patients likely represent broadly expressed, germ-line gene-encoded, polyreactive antibodies (i.e., the so-called natural antibodies) generated by an antigen-independent mechanism consisting of virus-driven, nonspecific (polyclonal) activation of B lymphocytes. The normal B-lymphocyte repertoire contains cells programmed to produce natural antibodies reactive with multiple self and heterologous epitopes on highly conserved molecules, such as nucleic acids, cytoskeletal proteins, the Fc portion of IgG, and cell-surface antigens.[77,81,82] An important feature of natural antibodies is that they are mostly secreted by CD5+ B cells in unmutated, germ-line configuration, by using a restricted set of V_H and V_L genes strongly conserved during evolution.[16,81,83] We demonstrated that enrichment in CD5+ B lymphocytes increases the proportion of EBV-transformed tonsillar lymphoid cell cultures that make antihistone IgM antibodies.[37]

Another indication that the EBV-induced antibodies seen in sera of IM patients might be unmutated copies of germ-line encoded antibodies comes from studies on anti-*i* autoantibodies that have been analyzed in detail due to their potential pathologic role. It has been observed that these antibodies, like antierythrocyte antibodies secreted by tumor B cells of monoclonal gammopathies,[84] bear an idiotope arising from heavy

TABLE III

Sequence of HY 5488 VH and Alignment with Gene V-A77(VH3)[a]

	Sequence
V-A77 (VH3)	GGA GGC TTG GTA CAG CCT GGG GGG TCC CTG AGA CTC TCC TGT
HY 5488 VH	*** *** *** *C* *** *** *** *** *** *** C** *** *** ***
V-A77 (VH3)	GCA GCC TCT GGA TTC ACC TTT AGC AGC TAT GCC ATG AGT TGG
HY 5488 VH	*** *** *** *** *** *** **T *** *** *** *G* *** *** ***
V-A77 (VH3)	GTC CGC CAG GCT CCA GGG AAG GGG CTG GAG TGG GTC TCA GCC
HY 5488 VH	*** *** *** C** *** *** *** *** *** *** *** *** *** CT*
V-A77 (VH3)	ATT AGT GGT AGT GGT GGT AGC ACA TAC TAC GCA GAC TCC GTG
HY 5488 VH	*** *** *** *AC *** *** **T *T* *** *** *** *** *** ***
V-A77 (VH3)	AAG GGC CGG TTC ACC ATC TCC AGA GAC AAT TCC AAG AAC ACG
HY 5488 VH	*** *** *** *** *** *** *** *** *** *** *** *** *** ***
V-A77 (VH3)	CTG TAT CTG CAA ATG AAC AGC CTG AGA GCC GAC ACG GCC
HY 5488 VH	*** *** *** *** *** *** *** *** *** **A *** *** ***
V-A77 (VH3)	GTA TAT TAC TGT GCG AGA A GGA TAT TGT AGT ACG AGT ACC
HY 5488 VH	*** *** *** *** *** A GGC *** TGG AGT GGC **A *T
V-A77 (VH3)	AGC TGC GAC GCC AAC TGG GGC GAC TCC TGG CAA
HY 5488 VH	CC* *** C** C** *G GCC TTC C** *G

[a] Asterisks indicate nucleotide identity.

TABLE IV
Sequence of HY 5488 Vκ and Alignment with Gene κVAC[a]

κVAC	CAG	TCT	CCA	GCC	ACC	CTG	TCT	GTG	TCT	CCA	GGG	GAA	AGA	GCC
HY 5488 k	***	***	***	***	***	***	***	***	***	***	***	***	***	***
κVAC	ACC	CTC	TCC	TGC	AGG	GCC	AGT	CAG	AGT	GTT	AGC	AGC	AAC	TTA
HY 5488 k	***	***	***	***	***	***	***	***	***	A**	***	*A*	C**	***
κVAC	GCC	TGG	TAC	CAG	CAG	AAA	CCT	GGC	CAG	GCT	CCC	AGG	CTC	CTC
HY 5488 k	***	***	***	***	***	***	***	***	*G*	***	***	***	***	***
κVAC	ATC	TAT	GGT	GCA	TCC	ACC	AGG	GCC	ACT	GGT	ATC	CCA	GCC	AGG
HY 5488 k	T**	***	A**	***	***	***	***	***	***	***	***	***	***	***
κVAC	TTC	AGT	GGC	AGT	GGG	TCT	GGG	ACA	GAG	TTC	ACT	CTC	ACC	ATC
HY 5488 k	***	***	***	***	*C*	***	***	***	***	***	***	***	***	***
κVAC	AGC	AGC	CTG	CAG	TCT	GAA	GAT	TTT	GCA	GTT	TAT	TAC	TGT	CAG
HY 5488 k	***	***	***	***	***	***	***	***	***	***	***	***	***	***
κVAC	CAG	TAT	AAT	AAC	TGG	CCT	CTC	ACC	TTC	GGC	CAA	GGG	ACA	CGA
HY 5488 k	***	***	***	***	***	***	***	***	***	***	***	***	***	***
κVAC														
HY 5488 k	CTG	GAG	ATT	AAA	CGA									

[a]Asterisks indicate nucleotide identity.

chains encoded by the V_H4–21 gene segments, which indicates a restriction of autoantibody specificity to a single V_H gene;[85] sequence analysis of EBV-induced antierythrocyte autoantibody showed remarkably little change from the germ-line sequence, thus ruling out an antigen selection process for the maturation of these antibodies.[85]

Although polyclonal activation of B lymphocytes encoding natural polyspecific antibodies in germ-line configuration seems to be the major cause of EBV-induced autoimmunity, alternative mechanisms are also possible. It has been suggested, for example, that certain EBV-induced IgM autoantibodies in IM are primarily made in response to a specific region of the virus-encoded *EBNA-1* protein, consisting of repeated sequences of glycine and alanine, and cross-react with homologous epitopes on several host proteins.[78] Very recently, two of these autoantigens have been partially cloned and expressed as fusion proteins; one of these, coded p542, was found to contain a 28-mer glycine/serine that appears to mimic the glycine/alanine repeat of *EBNA-1*; the other, coded p554, encodes a protein that is not cross-reactive with any EBV-encoded antigen, but forms complexes with other proteins. Interestingly, affinity-purified anti-p542 and anti-p554 IgM antibodies, contrary to natural antibodies, do not display polyreactivity and possess relatively high binding affinity. It has been suggested that anti-p542 IgM autoantibodies would be indeed produced by a mechanism of cross-reactive autoimmunity with *EBNA-1*, while anti-p554 IgM would be triggered by complexes between autologous and viral proteins with a consequent adjuvant effect favoring autoantibody production.[86] Moreover, based on the fact that upon gaining entry to the body, EBV infects and replicates in oropharyngeal epithelial cells, it has been suggested that some IM autoantibodies, such as those against the cytoskeleton filaments, are the result of autoimmunization by autoantigens released from virally damaged epithelial cells.[59,60]

6. CONCLUSIONS

Epstein-Barr virus infection has the potential to induce humoral autoimmunity, that is, reaction of antibodies with self antigens. Autoantibodies reactive with multiple autoantigens are in fact transiently produced in most, if not all, IM patients and are formed in general in the acute phase of the disease. Although most of these autoantibodies do not seem harmful to the host, some have been proposed to play a pathogenetic role. Thus, the anti-*i* and the anti-TPI antibodies have been suggested to be involved in the hemolytic anemia which is a severe though rare complication of IM; the anti-MnSOD antibody seems to be associated with clinical signs of the acute phase of the disease; and finally antineural antibodies have been proposed to contribute to the rare neurologic complications of IM.

The major mechanism responsible for the appearance of EBV-induced autoantibodies during IM seems to be the virus-driven, autoantigen-independent polyclonal (nonspecific) activation of B lymphocytes encoding natural, polyspecific antibodies. Such antibodies are secreted mostly in unmutated germ-line configuration by using a restricted set of V_H and V_L genes strongly conserved during evolution. Several lines of evidence point to this conclusion: 1) their transient duration and IgM nature, with no IgG class switching; 2) the available sequence data of cloned V_H and V_L antibody genes; and 3) the idiotype studies reported above. Molecular mimicry, especially with the glycine-rich sequence of *EBNA-1*, has also been proposed as a possible cause of autoimmunity to certain cellular antigens in IM; the lack of polyreactivity of the autoantibodies directed to such antigens and their relatively high affinities would support the existence of mechanisms other than polyclonal B-cell activation underlying EBV-induced autoimmunity.

Epidemiological, serological, and experimental observations have suggested an association between EBV and a variety of diseases of unknown etiology but with an autoimmune component, such as RA, SS, SLE, and MS. Molecular mimicry has often been invoked as evidence of a causal role played by the virus in the pathogenesis of autoimmune disease; however, in spite of the exciting and intriguing findings reported, the pathogenetic significance of the antigenic cross-reactions observed remains to be assessed. Cross-reactions, as well a viral reactivation and increased levels of EBV-specific antibodies or circulating EBV-infected B cells, might well be a result of the immunologic dysregulation that characterizes these patients, rather than primary pathogenetic events. Due to the ubiquitous nature of the virus, its lifelong persistence after primary infection, and the tendency to reactivate as a consequence of immune dysfunctions, great caution should be exercised in assigning EBV a primary direct role in the pathogenesis of these diseases. However, as EBV is a strong and lifelong immune stimulator, the possibility that it helps perpetuate an immune dysregulation in genetically predisposed individuals should be seriously considered.

ACKNOWLEDGMENTS. This work was supported by funds from Italian M.U.R.S.T. and C.N.R. I wish to thank Dr. F. K. Stevenson and Dr. C. J. Chapman (University of Southampton, United Kingdom) for sequence analysis of HY 5488 antibody.

REFERENCES

1. Liebowitz, D., and Kieff, E., 1993, Epstein-Barr virus, in: *The Human Herpesviruses* (B. Roizman, R. J. Whithly, and C. Lopez, eds.), Raven Press, New York, pp. 107–172.
2. Jondal, M., and Klein, G., 1973, Surface markers on human B and T lymphocytes. II. Presence of Epstein-Barr virus receptors on B lymphocytes, *J. Exp. Med.* **138**:1365–1378.
3. Sixbey, J. W., Nedrud, J. G., Raab-Traub, N., Hanes, R. A., and Pagano, J. S., 1984, Epstein-Barr virus replication in oropharyngeal epithelial cells, *N. Engl. J. Med.* **310**:1225–1230.
4. Jondal, M., Klein, G., Oldstone, M. B. A., Bokish, V., and Yefenof, E., 1976, An association between complement and Epstein-Barr virus receptors on human lymphoid cells, *Scand. J. Immunol.* **5**:401–410.
5. Miller, G., 1990, Epstein-Barr virus. Biology, pathogenesis, and medical aspects, in: *Virology*, 2nd Ed., Raven Press, New York, pp. 1921–1958.
6. Niederman, J., Miller, G., Pearson, H., and Pagano, J., 1976, Pattern of excretion of Epstein-Barr virus in saliva and other oropharyngeal sites during infectious mononucleosis, *N. Engl. J. Med.* **294**:1355–1359.
7. Lenoir, G. M., 1986, Role of the virus, chromosomal translocation and cellular oncogenes in the aetiology of Burkitt's lymphoma, in: *The Epstein-Barr Virus: Recent Advances* (M. A. Epstein and B. G. Achong, eds.), Wiley, New York, pp. 183–205.
8. Purtilo, D. T., 1983, Immunopathology of X-linked lymphoproliferative syndrome, *Immunol. Today* **4**:291–297.
9. Ernberg, I., and Altiok, E., 1989, The role of Epstein-Barr virus in lymphomas of HIV carriers. *APMIS (Suppl.)* **8**:63–66.
10. Greenspan, J. S., Greenspan, D., Lennette, E. T., Abrams, D. L., Conant, M. A., Petersen, V., and Freese, K., 1985, Replication of Epstein-Barr virus within the epithelial cells of oral "hairy" leukoplakia, an AIDS-associated lesion, *N. Engl. J. Med.* **313**:1564–1571.
11. Catalano, M. O., Carson, D. A., Niederman, J. C., Feorino, P., and Vaughan, J. H., 1980, Antibody to the rheumatoid arthritis nuclear antigen: Its relationship to *in vivo* Epstein-Barr virus infection, *J. Clin. Invest.* **65**:1238–1242.
12. Alspaugh, M. A., Henle, G., Lennette, E. T., and Henle, W., 1981, Elevated levels of antibodies to Epstein-Barr virus antigens in sera and synovial fluids of patients with rheumatoid arthritis, *J. Clin. Invest.* **67**:1134–1140.
13. Billings, P. B., Hoch, S. O., and Vaughan, J. H., 1984, Polymorphism of the EBNA-RANA antigen in Epstein-Barr virus positive cell lines, *Arthritis Rheum.* **27**:1423–1427.
14. Tosato, G., Steinberg, A. D., Yarchoan, R., Heliman, C. A., Pike, S. E., DeSeau, V., and Blaese, R. M., 1984, Abnormally elevated frequency of Epstein-Barr virus–infected B cells in the blood of patients with rheumatoid arthritis, *J. Clin. Invest.* **73**:1789–1795.
15. Slaughter, L., Carson, D. A., Jensen, F. J., Holbrook, T. L., and Vaughan, J. H., 1978, In vitro effects of Epstein-Barr virus on peripheral blood mononuclear cells from patients with rheumatoid arthritis and normal subjects, *J. Exp. Med.* **148**:1429–1434.
16. Burastero, S. E., Casali, P., Wilder, R. L., and Notkins, A. L., 1988, Mono-reactive high affinity and polyreactive low affinity rheumatoid factors are produced by CD5+ B cells from patients with rheumatoid arthritis, *J. Exp. Med.* **168**:1979–1992.
17. Rhodes, G., Carson, D. A., Valbrecht, J., Houghten, R., and Vaughan, J. H., 1985, Human immune responses to synthetic peptides from the Epstein-Barr nuclear antigen, *J. Immunol.* **134**:211–216.
18. Luka, J., Kreofsky, T., Pearson, G. R., Hennessy, K., and Kieff, E., 1984, Identification and

characterization of a cellular protein that cross-reacts with the Epstein-Barr virus nuclear antigen, *J. Virol.* **52**:833–838.

19. Fox, R. I., Sportsman, R., Rhodes, G. H., Luka, J., Pearson, G., and Vaughan, J. H., 1986, Rheumatoid arthritis synovial membrane contains a 62 kd protein that shares an antigenic epitope with the Epstein-Barr virus encoded EBNA-*1* antigen, *J. Clin. Invest.* **77**:1539–1547.

20. Roudier, J., Rhodes, G. H., Vaughan, J. H., and Carson, D. A., 1988, The Epstein-Barr virus glycoprotein gp110, a molecular link between HLA DR4, HLA DR1 and rheumatoid arthritic, *Scand. J. Immunol.* **27**:367–371.

21. Gregersen, P. K., Silver, J., and Winchester, R. J., 1987, Shared epitope hypothesis, *Arthritis Rheum.* **30**:1205–1213.

22. Klareskog, L., Forsum, U., Scheynius, A., Kabelitz, D., and Wigzell, H., 1982, Evidence in support of a self-perpetuating HLA-DR dependent delayed type cell reaction in rheumatoid arthritis, *Proc. Natl. Acad. Sci. U.S.A.* **79**:3632–3636.

23. Fox, R. I., Luppi, M., Pisa, P., and Kang, H.-I., 1992, Potential role of Epstein-Barr virus in Sjögren's syndrome and rheumatoid arthritis, *J. Rheumatol. (Suppl. 19)* **32**:18–24.

24. Akobashi, I., Jamamoto, J., Katsuki, T., and Matsuda, I., 1983, Unique pattern of EBV specific antibodies in recurrent parotitis, *Lancet* **2**:1049–1090.

25. Fox, R. I., Pearson, G., and Vaughan, J. H., 1986, Detection of Epstein-Barr virus associated antigens and DNA in salivary gland biopsies from patients with Sjögren's syndrome, *J. Immunol.* **137**:3162–3168.

26. Saito, I., Compton, S. B., and Fox, R. I., 1989, Detection of Epstein-Barr virus by DNA by polymerase chain reaction in blood and tissue biopsies from patients with Sjögren's syndrome, *J. Exp. Med.* **169**:191–198.

27. Lerner, M. R., Andrews, N. C., Miller, G., and Steitz, J. A., 1981, Two small RNAs encoded by Epstein-Barr virus and complexed with protein are precipitated by antibodies from patients with systemic lupus erythematosus, *Proc. Natl. Acad. Sci. U.S.A.* **78**:805–809.

28. Inoue, N., Harada, S., Miyasaka, N., Oya, A., and Yanagi, K., 1991, Analysis of antibody titers to Epstein-Barr virus nuclear antigens in sera of patients with Sjögren's syndrome and with rheumatoid arthritis, *J. Infect. Dis.* **164**:22–28.

29. Tsokos, G. C., and Balow, J. E., 1984, Cellular immune responses in systemic lupus erythematosus, *Prog. Allergy* **35**:93–161.

30. Nakamura, M., Burastero, S. E., Ueki, Y., Larrick, J. W., Notkins, A. L., and Casali, P., 1988, Probing the normal and autoimmune B cell repertoire with EBV. Frequency of B cells producing monoreactive high affinity autoantibodies in patients with Hashimoto disease and SLE, *J. Immunol.* **141**:4165–4172.

31. Sabbatini, A., Bombardieri, S., and Migliorini, P., 1993, Autoantibodies from patients with systemic lupus erythematosus bind a shared sequence of SmD and Epstein-Barr virus–encoded nuclear antigen EBNA I, *Eur. J. Immunol.* **23**:1146–1152.

32. Marchini, B., Dolcher, M. P., Sabbatini, A., Klein, G., and Migliorini, P., 1994, Immune response to different sequences of the EBNA I molecule in Epstein-Barr virus–related disorders and in autoimmune diseases, *J. Autoimmun.* **7**:179–191.

33. Incaprera, M., Bazzichi, A., Manunta, M., Chapman, C. J., Stevenson, P. K., and Garzelli, C., 1993, Epitope homology between Epstein-Barr virus nuclear antigen-2 (EBNA-2) and systemic lupus autoantigens, in: *Proceedings of International Symposium, The Impact of Biotechnology on Autoimmunity*, Florence, Italy, 1993, *Biotec* **8**:106.

34. Costa, O., and Monier, J. C., 1986, Antihistone antibodies detected by ELISA and immunoblotting in systemic lupus erythematosus and rheumatoid arthritis, *J. Rheumatol.* **13**:722–725.

216   CARLO GARZELLI

35. Shoenfeld, Y., and Segol, O., 1989, Anti-histone antibodies in SLE and other auto-immune diseases, *Clin. Exp. Rheumatol.* **7**:265–271.
36. Vainio, E., Lenoir, G. M., and Franklin, R. M., 1983, Autoantibodies in three populations of Burkitt's lymphoma patients, *Clin. Exp. Immunol.* **54**:387–396.
37. Garzelli, C., Incaprera, M., Bazzichi, A., Manunta, M., Rognini, F., and Falcone, G., 1994, Epstein-Barr virus–transformed human B lymphocytes produce natural antibodies to histones, *Immunol. Lett.* **39**:277–282.
38. Garzelli, C., Incaprera, M., Bazzichi, A., Manunta, M., and Falcone, G., 1995, Relationship of an idiotype expressed on a polyreactive monoclonal autoantibody induced by Epstein-Barr virus to anti-histone antibodies, *Immunol. Infect. Dis.* **5**:53–58.
39. Taub, J. W., Warrier, I., Holtkamp, C., Beardsley, D. S., and Lusher, J. M., 1995, Characterization of autoantibodies against the platelet glycoprotein antigens IIb/IIIa in childhood idiopathic thrombocytopenia purpura, *Am. J. Hematol.* **48**:104–107.
40. Xie, K., and Snyder, M., 1995, Two short autoepitopes on the nuclear dot antigen are similar to epitopes encoded by the Epstein-Barr virus, *Proc. Natl. Acad. Sci. U.S.A.* **92**:1639–1643.
41. Grose, C., Henle, W., Henle, G., and Feorino, P. M., 1975, Primary Epstein-Barr virus infection in acute neurologic diseases, *N. Engl. J. Med.* **292**:392–395.
42. Weinstein, M., 1991, Epstein-Barr virus and neurologic disease, *Neurol. Res.* **13**:199–200.
43. Bray, P. F., Culp, K. W., McFarlin, D. E., Panitch, H. S., Torkelson, R. D., and Schlight, J. P., 1992, Demyelinating disease after neurologically complicated primary Epstein-Barr virus infection, *Neurology* **42**:278–282.
44. Tsutsumi, H., Kamazaki, H., Nakata, S., Nagao, M., Chiba, S., Imai, S., and Osato, T., 1994, Sequential development of acute meningoencephalitis and transverse myelitis caused by Epstein-Barr virus during infectious mononucleosis, *Pediatr. Infect. Dis. J.* **13**:665–667.
45. Ito, H., Sayama, S., Irie, S., Kanazawa, N., Saito, T., Kowa, H., Haga, S., and Ikeda, K., 1994, Antineuronal antibodies in acute cerebellar ataxia following Epstein-Barr virus infection, *Neurology* **44**:1506–1507.
46. Lindberg, C., Andersen, O., Vahlne, A., Dalton, M., and Renmarker, B., 1991, Epidemiological investigation of the association between infectious mononucleosis and multiple sclerosis, *Neuroepidemiology* **10**:62–65.
47. Martyn, C. N., Cruddas, M., and Compston, D. A. S., 1993, Symptomatic Epstein-Barr virus infection and multiple sclerosis, *J. Neurol. Neurosurg. Psychiatry* **56**:167–168.
48. Haahr, S., Sommerlund, S., Moller-Larsen, A., Mogensen, S., and Andersen, H. M. K., 1992, Is multiple sclerosis caused by a dual infection with retrovirus and Epstein-Barr virus? *Neuroepidemiology* **11**:299–303.
49. Sumaya, C. V., Myers, L. W., and Ellison, G. W., 1980, Epstein-Barr virus antibodies in multiple sclerosis, *Arch. Neurol.* **37**:94–96.
50. Bray, P. F., Bloomer, L. C., Salmon, V. C., Bagley, M. H., and Larsen, P. D., 1983, Epstein-Barr virus infection and antibody synthesis in patients with multiple sclerosis, *Arch. Neurol.* **40**:406–408.
51. Larsen, P. D., Bloomer, L. C., and Bray, P. F., 1985, Epstein-Barr virus nuclear antigen and viral capsid antigen antibody titers in multiple sclerosis, *Neurology* **35**:435–438.
52. Sommerlund, M., Pallesen, G., Moller-Larsen, A., Hansen, H. J., and Haahr, S., 1993, Retrovirus-like particles in an Epstein-Barr virus–producing cell line derived from a patient with chronic progressive myelopathy, *Acta Neurol. Scand.* **87**:71–76.
53. Bray, P. F., Luka, J., Bray, P. F., Culp, K. W., and Schlight, J. P., 1992, Antibodies against

Epstein-Barr nuclear antigen (EBNA) in multiple sclerosis, CSF, and two pentapeptide sequence identities between EBNA and myelin basic protein, *Neurology* 42:1798-1804.

54. Holborow, E. J., Hemsted, E. H., and Mead, S. V., 1973, Smooth muscle autoantibodies in infectious mononucleosis, *Br. Med. J.* 3:323-325.

55. Sutton, R. N. P., Emond, R. T. D., Thomas, D. B., and Doniach, D., 1974, The occurrence of autoantibodies in infectious mononucleosis, *Clin. Exp. Immunol.* 17:427-436.

56. Andersen, P., and Faber, V., 1978, Antibodies to smooth muscle and other tissue components in infectious mononucleosis, *Scand. J. Infect. Dis.* 10:1-5.

57. Mead, G. M., Cowin, P., and Whitehouse, J. M. A., 1980, Antitubulin antibody in healthy adults and patients with infectious mononucleosis and its relationship to smooth muscle antibody (SMA), *Clin. Exp. Immunol.* 39:328-336.

58. Garzelli, C., Pacciardi, A., Basolo, F., and Falcone, G., 1989, Mechanisms other than polyclonal B cell activation possibly involved in Epstein-Barr virus–induced autoimmunity, *Clin. Exp. Immunol.* 76:412-416.

59. Linder, E., Kurki, P., and Andersson, L. C., 1979, Autoantibody to "intermediate filaments" in infectious mononucleosis, *Clin. Immunol. Immunopathol.* 14:411-417.

60. Kataaha, P. K., Mortazavi-Milani, S. M., Russel, G., and Holborow, E. J., 1985, Anti-intermediate filaments antibodies, antikeratin antibody, and antiperinuclear factor in rheumatoid arthritis and infectious mononucleosis, *Ann. Rheum. Dis.* 44:446-449.

61. Thomas, D. B., 1972, Antibodies to membrane antigen common to thymocytes and a subpopulation of lymphocytes in infectious mononucleosis sera, *Lancet* i:399-403.

62. Misra, R., Venables, P. J. W., Plater-Zyberk, C., and Watkins, P. F., 1989, Anti-cardiolipin antibodies in infectious mononucleosis react with the membrane of activated lymphocytes, *Clin. Exp. Immunol.* 75:35-40.

63. Garzelli, C., Manunta, M., Incaprera, M.., Bazzichi, A. Conaldi, P. G., and Falcone, G., 1992, Antibodies to histones in infectious mononucleosis, *Immunol. Lett.* 32:111-116.

64. Holborow, E. J., Asherson, G. L., Johnson, G. D., Barnes, R. D. S., and Carmichael, D. L., 1963, Antinuclear factor and other antibodies in blood and liver diseases, *Br. Med. J.* 1:656-658.

65. Wollheim, F. A., and Williams, R. C., 1966, Studies on the macroglobulins of human serum. I. Polyclonal immunoglobulin class M (IgM) increases in infectious mononucleosis, *N. Engl. J. Med.* 274:61-67.

66. Jenkins, W. J., Koster, H. G., Marsh, W. L., and Carter, R. L., 1965, Infectious mononucleosis: An unsuspected source of anti-i, *Br. J. Haematol.* 11:480-483.

67. Rosenfield, R. E., Schmidt, P. J., Calvo, R. C., and McGinniss, R. L., 1965, Anti-i, a frequent cold agglutinin in infectious mononucleosis, *Vox Sang.* 10:631-634.

68. Troxel, D. B., Innella, F., and Cohen, R. J., 1966, Infectious mononucleosis complicated by hemolytic anemia due to anti-i, *Am. J. Pathol.* 46:625-631.

69. Ritter, K., Brestrich, H., Nellen, B., Kratzin, H., Eiffert, H., and Thomssen, R., 1990, Autoantibodies against triosephosphate isomerase. A possible clue to pathogenesis of hemolytic anemia in infectious mononucleosis, *J. Exp. Med.* 171:565-570.

70. Geurs, F., Ritter, K., Mast, A., and Van Meale, V., 1992, Successful plasmapheresis in corticosteroid-resistant hemolysis in infectious mononucleosis: Role of autoantibodies against triosephosphate isomerase, *Acta Haematol.* 88:142-146.

71. Ritter, K., Kühl, R.-J., Semrau, F., Eiffert, H. D., Kratzin, H., and Thomssen, R., 1994, Manganese superoxide dismutase as a target of autoantibodies in acute Epstein-Barr virus infection, *J. Exp. Med.* 180:1995-1998.

72. Fong, S., Vaughan, J. H., Tsoukas, C. D., and Carson, D. A., 1982, Selective induction of

autoantibody secretion in human bone marrow by Epstein-Barr virus, *J. Immunol.* **129**:1941–1945.

73. Garzelli, C., Taub, F. E., Scharff, J. E., Prabhakar, B. S., Ginsberg-Fellner, F., and Notkins, A. L., 1984, Epstein-Barr virus–transformed lymphocyte produce monoclonal antibodies that react with antigens in multiple organs, *J. Virol.* **52**:722–725.

74. Robinson, J. E., and Stevens, K. C., 1984, Production of autoantibodies to cellular antigens by human B cells transformed by Epstein-Barr virus, *Clin. Immunol. Immunopathol.* **33**:339–350.

75. Garzelli, C., Taub, F. E., Jenkins, M. C., Drell, D. W., Ginsberg-Fellner, F., and Notkins, A. L., 1986, Human monoclonal autoantibodies that react with both pancreatic islets and thyroid, *J. Clin. Invest.* **77**:1627–1631.

76. Taub, F., Satoh, J., Garzelli, C., Essani, K., and Notkins, A. L., 1985, Human monoclonal autoantibodies reactive with multiple organs, in: *Human Hybridomas and Monoclonal Antibodies* (E. G. Engleman, S. K. H. Foung, J. Larrick, and A. Raubitchek, eds.), Plenum Press, New York, pp. 263–275.

77. Casali, P., and Notkins, A. L., 1989, Probing the human B-cell repertoire with EBV: Polyreactive antibodies and CD5+ lymphocytes, *Annu. Rev. Immunol.* **7**:513–535.

78. Rhodes, G., Rumpold, H., Kurki, P., Patrick, K. M., Carson, D. A., and Vaughan, J. H., 1987, Autoantibodies in infectious mononucleosis have specificity for the glycine-alanine repeating region of the Epstein-Barr virus nuclear antigen, *J. Exp. Med.* **165**:1026–1040.

79. Garzelli, C., Pacciardi, A., Carmignani, M., Conaldi, P. G., Basolo, F., and Falcone, G., 1989, A human monoclonal autoantibody isolated from a patient with infectious mononucleosis reactive with both self antigens and Epstein-Barr nuclear antigen (EBNA), *Immunol. Lett.* **22**:211–216.

80. Garzelli, C., Incaprera, M., Bazzichi, A., and Falcone, G., 1992, Detection of an idiotope on a human monoclonal autoantibody by monoclonal anti-idiotypic antibody and its relationship to Epstein-Barr virus–induced autoimmunity, *Autoimmunity* **11**:171–177.

81. Nakamura, M., Burastero, S. E., Notkins, A. L., and Casali, P., 1988, Human monoclonal rheumatoid factor–like antibodies from CD5 (Leu-1)+ B cells are polyreactive, *J. Immunol.* **140**:4180–4186.

82. Schwartz, R. S., 1988, Polyvalent anti-DNA autoantibodies: Immunochemical and biological significance, *Int. Rev. Immunol.* **3**:97–115.

83. Sanz, I., Dang, H., Takei, M., Talal, N., and Capra, D., 1989, V_H sequence of a human anti-Sm autoantibody. Evidence that autoantibodies can be unmutated copies of germ-line genes, *J. Immunol.* **142**:883–887.

84. Pasqual, V., Victor, K., Lelsz, D., Spellerberg, M. B., Hamblin, T. J., Thompson, K. M., Randen, I., Natvig, J., Capra, J. D., and Stevenson, F. K., 1991, Nucleotide sequence analysis of the V regions of two IgM cold agglutinins. Evidence that VH4-21 gene segment is responsible for the major cross-reactive idiotype, *J. Immunol.* **146**:4385–4391.

85. Chapman, C. J., Spellerber, M. B., Smith, G. A., Carter, S. J., Hamblin, T. J., and Stevenson, F. K., 1993, Autoanti–red cell antibodies synthesized by patients with infectious mononucleosis utilize the VH4-21 gene segment, *J. Immunol.* **151**:1051–1061.

86. Vaughan, J. H., Valbracht, J. R., Nguyen, M.-D., Handley, H. H., Smith, R. S., Patrick, K., and Rhodes, G. H., 1995, Epstein-Barr virus induced autoimmune responses. I. Immunoglobulin M autoantibodies to proteins mimicking and non mimicking Epstein-Barr virus nuclear antigen-1, *J. Clin. Invest.* **95**:1306–1315.

10

Retroviruses and Autoimmunity

KENNETH E. UGEN, LAURA FERNANDES,
H. RALPH SCHUMACHER, BIN WANG,
DAVID B. WEINER, and WILLIAM V. WILLIAMS

1. INTRODUCTION

Autoimmunity is an immune reaction against self constituents. Auto-immune phenomena are noted in a spectrum of clinical scenarios, ranging from the normal physiological processes and consequences of aging to severe pathological conditions which may be organ-specific or general.[1] Autoimmunity is likely to be involved in the etiology and/or pathogenesis of a variety of diseases such as rheumatoid arthritis, systemic lupus erythematosus, and diabetes. Under normal circumstances the immune system provides a response to insults such as neoplasm, injury, and infection. In pathological immune responses, however, the normal functioning of the immune system undergoes dysregulation resulting in an

KENNETH E. UGEN • Department of Medical Microbiology and Immunology, University of South Florida College of Medicine, Tampa, Florida 33612.   LAURA FERNANDES • Jefferson Medical College, Thomas Jefferson University, Philadelphia, Pennsylvania 19107. H. RALPH SCHUMACHER and WILLIAM V. WILLIAMS • Department of Medicine, University of Pennsylvania School of Medicine, Philadelphia, Pennsylvania 19104.   BIN WANG and DAVID B. WEINER • Department of Pathology and Laboratory Medicine, University of Pennsylvania School of Medicine, Philadelphia, Pennsylvania 19104.

Microorganisms and Autoimmune Diseases, edited by Herman Friedman *et al.* Plenum Press, New York, 1996.

autoaggressive immune response. The etiology and pathogenesis of many autoimmune disorders remains unsolved despite a tremendous growth in understanding of how the immune system works. The development of these autoimmune diseases appears to depend on a variety of genetic, hormonal, and neuroimmunological factors. Infectious agents and viruses, particularly, have been implicated in the etiology and pathogenesis of many autoimmune diseases.[2]

The purpose of this review is to summarize some of the evidence suggesting a role specifically for retroviruses in the etiology and pathogenesis of autoimmune diseases. Human immunodeficiency virus (HIV) will be used as the model for this review but other human as well as animal retroviruses will also be discussed.[3]

2. RETROVIRUSES AND AUTOIMMUNE DISEASE

Viruses have been the focus of attention in the investigation of autoimmune disease pathogenesis. There have been several mechanisms postulated by which viruses, including retroviruses, may induce autoimmunity. These include (1) the alteration of the expression or structure of host antigens, (2) generalized stimulation of the immune system (e.g., superantigens), and (3) molecular mimicry of host antigens by viruses resulting in inappropriate immune responses being directed against host antigen.[4,5] Most of the experimental evidence to support such theories has come not from human studies but experimental animals.[6] Therefore, based upon these studies, virus induction of host autoantigen expression may predispose an individual to autoimmunity. A generalized stimulation of the immune system can be caused by superantigens, such as the mouse mammary tumor virus envelope, also known as the Mls antigen. Some more recent investigations have indicated that endogenous murine superantigens may in fact represent molecules expressed by retroviruses. Immunological cross-reactivity has been implicated in many disorders such as rheumatic fever (cardiac myosin and streptococcal M proteins) and scleroderma (topoisomerase I and retroviral *gag* protein).

As indicated above, one group of infectious agents that has been implicated in the etiology and pathogenesis of autoimmune disorders is the family of retroviruses.[7,8] The Retroviridae is a large group of viruses which primarily infect vertebrates. There are three described subfamilies of retroviruses: (1) Oncovirinae, which includes cancer-causing viruses such as mammalian type B, C, and D retroviruses as well as the HTLV-BLV group; (2) Lentivirinae, which includes the human immunodeficiency viruses types 1 and 2 (HIV-1 and HIV-2), the simian and feline immunodeficiency viruses, the visna-maedi viruses, as well as the equine infectious anemia viruses (EIAV) and the caprine arthritis-encephalitis virus; and (3) Spumavirinae ("foamy" virus), which are endogenous viruses.[9] For a considerable period of time Spumavirinae had been thought to be benign. However, more recent evidence, as discussed in this review, suggests that members of this subfamily may be involved in the pathogenesis of several disorders of unknown etiology.[10]

3. GENERAL BIOLOGY OF RETROVIRUSES

Of the human retroviruses, the most intensively studied have been the human immunodeficiency viruses 1 and 2 (HIV-1 and HIV-2) and the human T-cell leukemia viruses (HTLV-1 and HTLV-2). Their overall structural organization is similar to that of other human and animal retroviruses although the human immunodeficiency viruses are considerably more complex. Structurally, HIV-1 consists externally of the envelope glycoproteins gp120 and gp41, which are associated with a lipid bilayer, and consists internally of the structural *gag* proteins and genomic viral RNA which is further associated with reverse transcriptase, the enzyme that is the hallmark of these viruses. HIV-1 interacts with most target cells by binding to a receptor on the cell surface known as the CD4 glycoprotein.[11] Figure 1A shows the general particle structure for retroviruses. Figure 1B shows the general retroviral genomic organization with the *gag*, *pol*, and *env* genes as well as the general retroviral genomic organization of the standard genes that comprise retroviruses. The designation *acc* refers to accessory (i.e., regulatory) genes which are a characteristic of lentiviruses such as HIV-1. Oncogenic retroviruses, except for HTLV-1, also express an oncogene.

Most of the retroviruses that are associated with persistent infection or oncogenesis are lymphotropic and particularly T cell-trophic. The following sequence illustrates the initial steps of the HIV life cycle and is also applicable to most retroviruses:

1. The retrovirus is internalized by cell fusion and/or receptor-mediated endocytosis and is uncoated in the cytoplasm.
2. At the cytoplasmic level, the vision-associated reverse transcriptase produces a hybrid RNA/DNA molecule and converts it to a double-stranded linear DNA molecule. This DNA molecule contains two copies of the long terminal repeat (LTR), a sequence of DNA that contains codons important in the process of viral integration and transcription.

(a)

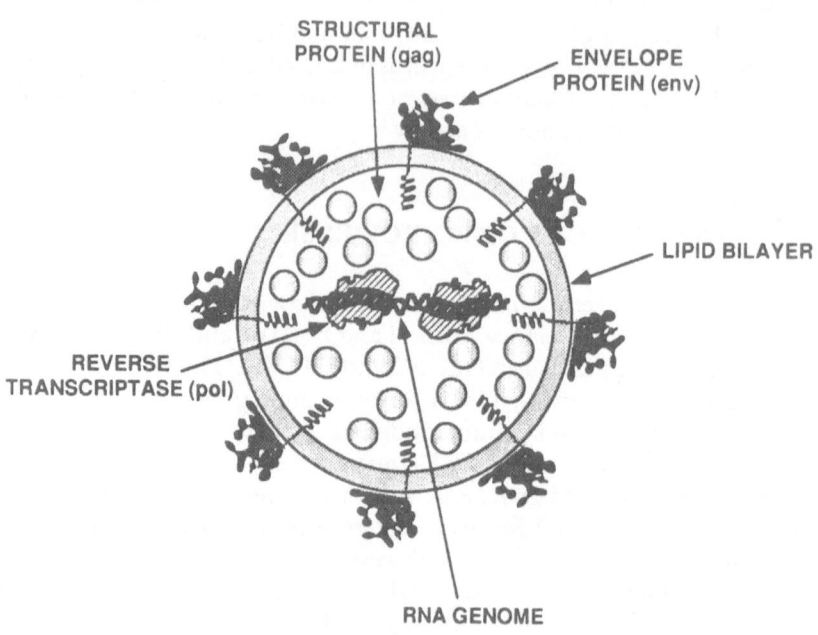

(b)

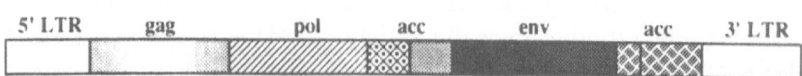

FIGURE 1. Top (a): Characteristic structure of a retroviral particle with a basic lipid bilayer membrane surrounding the viral RNA genome and the requisite enzyme reverse transcriptase (*pol*) as well as the structural protein *gag*. On the surface of the particle is the envelope protein (*env*), which is important for binding and fusion to infectible target cells. Bottom (b): Basic organization of the retroviral genome. This genome includes two long terminal repeats (LTR), one at the 5' end and the other at the 3' end as well as the *gag, pol,* and *env* genes described above. Also included are the accessory (*acc*) genes which are involved in the control of viral replication and pathogenesis.

3. Following this, there is transport of the viral DNA from the cytoplasm to the nucleus with its integration into the host cell chromosome, forming the provirus. This integrated provirus is the template for viral RNA formation, virion protein synthesis, assembly, and release of viral particle from the cell by budding.

Retroviral genomes range from 8000 to 10,000 base pairs. In addition to the structural *gag, pol,* and *env* genes, which are essential for virus replication, there are also noncoding repeated sequences at each end of the genome called long terminal repeats. The retroviral LTRs contain the promoter and regulatory sequences which regulate transcription of viral genes.

The *gag* gene encodes a protein precursor that following viral protease digestion forms the internal capsid proteins. These proteins surround the RNA genome and constitute the core of the virus particle. The *pol* gene also encodes a precursor protein that during the maturation process releases three important viral enzymes which are observed in the HIV-1 provirus: protease, reverse transcriptase (RT), and integrase. In the HTLV-1 provirus, the protease is in a reading frame distinct from the *gag* and *pol* gene reading frames. Protease acts on the *gag* and *pol* precursor and is responsible for cleavage observed in the protein maturation process. Reverse transcriptase reverses the typical pathway of transcription from DNA to RNA, catalyzing instead DNA synthesis from the RNA genome. The integrase functions to integrate the viral DNA at the LTRs into the host cellular DNA, and this process occurs at random insertion points in the host genome.

The last structural gene, the envelope, encodes an immature protein that under cleavage forms a receptor-binding protein, located at the virus particle surface, and a transmembrane protein that binds to the viral lipoprotein membrane and to the receptor-binding protein. Antigenic sites capable of inducing neutralizing antibodies are found in the envelope protein, including the receptor-binding site identified in the surface protein gp120.[12]

In addition to structural genes, regulatory genes are present.[13,14] HTLV-1 and 2 encode *tax* and *rex* as regulatory genes which are analogous to *tat* and *rev* in HIV-1.[15] HIV-1 also encodes the additional regulatory genes *vif, nef, vpr,* and *vpu* or *vpx* for HIV-1 and HIV-2, respectively. While in general still poorly understood, the regulatory gene products function at least in part to regulate virus gene expression. They also can have specific effects on host cell growth and differentiation. Thus, while retroviruses are relatively simple in organization, their ability to integrate into the host chromosome and induce these host cells to produce foreign proteins, alter the differentiation state of cells, and stimulate immune responses all suggest mechanisms for autoimmunity induction. Specific examples of retroviral-induced autoimmunity and inflammatory disease will be considered next.

3.1. Caprine Arthritis-Encephalitis Virus

Interest in the autoimmune and particularly rheumatologic aspects of retroviruses was stimulated in the early 1980s with the isolation and characterization of caprine arthritis-encephalitis virus (CAEV).[16] This virus affects the joints, lungs, and brain of chronically infected goats. Symptoms start as early as one week after experimental infection, and histopathologic aspects of the disease are still present at 8 months follow-up. In the case of natural infection, the first signs of disease appear over 6 months or more. Clinically the most common manifestation is an oligo-arthritis, affecting frequently the carpal joints and characterized by remissions and relapses. Occasionally progression to a rapidly erosive-destructive arthritis can be observed. In addition, interstitial pneumonia, demyelinating encephalomyelitis, and wasting are clinical manifestations in severely affected animals.

Even though this lentivirus can be isolated from peripheral blood monocytes from chronically infected goats, other important sources of virus are synovial macrophages or macrophage-like cells. Furthermore, the histopathology of involved joints shows similarities with human rheumatoid arthritis where synovial lining cell hypertrophy and hyperplasia, villi formation, focal necrosis of synovial cells, mononuclear perivascular infiltration, and occasionally lymphoid follicle–like structures can be observed. Recent studies investigating the immunological role for persistent arthritis have raised the hypothesis that antigenic variants of CAEV that are resistant to neutralizing antibodies can have an important function in perpetuating the inflammatory response. This is supported by the observation that neutralizing antibodies for CAEV do not arrest viral replication or virus-induced arthritis.[17] In addition, the severity of arthritis presentation correlates with the amount of virus isolated from affected joints. Thus, while CAEV infection results in a clinical syndrome with striking similarity to human autoimmune diseases, it may in fact represent a chronic infection. These observations have heightened interest in the potential role retroviruses play in human diseases currently considered to be autoimmune syndromes.

3.2. Human T-Cell Leukemia Virus

Concomitant with chronic arthritis reported in farm animals by the CAEV lentivirus, the HTLV-1 oncovirus has also been described as responsible for arthritis-related symptoms in chronically infected individuals.[17-20]

Recently, in transgenic mouse studies, Iwakura et al. demonstrated that transgenic mice carrying the pX region of the HTLV-1 genome developed polyarthritis at a high rate.[21] Moreover, further histopathologic evaluation of the transgenic mouse synovial tissue demonstrated many aspects similar to human rheumatoid arthritis.[22] There was synovial lining cell proliferation, villous hyperplasia, inflammatory cell infiltration, lymphoid follicle formation, vascular alterations, pannus formation, and cartilage and bone destruction. The authors strongly support HTLV-1 transgenic mice as an important model for rheumatoid arthritis. In fact, they raise the hypothesis that HTLV-1 is one of the causative agents of rheumatoid arthritis. Interestingly, the HTLV-1 pX region encodes both the *tax* and *rex* genes. The *tax* protein is well known not only as a transactivator of viral transcription, but also as a transactivator for interleukin-2 (IL-2), IL-2 receptor, and several cellular oncogenes.[23] In adult T-cell leukemia (ATL), stimulation of interleukin-2 production by *tax* protein is essential for autonomous cell growth.[24] Similarly, in arthritis induced by the HTLV-1 pX transgene, the *tax* protein could also transactivate growth factors at a synovial level. In addition to chronic arthritis, Sjögren's syndrome, Bechet's disease, polymyositis, dermatomyositis, uveitis, and progressive myelopathy have also been diagnosed in HTLV-1-infected individuals.[25-30]

3.3. Human Immunodeficiency Virus

After a decade into the AIDS pandemic, knowledge of the syndrome induced by HIV infection has been markedly expanded. Numerous publications have addressed the autoimmune aspects of HIV-1 infection. The first clinical study addressing some of the rheumatologic aspects was published by Winchester et al. and described an association between Reiter's syndrome and HIV infection.[31] Following this, Espinoza and collaborators reported a 70% prevalence of rheumatologic manifestations among HIV-infected individuals.[32] These studies focused mainly on a homosexual HIV-infected population.

In addition to these clinical pathologic aspects, unique autoimmune phenomena triggered by HIV-1 have been observed over the last few years.

HIV-1 can induce autoimmunity by many different pathways: T-cell dys-regulation is responsible for a polyclonal B-cell proliferation with resultant antibody production and the hypergammaglobulinemia frequently observed in these patients. This polyclonal response is triggered by the *tat* regulatory protein, which transactivates the IL-6 and tumor necrosis factor-alpha (TNF-α) genes in HIV-infected cells and results in cytokine production and B-cell growth. Autoantibodies produced by this mechanism may be involved in some of the clinicopathologic manifestations of HIV-1 infection.

In terms of the immunopathogenic mechanism involved in the development of AIDS it appears that the disease manifestations do not result solely from a cytopathic effect of the virus.[33] A number of aspects of autoimmunity have been proposed to contribute to the development of AIDS. These include: (1) the elimination of CD4+ cells by immune responses directed against either HIV[34] or CD4+ cell antigens,[35] (2) a self major histocompatibility complex (MHC)-stimulated cross-reactive recognition with a secondary antiidiotypic response to CD4,[36] (3) HIV components acting as superantigens leading to widespread depletion of CD4+ cells,[37] (4) immunosuppressive effects mediated by *tat*[38] and gp120,[39] and (5) apoptosis of mature CD4+ cells.[40]

Based on the experiments of a number of investigators there has been considerable debate concerning the mechanisms involved in the death of CD4+ cells in HIV-1 infection and AIDS. For example, HIV-1 DNA has not been detected in a sufficient number of CD4+ cells to directly correlate CD4+ cell death with direct infection of these cells by HIV-1, even though the technique of PCR has allowed detection of HIV-1 DNA in a larger number of cells.

4. CLASSIC AUTOIMMUNE DISORDERS AND ASSOCIATIONS WITH RETROVIRUSES

Rheumatoid arthritis (RA) is an example of the classic "autoimmune" disease. It is characterized by a symmetric, inflammatory poly-arthritis which progressively destroys the joints of affected individuals. Immunologically, patients with RA typically have high titers of autoantibodies (usually of the IgM isotype) which bind the Fc portion of the IgG molecule (rheumatoid factor). The presence of rheumatoid factor in the majority of RA patients led to its classification as an autoimmune disease. Type C retrovirus particles have been isolated from rheumatoid synovial fluid. A recent publication by Stephan Gay and cólleagues describes the presence of spherical virus particles, 200 nm in diameter in five out of eight synovial fluids from rheumatoid arthritis patients.[41] Also inoculation of the HUT 78 cell line with the synovial fluids of these patients revealed the presence of virus-like particles (VLP) 1 to 2 weeks after inoculation. Cultured fibroblasts derived from rheumatoid synovium also showed VLP morphologically similar to those observed in synovial fluid cells. In contrast to these studies, immunocytochemistry and electron microscopy have been negative against HTLV-1 p19 and HIV p24 antigens in this study. These observations, together with Schumacher's earlier publication on VLP in synovial membrane[42] and fluid from early rheumatoid arthritis patients and the recently developed HTLV-1 transgenic mice arthropathy model, have raised a strong possibility that a retrovirus is the trigger for rheumatoid arthritis.[21] However, definitive evidence for this association remains elusive.

Sjögren's syndrome is an inflammatory exocrinopathy which affects the salivary and lacrimal glands, producing dry eyes and dry mouth in affected individuals. While Sjögren's syndrome most frequently occurs in association with other rheumatic diseases (e.g., rheumatoid arthritis, systemic lupus erythematosus, and scleroderma), primary Sjögren's syndrome is also a well-described entity. In cases of Sjögren's syndrome the degeneration of lacrimal and salivary gland function is accompanied by lymphocytic infiltration. It has been noted that both glandular pathology and other symptoms of Sjögren's syndrome were noted in certain individuals infected with HIV-1. Based upon these observations several laboratories have investigated the possible retroviral etiology of this syndrome. Most notably, this syndrome has been shown to occur in association with infection by HTLV-1 and HIV-1. Although the etiological basis for Sjögren's syndrome remains unknown, an infectious basis has been postulated by some investigators. A candidate etiological agent is the Epstein-Barr virus (EBV), since it has been shown to be trophic for salivary epithelium as well as capable of stimulating a number of autoimmune phenomena such as polyclonal hypergammaglobulinemia and rheumatoid factor production.[43]

Even though the recognition of retroviruses as important agents in triggering immunological events and inducing autoimmune manifestations has been established, a specific therapeutic approach toward virus pathogenicity has been delayed by numerous difficulties in dealing with these complex viruses. The integration process that underlies the retroviral life cycle makes eradication of infection a difficult problem to solve. Antiviral therapy is likely to be the major approach in decreasing virus load and the concomitant continuous stimulation of the immune system.

Perhaps a decrease in circulating antivirus antibodies, antiidiotypic antibodies, and viral transactivators can contribute to the control of the many immunological manifestations and rheumatological syndromes developed by retrovirus-infected patients. Although azidothymidine (AZT) has been used to treat psoriasis and Reiter's disease in HIV-infected individuals with some promising outcomes, the drug has not been accepted as an effective agent in managing arthritis and other autoimmune manifestations in these populations.

The development of gene therapy, in conjunction with extensive studies in retrovirus biology, has suggested new therapeutic approaches. These are being evaluated *in vitro* with the eventual goal of performing clinical trials. Somatic gene therapy with antisense oligonucleotides directed against HIV and HTLV regulatory genes has been developed. These oligonucleotides are designed in the reverse sequence of the target mRNA for the specific gene under study. While in contact with the specific mRNA it blocks translation and virus protein production. Inhibition of HIV-1 replication has been shown by intracellular expression of *tat* and *rev* antisense RNA.[44] In addition, HTLV-1 pX region antisense RNA has also shown inhibition of virus replication as well as inhibition of virus-mediated cell immortalization.[45] Another approach to gene therapy developed by Marasco *et al.* uses a single chain antibody that recognizes the CD4 binding site of the HIV-1 envelope protein.[46] When expressed intracellularly, these antibodies act by binding to the immature envelope proteins and disturbing their maturation process. The results of these interactions is a delay in virus replication and a decrease in virus infectivity.

5. SUMMARY

It is clear that infection with retroviruses can result in autoimmune phenomena in both animals and humans. Manifestations of autoimmune disease have been most extensively studied in HIV-1 where the loss of $CD4^+$ lymphocyte populations is likely to be due to multiple causes. The relative contributions of these mechanisms to the autoimmune phenomena, however, remain to be determined. Some of the most tenable mechanisms for autoimmunity in AIDS include (1) aberrant production of autoantibodies probably due to molecular mimicry between regions of the HIV envelope and class II MHC molecules, (2) nonspecific molecules such as tumor necrosis factor as well as immune complexes, and (3) the suppression of the production of cytokines necessary for T-cell maturation.

Therapeutic approaches to retroviral diseases, specifically HIV and

HTLV, are being developed based on recent knowledge of envelope–receptor binding, viral regulatory proteins, and virus enzymes. An effort to decrease the amount of circulating virus is the key to decreasing immune system dysfunction and autoimmune phenomena. This should decrease autoimmune manifestations of these infections.

REFERENCES

1. Roitt, I. M., Hutchings, P. R., Dawe, K. I., Sumar, N., Bodman, K. B., and Cooke, A., 1992, The forces driving autoimmune disease, *J. Autoimmun.* **5**(Suppl. A):11.
2. Nakamura, M. C., and Nakamura, R. M., 1992, Contemporary concepts of autoimmunity and autoimmune diseases, *J. Clin. Lab. Anal.* **6**:275.
3. O'Neill, H. C., 1992, The diversity of retroviral diseases of the immune system, *Immunol. Cell Biol.* **70**:193.
4. Fujinami, R. S., and Oldstone, M. B. A., 1985, Amino acid homology between the encephalitogenic site of myelin basic protein and virus: Mechanism for autoimmunity, *Science* **230**:1043.
5. Fujinami, R. S., 1992, Molecular mimicry, in: *The Autoimmune Diseases II* (N. R. Rose and I. R. Mackay, eds.), Academic Press, Inc., San Diego.
6. Bernard, C. C. A., and Mandel, T. E., 1992, Experimental models of human autoimmune disease, in: *The Autoimmune Diseases II* (N. R. Rose and I. R. Mackay, eds.), Academic Press, Inc., San Diego.
7. Krieg, A., 1990, Retroviruses and autoimmunity, *J. Autoimmun.* **3**:137.
8. Ugen, K. E., Wang, B., Ayyavoo, V., Agadjanyan, M., Boyer, J., Li, F., Kudchodkar, S., Lin, J., Merva, M., Fernandes, L., Williams, W. V., and Weiner, D. B., 1994, DNA inoculation as a novel vaccination method against human retroviruses with rheumatic disease associations. *Immunol. Res.* **13**:154.
9. Doolittle, R. F., Feng, D. F., McClure, M. A., and Johnson, M. S., 1994, Retrovirus phylogeny and evolution, *Curr. Top. Microbiol. Immunol.* **157**:1.
10. Rasmussen, H. B., Perron, H., and Clausen, J., 1993, Do endogenous retroviruses have etiological implications in inflammatory and degenerative nervous system diseases? *Acta Neurol. Scand.* **88**:190.
11. Capon, D. J., and Ward, R. H., 1991, The CD4-gp120 interaction and AIDS pathogenesis, *Annu. Rev. Immunol.* **9**:649.
12. Gorny, M., Pinter, A., and Zolla-Pazner, S., 1972, Specific immunity to HIV and other retroviral infections, in: *Progress in AIDS Pathology* (H. Rotterdam, ed.), Field and Wood Publications, New York.
13. Arya, S. K., and Gallo, R. C., 1986, Three novel genes of human T-lymphotropic virus type III: Immune reactivity of their products with sera from acquired immune deficiency syndrome patients, *Proc. Natl. Acad. Sci. U.S.A.* **83**:2209.
14. Salfeld, J., Gottlinger, H., Sia, R., Park, R., Sodroski, J., and Haseltine, W., 1990, A tripartite HIV-1 tat-env-rev fusion protein, *EMBO J.* **9**:965.
15. Greene, W. C., Ballard, D. W., Bohnlein, E., Rimsky, L. T., Hanly, S. M., Kim, J. H., Malim, M. H., and Cullen, B. R., 1990, The trans-regulatory proteins of HTLV-I: Analysis of tax and rex, in: *Human Retrovirology* (W. A. Blattner, ed.), Raven Press, New York, pp. 35–43.
16. Crawford, T. B., Adams, D. S., Sandle, R. D., Gorham, J. R., and Henson, J. B., 1980, The

connective tissue component of the caprine arthritis-encephalitis syndrome, *Am. J. Pathol.* **100**:443.

17. Cheevers, W. P., Knowles, D. P. J., and Norton, L. K., 1991, Neutralization-resistant antigenic variants of caprine arthritis-encephalitis lentivirus associated with progressive arthritis, *J. Infect. Dis.* **164**:679.

18. Taneguichi, A., Takenaka, Y., Noda, Y., Ueno, Y., Shichikawa, K., Sato, K., Miyasaka, N., and Nishioka, K., 1988, Adult T cell leukemia presenting with chronic proliferative synovitis, *Arthritis Rheum.* **31**:1076.

19. Sato, K., Maruyama, I., Maruyama, Y., Kitajima, Y., Higaki, M., Yamamoto, K., Miyasaka, N., Osame, M., and Nishioka, K., 1991, Arthritis in patients infected with human T lymphocytic virus type I. Clinical and immunopathologic features, *Arthritis Rheum.* **34**:714.

20. Nishioka, K., Maruyama, I., Sato, K., Kitajima, I., Nakajima, Y., and Osame, M., 1989, Chronic inflammatory arthropathy associated with HTLV-I (Letter), *Lancet* **1**:441.

21. Iwakura, Y., T.su, M., Yoshida, E., Takiguchi, M., Sato, K., Kitajima, I., Nishioka, K., Yamamoto, K., Takeda, T., Hatanaka, M., Yamamoto, H., and Sekiguchi, T., 1991, Induction of inflammatory arthropathy resembling rheumatoid arthritis in mice transgenic for HTLV I, *Science* **253**:1026–1028.

22. Yamamoto, H., Sekiguchi, T., Itagaki, K., Saijo, S., and Iwakura, Y., 1993, Inflammatory polyarthritis in mice transgenic for human T cell leukemia virus type I, *Arthritis Rheum.* **36**:1612.

23. Ballard, D. W., Bohnlein, E., Lowenthal, J. W., Wano, Y., Franza, B. R., and Greene, W. C., 1988, HTLV-I tax induces cellular proteins that activate the kappa B element in the interleukin 2 receptor alpha gene, *Science* **241**:1652.

24. Wano, Y., Feinberg, M., Hosking, J. B., Bogerd, H., and Greene, W. C., 1988, Stable expression of the tax gene of type I human T-cell leukemia virus in human T cells activates specific cellular genes involved in growth, *Proc. Natl. Acad. Sci. U.S.A.* **85**:9733.

25. Kuroda, Y., Fukuoka, M., Endo, C., Matsui, M., Kurohara, K., Kakigi, R., and Tokunaga, O., 1993, Occurrence of primary biliary cirrhosis, CREST syndrome and Sjögren's syndrome and Sjögren's syndrome in a patient with HTLV-I associated myelopathy, *J. Neurol. Sci.* **116**:47.

26. Matsumoto, Y., Muramatsu, M. O., and Sato, K., 1993, Mixed connective tissue disease and Sjögren's syndrome, accompanied by HTLV-I infection, *Intern. Med.* **32**:261.

27. Kanazawa, H., Ijichi, S., Eiraku, N., Igakura, T., Higuchi, I., Nakagawa, M., Kuriyama, M., Tanaka, S., and Osame, M., 1993, Bechet's disease and Sjögren's syndrome in a patient with HTLV-I associated myelopathy, *J. Neurol. Sci.* **119**:121.

28. Mochizuki, M., Yamaguichi, K., Takatsuki, K., Watanabe, T., Mori, S., and Tajima, K., 1992, HTLV-I and uveitis, *Lancet* **339**:1110.

29. Osame, M., Usuku, K., Izumo, S., Ijichi, N., Amitani, H., Igata, A., Matsumoto, M., and Tara, M., 1986, HTLV-I associated myelopathy: A new clinical entity, *Lancet* **1**:1031.

30. Nishikai, M., and Soat, A., 1991, Human T lymphotropic virus type I and polymyositis and dermatomyositis in Japan, *Arthritis Rheum.* **34**:791.

31. Winchester, R., Bernstein, D. H., Fischer, H. D., Enlow, R., and Solomon, G., 1987, The co-occurrence of Reiter's syndrome and acquired immunodeficiency, *Ann. Intern. Med.* **106**:19.

32. Espinoza, L. R., Aguilar, J. L., Berman, A., Gutierrez, F., Vasey, F. B., and Germain, B. F., 1989, Rheumatic manifestations associated with human immunodeficiency virus infection, *Arthritis Rheum.* **32**:1615.

33. Ziegler, J. L., and Stites, D. P., 1986, Hypothesis: AIDS is an autoimmune disease directed

at the immune system and triggered by a lymphocytic retrovirus, *Clin. Immunol. Immunopathol.* **41**:305.

34. Morrow, W. J., Isenberg, D. A., Sobol, R. E., Stricker, R. B., and Kieber-Emmons, T., 1991, AIDS virus infection and autoimmunity: A perspective of the clinical, immunological, and molecular origins of the alloallergic pathologies associated with HIV disease, *Clin. Immunol. Immunopathol.* **58**:163.

35. Zarling, J. M., Ledbetter, J. A., Sias, J., Fultz, P., Eichberg, J., Gjerset, G., and Moran, P. A., 1990, HIV-infected humans, but not chimpanzee, have circulating cytotoxic T lymphocyte that lyse uninfected CD4+ cells, *J. Immunol.* **144**:2992.

36. Ascher, M. S., and Sheppard, H. W., 1988, AIDS as immune system activation: A model for pathogenesis, *Clin. Exp. Immunol.* **73**:165.

37. Imberti, L., Sottini, A., Bettinardi, A., Puotti, M., and Primi, D., 1991, Selective depletion of HIV infection of T cells that bear specific T cell receptor V beta sequences, *Science* **254**:860–862.

38. Viscidi, R. P., Mayur, K., Lederman, H. M., and Frakel, A. D., 1989, Inhibition of antigen-induced lymphocyte proliferation by tat protein from HIV-1, *Science* **246**:1606–1608.

39. Weinhold, K. J., Lyerly, H. K., Stanley, S. D., Austin, A. A., Matthews, T. J., and Bolognesi, D. P., 1989, HIV-1 gp120-mediated immune suppression and lymphocyte destruction in the absence of viral infection, *J. Immunol.* **142**:3091.

40. Groux, H., Torpier, G., Monte, D., Moulton, Y., Capron, A., and Ameisen, J. C., 1992, Activation-induced death by apoptosis in CD4+ T cells from human immunodeficiency virus–infected asymptomatic individuals, *J. Exp. Med.* **175**:331.

41. Stransky, G., Vernon, J., Aicher, W. K., Moreland, L. W., Gay, R. E., and Gay, S., 1993, Virus-like particles in synovial fluids from patients with rheumatoid arthritis, *Br. J. Rheumatol.* **32**:1044.

42. Schumacher, H. R., 1975, Synovial membrane and fluid morphologic alterations in early rheumatoid arthritis: Microvascular alterations and virus-like particles, *Ann. N.Y. Acad. Sci.* **256**:39.

43. Slaughter, L., Carson, D. A., and Jensen, F., 1978, In vivo effects of Epstein Barr virus on peripheral blood mononuclear cells from patients with rheumatoid arthritis and normal subjects, *J. Exp. Med.* **148**:1429.

44. Sczakiel, G., Oppendander, M., Rittner, K., and Pawlita, M., 1992, Tat-and-Rev directed antisense RNA expression inhibits and abolishes replication of human immunodeficiency virus type I: A temporal analysis, *J. Virol.* **9**:5576.

45. von Ruden, T., and Gilboa, E., 1989, Inhibition of human T cell leukemia virus type I replication in primary human T cells that express antisense RNA, *J. Virol.* **63**:677.

46. Marasco, W. A., Haseltine, W. A., and Chen, S. Y., 1993, Design, intracellular expression, and activity of a human anti-human immunodeficiency virus type I gp120 single-chain antibody, *Proc. Natl. Acad. Sci. U.S.A.* **90**:7889.

11

Autoimmunity in Chagas Disease

ANTONIO R. L. TEIXEIRA, CARLOS M. RIPOLL, and CHARLES A. SANTOS-BUCH

1. INTRODUCTION

Two conflicting theories have interpreted the relationship between *Trypanosoma cruzi* infections and Chagas disease. On the one hand, there is the concept that the disease is a direct consequence of the protozoan infection. This theory originated from the early descriptions of chagasic myocarditis and the gastrointestinal syndromes associated with megaesophagus and/ or megacolon, whose clinical manifestations characterize some of the chronic infections. Despite the early descriptions showing a paucity of parasite forms in the chronic lesions of human and experimental Chagas disease,[1-3] the notion that its pathogenesis is associated with the release of metacyclic trypomastigotes from the rupture of parasitized cells followed by a reactive localized inflammatory response is still strongly favored by some.[4] However, when compared to the cells in the heart rejection process, there is a persistent lower number of CD4+ cells in chronic

ANTONIO R. L. TEIXEIRA • Laboratory of Multidisciplinary Research on Chagas Disease, Department of Pathology, Faculty of Health Sciences, University of Brasilia, Brasilia, 70-919-970 Brazil.   CARLOS M. RIPOLL • Department of Regional Pathology, Laboratory of Public Health, San Salvador de Jujuy, Jujuy, 4600 Argentina.   CHARLES A. SANTOS-BUCH • Department of Pathology, Cornell University Medical College, New York, New York 10021.

Microorganisms and Autoimmune Diseases, edited by Herman Friedman *et al.* Plenum Press, New York, 1996.

Chagas myocarditis, suggestive of an immunological imbalance.[4] Recently Jones *et al.* demonstrated *T. cruzi* genomic DNA in the inflammatory exudate of chagasic hearts by the polymerizing chain reaction technique.[5] These findings have given some credence to the criticisms leveled a decade ago[6] against the data which supported the autoimmune basis of Chagas heart disease.[7-11] On the other hand, there is strong evidence that the disease stems from a much more complex immunobiological mechanism, partly resulting from antigen-driven interactions. These mechanisms defy narrow and somewhat archaic definitions of autoimmunity.

In favor of this broader concept are several observations made in nature: 1) Over one hundred species of wild mammals serve as reservoirs of *T. cruzi* and do not develop Chagas disease as a result of the infection. The parasite circulates among these different species and they carry the infection to the Triatomine insect vector, and, thus, American trypanosomiasis perpetuates in nature; 2) The clinical manifestations of Chagas disease are usually seen more often among the descendants of Europeans and Africans and in their domestic animals introduced in post-Columbian America[12] than in the Amerindian population; 3) The acute infection in infants and children goes unrecognized and subsides spontaneously; that is, the *T. cruzi*-infected child usually does not become sick at a time when tissue parasitism is fairly high but may develop striking chagasic lesions 25 to 30 years later when the parasite is difficult to find; 4) Superinfections do not appear to accelerate the development of chronic Chagas disease lesions. None of these natural situations can be explained by the conventional theory which maintains that parasitism and the inflammatory response to the parasite are fundamental to the development of the lesions found in Chagas disease.

Since "natural experiments" cannot be reasonably explained by conventional theory, we have chosen to describe in the following pages the broad diversity of immunobiological responses in *T. cruzi*—infected, genetically susceptible mammalian populations. These observations have shed light on those features related to genetically controlled immunological phenomena. In fact, the data accumulated in the literature over the last two decades show that autoimmunity indeed plays an important role in the pathogenesis of Chagas disease.

2. UNCOMMON MOLECULAR FEATURES OF *TRYPANOSOMA CRUZI*

Trypanosoma cruzi is a flagellated protozoan of the order Kinetoplastid. The single kinetoplast is known to carry out all known functions characterized in mitochondria. The extranuclear DNA in the kinetoplast is suspected to be of bacterial origin.[13] It has been argued that bacteria serve as vectors for horizontal transmission of genes between eukaryotes.[14] kDNA is a disk-like structure within the single kinetoplast matrix which makes up 15–25% of the total parasite DNA. Intact kDNA contains about 10,000 minicircles and 50 maxicircles catenated in a large network.[15] Maxicircles contain genes for mitochondrial proteins and RNA, whereas minicircles are associated with RNA editing.[16] The heterogeneity of minicircles is apparently required to provide sufficient genetic information for extensive editing.[17,18] It has also been shown that division and recombination contributes to minicircle heterogeneity.[19]

Isoenzyme analysis has shown a broad genetic diversity among different *T. cruzi* isolates which have been classified into three distinct zymodemic groups.[20] Despite the fact that some features of biological behavior appear to have an association with enzymatic profiles, a definitive link is lacking.[21,22] In this regard, it has not been possible to tie a particular zymodeme with a specific clinical form of Chagas disease. It has been shown, however, that the parasite isolates from Venezuela belong to zymodemes 1 and 3, whereas those of Brazil belong to zymodeme 2. Epidemiological and clinical data appear to show that chagasic cardiomyopathies predominate in Venezuelan patients in contrast to the more diverse clinical manifestations of cardiomyopathies and megaesophagus and megacolon in Brazil.[23] However, it has been shown that isolates from patients with acute infections show all three zymodeme groups, and, therefore, a strong link of zymodeme profiles with the clinical form of Chagas disease awaits further confirmation.[24] The absence of stable molecular and biochemical markers and prevalent broad intraspecific variations have hampered classification of isolates and clones of *T. cruzi* in studies which attempt to tie a particular strain or clone to clinical and pathological manifestations of chronic Chagas disease.

Polymorphism of *T. cruzi* isolates have been studied by pulse field gel electrophoresis (PFGE), restriction fragments length polymorphism (RFLP), and random amplified polymorphic DNA (RAPD). The protozoan molecular karyotype shows at least 20 chromosomes ranging from 300 to 2000 kilobases.[25,26] Studies using RFLP have shown enormous heterogeneity among isolates and clones. The profound plasticity in *T. cruzi* genome organization shows that a single gene can be present in chromosomes of different sizes. The restriction map sites corroborate the hypothesis that *T. cruzi* is diploid with different homozygous and heterozygous alleles present in different isolates and clones.[27]

The infective trypomastigote stage of *T. cruzi* is the only form capable of entering mammalian host cells. The trypomastigote evades antibody-directed immune destruction, and this property is attributed to its ability to inhibit complement-mediated lysis.[28] Experimental observations have shown that *T. cruzi* parasitemia is often followed by myotropic infection of type I striated muscle fibers, like those prevalent in the heart.[29] Recent investigation has shown that myotropic host cell recognition is driven by the chemoaffinity of the complementary structures of *T. cruzi* attachment molecules to surface muscarinic cholinergic and beta adrenergic receptors of the surface of the host cell.[30,31] The host membrane interaction of the parasite attachment molecules trigger a signal to a protein Gi–regulated adenylyl cyclase complex of the plasma membrane, which results in the attenuation of parasite cAMP cytoplasmic levels in a matter of seconds.[32] At sites of parasite entry recruitment of host cell lysosomes may occur and a parasitophorous vacuole may be formed. Escape from the vacuole depends on the coordinated action of parasite-derived plasma membrane neuraminidase/trans-sialidase and a parasite-derived transmembrane pore-forming protein in acid medium.[33] The trypomastigote form has stage-regulated surface ATP receptors with very high molecular affinities capable of binding the low concentrations of free ATP within the host cell cytoplasm, and exogenous ATP is enzymatically transported and rapidly utilized by the activation of a cascade of parasite calcium-dependent, staurosporine-sensitive phosphorylations. It is believed that ATP receptors play an important role in the transport of the exogenous nucleotide (J. Inverso, Y. Song, and C. A. Santos-Buch, unpublished observations).

3. THE PARADOX OF ANTIGEN-DRIVEN DELAYED-TYPE HYPERSENSITIVITY

In view of the experiments of nature described above, which appear to show that morbidity in Chagas disease is not directly related to the presence of the parasite in blood and tissues of the mammalian host, we decided to standardize the rabbit model of human Chagas disease. We considered that the severe tissue damage that accompanies Chagas myocarditis is probably related to delayed-type hypersensitivity.[8] The rabbit was chosen for these experiments since several diseases associated with delayed hypersensitivity, such as tuberculosis, syphilis, and rheumatic fever, have been studied in this animal model. The animals survived acute *T. cruzi* infection but they died of chronic Chagas disease with the same features seen in the human disease.[34] This observation led to other experiments in which we evaluated the benefits of treating *T. cruzi* infections with the trypanocidal nitroarene benznidazole.[35] We have shown that the administration of benznidazole in acute and chronic *T. cruzi* infections of rabbits does not stop the progressively destructive myocarditis, the hallmark of Chagas disease, although the parasite was no longer demonstrable in the body of the treated animals. What could be sustaining such an active, self-destructive host immune response in benznidazole-treated rabbits? We postulated that *T. cruzi* DNA may have been retained in the host cells after the overt infection was ended by treatment with the trypanocidal drug.

Although current notions of biology indicate that natural barriers would not allow transkingdom gene transfer, it has been shown that horizontal DNA conjugation is common among prokaryotes.[36] Further, Heinemann and Sprague[37] have shown that conjugative plasmids of *Escherichia coli* can mobilize DNA transmission from this bacterium to the yeast *Saccharomyces cerevisiae*. In fact, bacterium-yeast conjugation is not the only example of prokaryote-eukaryote gene transfer. Bakkeren *et al.*[38] have shown direct DNA transfer from the soil-borne phytopathogen *Agrobacterium tumefaciens* into plant cells. During this cell-cell interaction, a specific transfer DNA (tDNA) is mobilized from a donor bacterium to the plant cell nucleus and stabilized by chromosomal integration. Expression of tDNA genes in the plant cell leads to the transformation and the production of specialized nutrients.[14] Furthermore, in certain cases of virus integration in human and bird liver cells, insertional activation of genes involved in specific cell functions have been observed.[38-41] More recently, the horizontal transfer of *Drosophila* genes by the mite *Proctolaelaps regalis* has been recorded.[42] If ectoparasites can accomplish interspecies gene transfer, there seemed no reason why the endoparasite *T. cruzi* should not accomplish a similar transfection.

Investigation into the possible transfection of *T. cruzi* DNA into the host genome was begun with macrophages because these phagocytic cells are readily parasitized.[43] Furthermore, macrophages play a central role in the host immune response as antigen-presenting cells to helper (CD4+) T lymphocytes associated with delayed hypersensitivity and rejection of syngeneic heart tissues.[9,10,44,45] Cytogenetic analysis of metaphase chromosomes obtained from peritoneal macrophages of *T. cruzi*-infected Balb/c mice showed an association between parasite chromatin and macrophage chromosomes. Free amastigote DNA was identified among the chromosomes, and these associations occurred at all stages of the experimental infection. *T. cruzi* chromatin was observed in macrophage chromosome spreads obtained at different periods of the infection (Fig. 1).[43] These associations were seen in several metaphase plates obtained

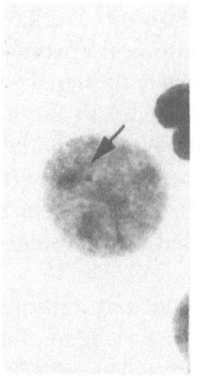 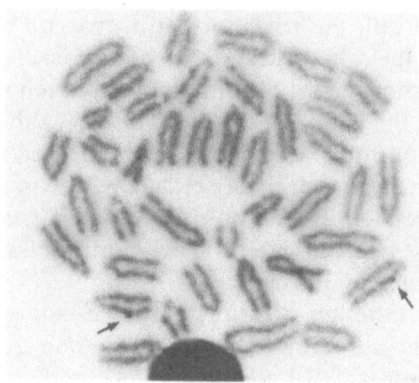

FIGURE 1. Autoradiograph of a peritoneal macrophage chromosome in metaphase of a Balb/c mouse infected with 10⁵ ³H-thymidine-labeled trypomastigotes of the Berenice strain. The arrows show silver granules of *T. cruzi* ³H-DNA inserted in the chromosome (Giemsa, 1000×).

from *T. cruzi*–infected mice and were never observed with mock-infected control mice.[43]

Analysis of autoradiograph spreads of peritoneal macrophages collected from ³H-DNA-labeled *T. cruzi*–infected mice showed ³H-DNA silver grains in macrophage chromosomes, including metaphase chromosomes. G-banding and karyotyping of metaphase macrophages from *T. cruzi*–infected mice were used to determine whether the parasite DNA associated preferentially with a particular host chromosome. We observed that there was indeed a preferential association of the parasite DNA with chromosomes 3, 6, and 11. In sharp contrast, the cytogenetic analysis of metaphase macrophage spreads of noninfected control mice showed none of these preferences.

The presence of *T. cruzi* DNA in the host genome was also shown by *in situ* hybridization of *T. cruzi* DNA probes with macrophage chromosomes obtained from acute and chronic infections in mice.[46] Human U-937 transformed macrophages that had been infected with *T. cruzi* trypomastigotes and treated with trypanocidal doses of benznidazole were also used. Results indicated that the infection had been eradicated because the parasite was not recovered from cell cultures or from cultures of the supernatant medium after several passages, for as long as 4 months. In these experiments specific hybridization signals were obtained with biotinylated DNA probes (Fig. 2). In sharp contrast, there was a total absence

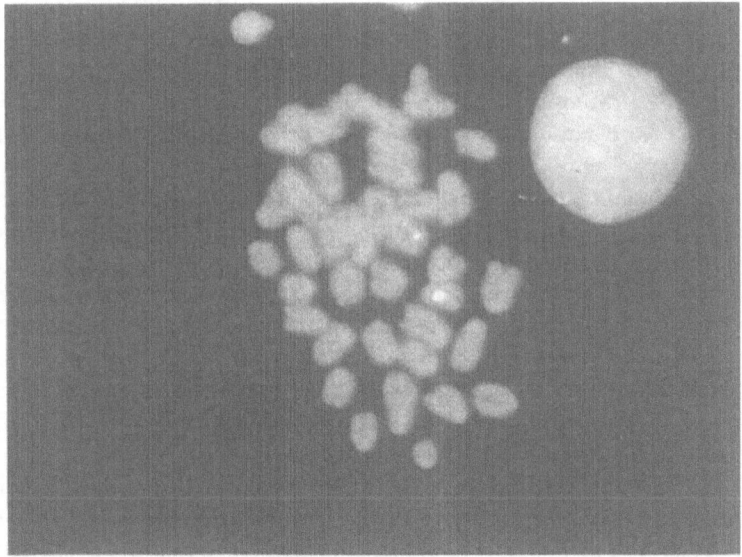

FIGURE 2. Localization of biotinylated *T. cruzi* DNA in chromosomes of murine macrophages at 11 days of inoculation (*in situ* hybridization).

of hybridization signals in host cells derived from noninfected control mice or noninfected U-937 macrophages.[46]

Southern blot analyses were done in order to investigate whether *T. cruzi* DNA was merely associated with the host genome or whether it had been integrated into the host chromosomes. Positive hybridization signals were obtained with genomic DNA extracted from *T. cruzi*–infected host cells. A 1.7-kb band was consistently observed when *T. cruzi*–infected macrophage or infected and treated macrophage DNA was hybridized with a kDNA probe. Furthermore, control macrophage genome showed none of these findings with the same experimental conditions. It was further shown that the genomic host DNA from *T. cruzi*–infected and treated L-929 fibroblasts did not cross hybridize to the cloned kDNA minicircle probe (kindly donated to us by Dr. Samuel Goldenber, Instituto Oswaldo Cruz, Brazil). The possibility may exist that genomic *T. cruzi* kDNA integration is fairly specific to antigen-processing cells such as macrophages and Kupfer cells of the liver and skin.

Restriction enzyme digest analysis of genomic DNA from *T. cruzi* and

DNA from parasite-infected macrophages showed the 1.7-kb band when *Eco* RI, *Bam* HI, and *Taq* I restriction fragments of host DNA from infected cells were hybridized with the kDNA minicircle probe. In sharp contrast, several bands, ranging from 0.8 to 3.6 kb were formed when *Alu* I and *Hinf* I digests of the same host DNA were hybridized with the same kDNA minicircle probe.

These results strongly indicated to us that kDNA minicircles from *T. cruzi* were integrated into the genome of host cell macrophages following infection.[46] These findings also suggested to us that the antigen-driven signals to sustain the progressive Chagas myocarditis seen in benz-nidazole-treated *T. cruzi*–infected rabbits may have been a result of the transfection of kDNA minicircles into host cell macrophages through complex and presently unknown steps. We suspect that similar mechanisms of *T. cruzi* antigen-driven disease may be operative in human infections with chronic Chagas myocarditis.

Even though kDNA minicircles may encode for kinetoplast proteins, a complex mechanism of guide RNA editing associated with kDNA has been described in *T. cruzi*.[16–18] We suggest therefore that the preferential integration of *T. cruzi* minicircle DNA into chromosomes 3, 6, and 11 of the host cell genome may alter the regulation of host gene expression by editing its RNA, perhaps by modifying ever so slightly critical cross-reactive epitopes which play a role in autoimmunity through expressed proteins of the host cell surface. Recently, a lytic monoclonal antibody to *T. cruzi* trypomastigotes that recognizes an epitope expressed in tissues affected in Chagas disease has been described.[47]

In this respect, it is important to emphasize that the gene encoding the β chain of the T-lymphocyte receptor is in locus B of chromosome 6. Interestingly, striated muscle α-actin and heavy chain myosin genes are also present in chromosomes 3 and 11, respectively.[48] It is noteworthy that some autoimmune lesions in Chagas disease appear to be related to T-lymphocyte cytotoxicity against heart and skeletal muscle myosin (see Section 6).[9,11]

4. THE PARADOX OF THE SEVERE FORMS OF ACUTE CHAGAS DISEASE IN SANTIAGO DEL ESTERO INFANTS

Santiago del Estero province is in north-central Argentina. It has a tropical, dry climate and is populated by 650,000 people. Two thirds of the population live in rural areas in "ranchos," which are infested by *Triatoma infectans* (84.2%), and nearly two thirds of the vectors carry *T.*

cruzi.[49,50] The vast majority of the rural-based, asymptomatic pregnant women are seropositive for *T. cruzi* infection.

In endemic areas of Santiago del Estero, it is estimated that 3–5% of all symptomatic acute infections develop severe clinical forms with dramatic atypical effects.[49,50] Infants with the life-threatening, severe forms of acute Chagas disease are a medical emergency and require prompt and effective treatment by experienced clinicians.

In this particular group of very sick, infected infants there is at times no apparent portal of entry. In a recent study, *cutaneous inoculation chagomas*, which are clearly related to a profound but localized lesion of delayed-type hypersensitivity in the probable portal of entry, were found in 17.2% of infants 1 year old or less (mean age = 6.2 months), in 5.5% of infants 1 to 4 years old (mean age = 2.4 years), and in 6.7% of children 5 to 9 years (mean age = 6.7 years); none were observed in infected children older than 10 years.[51] Similarly, *lipochagomas*, which histologically show marked fat necrosis and profound delayed-type hypersensitivity and clinically are considered pathognomonic of Chagas disease and are found in Bichat's space, deep in the cheeks of the face, and more rarely in the subcutaneous tissue of the trunk and extremities, are almost exclusively found in infants who are less than 1 year of age.[51] *Hematogenous, metastatic chagomas*, which histologically show blood vessel and capillary fibrinoid necrosis and a profound lymphoid inflammatory reaction, were also found predominantly in the group of infants 1 year old or less (Fig. 3).[51] This observation confirmed the findings of earlier clinical studies[52-56] which showed that metastatic chagomas are almost exclusively found in breast-fed infants. Interestingly, congestive heart failure was also associated with these severe cutaneous and vascular lesions in a small proportion of the infants 1 year old or less.[51] The ultrastructural changes of microangiopathy and myocardial lesions have been described recently in experimental acute Chagas disease.[57]

Why should breast-fed infants of chronically infected asymptomatic mothers develop these life-threatening, antigen-driven, systemic lesions of rapid onset in the postnatal period, which is normally characterized by maternal antibody immune protection from other infections? Why should the risk of developing these severe forms of acute Chagas disease in *exactly the same endemic zones* greatly diminish with age as circulating maternal antibody titers normally drop?

The answer may lie in the nature of the idiotype-antiidiotype network and in the expression of surface cross-reacting epitopes with molecular mimicry in antiidiotypes. Data show that titers of IgGαF(ab')$_2$α *T. cruzi* surface epitopes are significantly increased in patients with active chronic

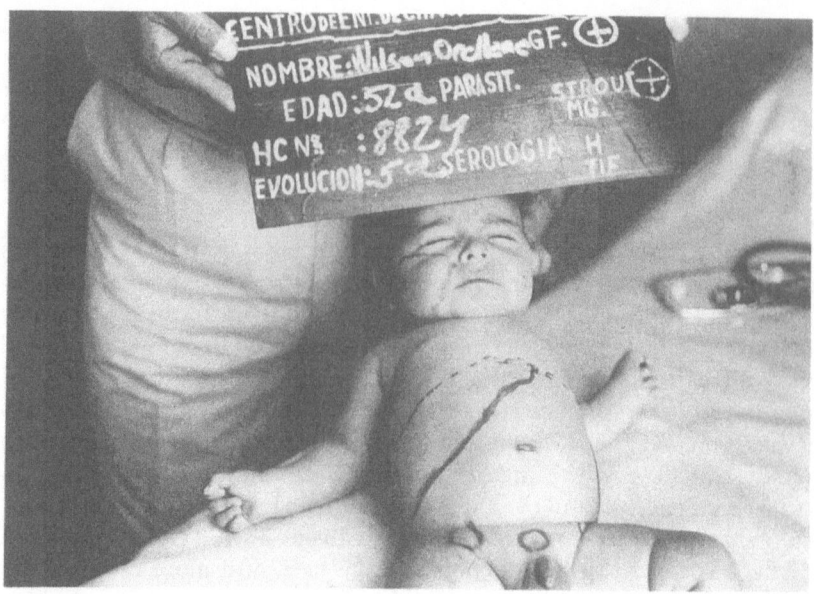

FIGURE 3. *Metastatic chagomas* in a 52-day-old Santiago del Estero baby with the severe form of acute Chagas disease. Note the *inoculation subcutaneous chagoma* of the right cheek, the portal of entry of the primary infection, 5 days before photography. The outline of the enlarged liver is outlined with a marker. Trypomastigotes were detected in the peripheral blood by the method of Stroud. (Photograph kindly supplied by Dr. Oscar Ledesma Patino, Centro de Chagas y Patologia Regional de Santiago del Estero, Argentina.)

Chagas myocarditis[58] and not in asymptomatic chronically infected populations. Interestingly, these antiidiotypic antibodies also showed antimyocardial cell membrane activity in the group of patients with Chagas heart disease.[58] In this regard, it has been shown that peripheral blood T lymphocytes from patients with the cardiac form of Chagas disease are stimulated to proliferate by monospecific antiflagellate antibody, raising the possibility that these patients have autoantiidiotypic T cells.[59] We have recently shown that inbred mice develop anti–muscle cell T-lymphocyte cytotoxicity when immunized with killed Lyt2+ lymphocytes derived from *T. cruzi* plasma membrane–immunized syngeneic mice.[60]

These data led us to conclude that primary anti–*T. cruzi* Lyt2+ cells express host cell–like surface epitopes by molecular mimicry and that primary IgGα *T. cruzi* also expressed similar mimes.[60] The antiidiotypic response, therefore, may result in autoreactive, anti–host cell antibody and immune cell-mediated activity.

An explanation of the severe delayed-type hypersensitivity lesions seen in breast-fed infants may be found in anti– *T. cruzi* idiotypic immune reactions (Fig. 4). The placental passage of maternal anti– *T. cruzi* idiotypic antibody and, in special cases, the placental passage of anti– *T. cruzi* idiotypic lymphocytes may prime the infant's antigen-processing cells of the skin, subcutaneous tissues, and lymphatic-rich fibrodipose tissues of the face. Subsequent vectorial *T. cruzi* inoculation may result in the abrupt development of severe, delayed hypersensitivity lesions in sites of host cell parasitism like those found in the subpopulation of breast-fed infants of Santiago del Estero. As the maternal antibody pool is metabolized in the postnatal period of 7–9 months, the pool of autoantigens expressed in the maternal antiidiotypic antibodies and antiidiotypic T cells reduces the antigen-driven reactions of the infant's immune system and the risk of developing the severe forms of acute Chagas disease.

A similar pathogenesis may be operative in the ophthalmic-dacro-adenitis-ganglion complex formed when the conjunctiva or the face is the portal of entry of *T. cruzi* (Romana's sign; Fig. 5).

It is unlikely that maternal passage of freely swimming *T. cruzi* trypomastigotes happens with sufficient frequency to prime the infant's tissue antigen-processing cells because congenital Chagas disease is rather rare (about 4% of asymptomatic infected mothers),[61,62] and usually congenital Chagas disease babies with parasitemia have a *benign* clinical course and do not develop the severe forms of delayed-type hypersensitivity lesions seen in breast-fed infants at a later age (6 months).[61,62] More relevant to

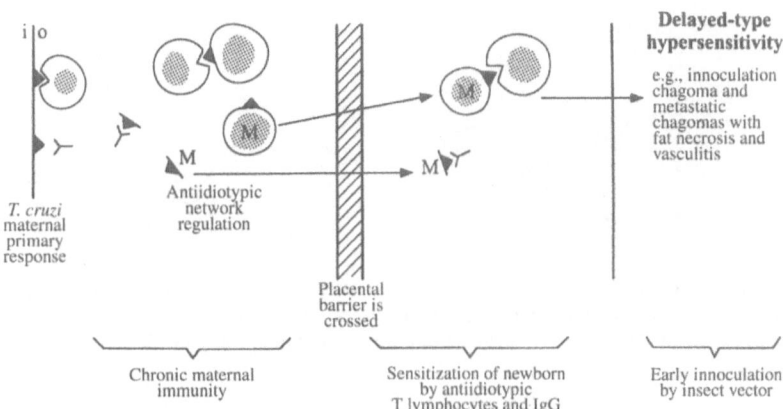

FIGURE 4. Idiotype-antiidiotype transplacental sensitization.

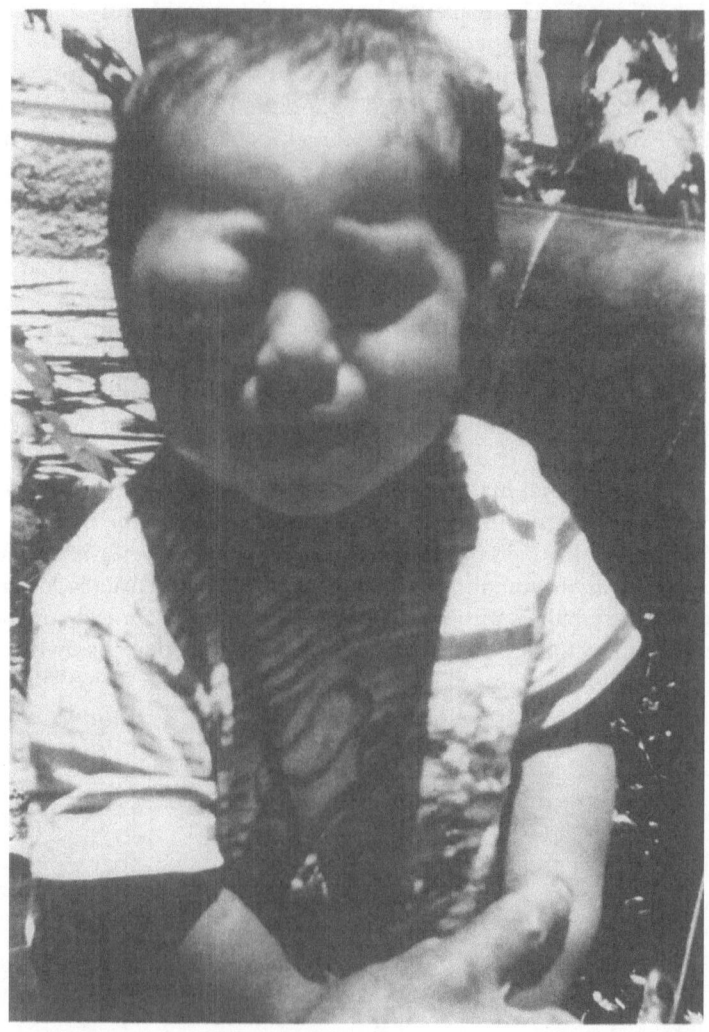

FIGURE 5. *Ophthalmic-dacroadenitis-ganglion complex* (Romana's sign) in infant approximately 2 years of age of the Departamento de Potosi, Bolivia. (Photograph kindly supplied by Dr. Carlos M. Ripoll and Dr. Justo Chungora of the Public Health Unit of Tupiza-Potosi, Bolivia.)

the issue is the distinct possibility that specific *T. cruzi* strains infecting the infants in great numbers (superinfection) play a role in the induction of the severe forms of acute Chagas disease in susceptible breast-fed infants.

The role that *T. cruzi* superinfections play in lesion induction was tested in experimental rabbits using six *T. cruzi* clones derived from the cardiac disease isolate SLU-239 and from the megacolon syndrome isolate SLU-142.[63] Discrete allelic differences were shown by isoenzyme and RFLP profiles (data not shown). Superinfections were created by massive systemic reinoculations (1×10^6 trypomastigotes), 10 months after a similar primary infection. Delayed-type skin reactions to homologous *T. cruzi*–soluble antigen were assayed and the weakest reactions were seen in superinfected rabbits with long-lasting parasitemia. Of great interest was the observation that the superinfected rabbits with long-lasting parasitemia and negative delayed-type skin hypersensitivity reactions showed a significantly lower incidence of characteristic Chagas heart disease electrocardiograph alteration[63,64] than rabbits that had been infected with a single inoculum of the same *T. cruzi* clone and had a short-lasting parasitemia in the course of the two-year study. The data derived from this experimental study appear to suggest that superinfections increased neither morbidity nor mortality in rabbits. In fact, morbidity and mortality were significantly higher in rabbits that had received a single *T. cruzi* inoculum. In these experiments, regardless of the *T. cruzi* clone used [SLU-239, (cardiac disease isolate) or SLU-142 (megacolon disease isolate)], only chronic Chagas myocarditis was observed at autopsy.

5. DOWNREGULATION OF THE IMMUNE RESPONSE

Polyclonal activation and suppression of the immune system in *T. cruzi*–infected experimental animals have been studied extensively.[65,66] CD5+ B lymphocytes appear to be associated with polyclonal activation.[67] These cells, which produce interleukin-10 (IL-10), appear to activate Th1 lymphocytes, which, in turn, impair IL-2 and interferon-γ production and decrease Th2-lymphocyte activity. Polyclonal activation of CD4αβ, CD4αβ, and CD8γδ T cells has been reported.[68] The role these T-lymphocyte subpopulations play in the pathogenesis of chagasic lesions is not clear.

Experimental observations suggest that cytotoxic T lymphocytes are the most prevalent cell type in the heart inflammatory cells of *T. cruzi*–infected mice.[69] Mice depleted of cytotoxic T lymphocytes with an otherwise intact T-cell repertoire showed parasitemias similar those seen in mice with a complete γδ repertoire.[68] The mortality of the former animals, however, was lower presumably because myocarditis was not found at autopsy. These data would suggest that γδ T cells may not control parasite growth but may produce tissue damage.[68]

The role that T lymphocytes play in the pathogenesis of chagasic lesions has been shown by passive transfer experiments.[10,70] In these adoptive transfer experiments, CD4+ immune lymphocytes were used to obtain lesions in naive recipients. Induction of cardiac lesions have also been induced by passive transfer of CD8+ T lymphocytes to naive syngeneic recipients.[69] Despite the fact that CD8+ T lymphocytes are abundant in Chagas heart lesions,[4] it has not been possible to clone and expand this cell population,[69] and, therefore, its stimulation with autoantigens remains unexplored.

The interaction of *T. cruzi* plasma membrane attachment molecules with naive lymphocytes transduced second messenger signals in T helper and T suppressor lymphocytes.[71] The attachment of *T. cruzi* plasma membranes onto helper cells triggered signals that resulted in increased lymphocyte cAMP levels and decreased cGMP levels. The assays showed that β-adrenoreceptors mediated the stimulation of T helper cell adenyl cyclase. Furthermore, the attachment of *T. cruzi* plasma membranes onto T suppressor lymphocytes resulted in the transduction of signals which induced the secretion of the effector prostaglandin E_2 (a known suppressor), an increase in cGMP, and a decrease in cAMP levels by triggering suppressor T-lymphocyte muscarinic cholinergic receptors.[71] It can be argued that persistent parasitemia provides suppressor factors like PGE_2 by its action on T suppressor cell muscarinic cholinergic receptors or by parasite interactions with the β-adrenergic receptors of T helper lymphocytes which trigger signals that result in the inhibition of T helper cell function.[71] This notion presupposes that there is a background of nonspecific decreased dysfunctional immune regulation in subjects with *T. cruzi* parasitemia.

It is not clear, however, why immune suppression appears to be specific for delayed-type hypersensitivity reactions in *T. cruzi*–infected animals or why there is immune dysregulation in asymptomatic chronically infected patients with a low parasite load and transient, evanescent parasitemia.

An alternative explanation is that *failure to trigger normal homeostatic immune suppressor mechanisms* could be the primary defect in *T. cruzi* infections.[9,72,73] There are data to support the notion that a specific breakdown of tolerance to myosin is critical in the development of autoimmune myocarditis in chagasic mice.[9] The exact mechanisms of how this breakdown in tolerance to myosin in *T. cruzi*–infected mice comes about are now being actively investigated.

6. AUTOREACTIVE ORGAN-SPECIFIC DELAYED-TYPE HYPERSENSITIVITY AND AUTOANTIBODIES

The general features of Chagas disease led us to postulate that autoimmunity may be involved in lesion production.[7] In a series of experiments we showed that immune lymphocyte reactivity against parasitized and nonparasitized rabbit heart cells was the basis of organ-specific injury in Chagas disease.[3,7] Independent and simultaneous publications by Cossio and his group described the presence of heterophile antibodies in sera of chronic Chagas disease patients reactive against endocardium, sarcolemma of myofibers, and the interstitium.[74] In the following years Cossio et al.[75] and Teixeira et al.[8] showed positive inhibition migration indices of *T. cruzi* immune peripheral blood mononuclear cells from chagasic patients in the presence of heart cell antigens, thus confirming previous experimental data.[3,7] We further showed that immunization of animals representative of the evolutionary scale developed antimuscle antibody activity when immunized with actomyosin-extracted membranes of *T. cruzi* flagellates.[76] Further experiments also showed that monospecific polyclonal antibody against sarcolemmal myocardial calcium ATPase cross-reacted with epitopes of the flagellate plasma membrane and that epitopes of the calcium transport enzyme were present in the sarcolemma membrane of heart muscle cells,[77] possibly indicating molecular mimicry of mammalian cell proteins in expressed parasite epitopes.

Very little is known of basic mechanisms associated with the pathogenesis of the "mega" conditions because scientific research in this field has been neglected. Over a decade ago it was reported[78] that neuronal and *T. cruzi* epitopes could be recognized by a monoclonal antibody generated against mammalian dorsal root ganglia, which may be important to the study of the pathogenesis of this disorder. This observation may explain, for instance, the reported affinity and adhesion of *T. cruzi*–immunized lymphocytes to parasympathetic neurons.[79] The possible involvement of autoimmune mechanisms in the destruction of neuronal cells of the digestive tract should be considered based on the observation of lymphocytic adherence and lysis of ganglion cells when preparations of intestinal parasympathetic ganglia are incubated with mononuclear cells of *T. cruzi*–infected rabbits.[79] Recently, it has been reported that human chagasic IgG inhibited in a noncompetitive manner the binding of ^{3}H-quinuclidinyl benzilate to the cardiac membrane and that the IgG antibody behaved as a partial agonist of muscarinic cholinergic receptors.[80] The prevalence of the cholinergic antibody was higher in sera from *T. cruzi*–infected asymptomatic subjects with a demonstrable dysfunctional autonomic system[81] than in those without measurable autonomic nervous system alterations. The presence of these antibodies could explain the receptor blockade of the parasympathetic autonomic system sometimes seen in Chagas disease.[80]

In earlier experiments,[3,7,8] it was shown that reported effector cell cytotoxicity was not under human lymphocyte antigen (HLA) restriction, because the cytotoxic T lymphocytes that lysed target rabbit and human heart cells did not destroy liver and kidney cells. The reported *in vitro* destruction of target cell by immune lymphocytes paralleled the ultra-structural descriptions of cardiac cell lysis by lymphocytes which infil-trated the myocardial biopsies of patients with Chagas myocarditis.[74,82] In other experiments[44] direct cytotoxicity of immune lymphocytes from isogeneic chagasic III/J rabbits to heart cells was shown *in vitro*. The high degree of cytotoxicity observed in this target cell–effector cell interaction paralleled the strong delayed-type cutaneous reactivity elicited in iso-geneic chagasic III/J rabbits against a parasite antigen.[12,44]

Infection of mice with *T. cruzi* induced the expansion of a CD4+ heart-reactive T-lymphocyte population that is able to mediate the rejec-tion of syngeneic heart implants, and this data strongly suggests the induction of an autoreactive delayed-type hypersensitivity reaction in experimental infections of inbred mice.[10] Autoimmune myocarditis with demonstrable cytotoxicity to syngeneic heart cells was shown in infected mice with *T. cruzi* parasitemia, and a similar observation was made in mice repeatedly immunized with enriched preparations of *T. cruzi* plasma membranes.[83] In all of these experimental models, an antigen-driven, autoreactive, cell-mediated immune reaction was required, because no autoreactivity was observed when *T. cruzi* antigens were not provided in the absence of parasitemia and parasitosis or when multiple adjuvant immunizations with cross-reacting antigens were not used throughout the experimental protocol.

When the polymerase chain reaction was used to amplify each known T-cell receptor (TCR) Vα and Vβ subfamily-specific sequence in tran-scripts derived from heart samples obtained from patients with chronic Chagas myocarditis, the number of TCR Vα was significantly lower than the number of TCR Vβ.[84] In normal hearts, the diversity of Vα and Vβ was similar in the scanty T cell population. Such data of restricted TCR Vα region repertoire have been described in experimental and human auto-immune diseases.[84]

T lymphocyte–mediated cytotoxicity may not be the only mechanism engineering cell destruction in Chagas heart disease. Laguens *et al.*[70] showed that antiheart autoantibodies can mediate spleen cell cytotoxicity against syngeneic heart cells, and antiheart autoantibody was demonstrated in scattered B lymphocytes in the myocarditis of infected susceptible mice.[70] Antisera of *T. cruzi*–infected mice that had developed heart disease histologically similar to that of humans showed reactivity with three cardiac antigens, and two of these were identified as the contractile protein myosin and the intermediate filament protein desmin.[9,85] Antibodies cross-reactive with human and *T. cruzi* ribosomal proteins (23 kDa) have been characterized,[86] and the antibodies have a strong reactivity against the cloned C-terminal portions of four *T. cruzi* ribosomal proteins, TcP1, TcP2α (TcP2b), TcP2β (TcPJL5), and TcPo.[86] Anti-β-adrenergic receptor and anti–muscarinic cholinergic receptor IgG in patients with chronic Chagas disease are partial agonists and modulate heart and parasympathetic system function in these patients.[80,87]

To show organ-specific autoimmune reactions in Chagas heart disease unequivocally would require the demonstration of organ-specific autoantigens. In this regard Rizzo et al.[9] assessed the proliferative response of T lymphocytes from mice chronically infected with *T. cruzi* and showed that CD4+ immune lymphocytes proliferated in response to cardiac myosin but not actin. In order to determine whether cardiac antigen-specific autoantibodies are produced in *T. cruzi*–infected mice, Tibbetts et al.[11] infected mice with the pathogenic Brazil strain. Antisera from these mice were found to react with three main cardiac antigens. Two of these antigens were identified as the contractile protein myosin (200 kDa) and the intermediate filament protein desmin (5 kDa). These data suggest that *T. cruzi* infection in susceptible hosts may lead to the development of an organ-specific autoantigen-defined autoimmune disease. Rizzo et al.[9] and Tibbetts et al.[11] consider these findings to be of special interest because autoantibodies specific to contractile proteins have been described in other states of heart-specific autoimmunity resulting from Coxsackie B infections.

Very recently, Cunha Neto et al.[88] eluted T lymphocytes from endomyocardial biopsies of Chagas heart disease patients. These cells could be cultivated in the presence of PHA, 40 U/ml IL-2, and irradiated peripheral blood mononuclear cells. Three CD4+ T-cell clones reacted against *T. cruzi* soluble antigen and the recombinant CRA, FRA, B13, and JL5 antigens. Of interest, the T-cell clone E205 proliferated in the presence of the parasite recombinant antigens and *in the presence of cardiac myosin.* These investigators also showed that autoantibodies in sera of chagasic patients reacted against the *T. cruzi* 140/116-kDa proteins.[85] In addition, it was shown that the recombinant (MXT) protein reacted with anti–*T. cruzi* antibodies and with anti–cardiac myosin antibodies in the ELISA assay.

Homology between the MXT protein and the human cardiac myosin used was shown in two sites, each formed by six amino acids residues. Synthetic peptides derived from the homologous sequences inhibited the binding of the antibodies to the MXT protein (Cunha-Neto and J. Kalil, unpublished observations). This finding would suggest the presence of cross-reactive epitopes between the *T. cruzi* MXT protein and human cardiac myosin.

The results of these experiments suggest that cross-reacting molecular mimicry between the protozoan parasite and the target organ epitopes may initiate autoreactive humoral and cell-mediated, organ-specific immune responses.

ACKNOWLEDGMENTS. We are grateful to Dr. Humberto Lugones and Dr. Oscar Ledesma Patino of El Centro de Chagas y Patologia Regional de Santiago del Estero, Argentina, for their enormous contributions on the discussion of the severe forms of acute Chagas disease. We are greatly indebted to over 50 collaborators, students, and fellows who have added to our knowledge gained in Chagas disease research from 1970–1995. This work was supported by numerous private and public entities in Argentina, Brazil, and the United States.

REFERENCES

1. Viana, G. O., 1911, Contribuicao para o estudo da anatomia patolojica da Molestia de Carlos Chagas, *Mem. Inst. Oswaldo Cruz* **3**:276–294.
2. Torres, C. B. M., 1929, Patogenia de la miocarditis cronica en la enfermedad de Chagas, *Sociedad Argentina de Patologia Regional del Norte, Quinta Reunion* **2**:902–916.
3. Teixeira, A. R. L., and Santos-Buch, C. A., 1975, The immunology of experimental Chagas disease. IV. Production of lesions in rabbits similar to those of chronic Chagas disease in man, *Am. J. Pathol.* **80**:37–46.
4. Higuschi, M. L., Gutierrez, P. S. Aiello, V. D., Palomino, S., Bocchi, E., Kalil, J., Belloti, G., and Pileggi, F., 1993, Immunohistochemical characterization of infiltrating cells in human chronic chagasic myocarditis: Comparison with myocardial rejection process, *Virchows Arch. A* **423**:157–160.
5. Jones, E. M., Coley, D. G., Tostes, S., Lopes, E. R., Vnecak-Jones, C. L., and McCurley, T. L., 1993, Amplification of a *Trypanosoma cruzi* DNA sequence from inflammatory lesions in human chagasic cardiomyopathy, *Am. J. Trop. Med. Hyg.* **48**:348–357.
6. Kierszenbaum, F., 1986, Autoimmunity in Chagas disease, *Parasitol. Today* **1**:5–9.
7. Santos-Buch, C. A., and Teixeira, A. R. L., 1974, The immunology of experimental Chagas disease. III. Rejection of allogeneic heart cells in vitro, *J. Exp. Med.* **140**:38–53.
8. Teixeira, A. R. L., Teixeira, G., Macedo, V., and Prata, A., *Trypanosoma cruzi*–sensitized T lymphocyte ^{51}Cr release from human heart cells in Chagas disease, *Am. J. Trop. Med. Hyg.* **27**:1097–1107.

9. Rizzo, L. V., Cunha-Neto, E., and Teixeira, A. R. L., 1989, Autoimmunity in Chagas disease. Specific inhibition of reactivity of CD4+ T cells against myosin in mice chronically infected with *Trypanosoma cruzi*, *Infect. Immun.* **57**:2640–2644.

10. Ribeiro dos Santos, R., Rossi, M. A., Laus, J. L., Silva, j. S., Savino, W., and Mengels, J., 1992, Anti-CD4R+ abrogates rejection and reestablishes long-term tolerance to syngeneic newborn hearts grafted in mice chronically infected with *Trypanosoma cruzi*, *J. Exp. Med.* **175**:29–39.

11. Tibbetts, R. S., McCormick, T. S., Rowland, E. C., Miller, S., and Engman, D. M., 1994, Cardiac antigen-specific autoantibody production is associated with cardiomyopathy in *Trypanosoma cruzi* infected mice, *J. Immunol.* **152**:1493–1499.

12. Teixeira, A. R. L., 1987, The stercorian trypanosomes, in: *Immune Responses in Parasitic Infections: Immunology, Immunopathology and Immunoprophylaxis*, Vol. III (E. J. L. Soulsby, ed.), CRC Press, Boca Raton, FL, pp. 25–118.

13. Margulis, L., 1981, Several prokaryotes make a eukaryote, in: *Symbiosis in Cell Evolution*, Freeman, San Francisco.

14. Stachel, S. E., and Zambriski, C., 1989, Genetic trans-kingdom sex? *Nature* **340**:190–191.

15. Borst, P., 1991, Why kinetoplast network? *Trends Genet.* **7**:139–141.

16. Sturm, N., and Simpson, L., 1990, Kinetoplast DNA minicircles encode guide RNAs for editing of cytochrome oxidase subunit III mRNA, *Cell* **61**:879–884.

17. Weiner, A. M., and Maizeles, N., 1990, RNA editing: Guided but not templated? *Cell* **61**:917–920.

18. Hadjuk, S. E., Hanis, M. E., and Pollard, V. W., 1993, RNA editing in kinetoplastid mitochondria, *FASEB J.* **7**:54–63.

19. Gibson, W., and Garside, L., 1990, Kinetoplast DNA minicircles are inherited from both parents in genetic hybrids of *Trypanosoma brucei*, *Mol. Biochem. Parasitol.* **42**:45–54.

20. Tibayrenc, M., and Miles, M., 1983, A genetic comparison between Brazilian and Bolivian zymodemes of *Trypanosoma cruzi*, *Trans. R. Soc. Trop. Med. Hyg.* **77**:76–83.

21. Garcia, E. S., and Dvorak, J. A., 1982, Growth and development of two *Trypanosoma cruzi* clones in the arthropod *Dipelogaster maximus*, *Am. J. Trop. Med. Hyg.* **31**:259–262.

22. Flint, V. E., Schecter, M., Chapman, M. D., and Miles, M. A., 1984, Zymodeme and species specificities of monoclonal antibodies raised against *Trypanosoma cruzi*, *Trans. R. Soc. Trop. Med. Hyg.* **78**:192–202.

23. Miles, M. A., Povoa, M. M., DeSouza, A. A., Lainson, R., Shaw, J. J., and Ketteridge, D. S., 1981, Chagas disease in the Amazon Basin: II. The distribution of *Trypanosoma cruzi* zymodemes 1 and 3 in Para State north Brazil, *Trans. R. Soc. Trop. Med. Hyg.* **7**:667–674.

24. Luquetti, A. O., Miles, M. A., Rassi, A., Rezende, J. M., Souza, A. A., Povoa, M. M., and Rodrigues, I., 1986, *Trypanosoma cruzi*: Zymodemes associated with acute and chronic Chagas disease in Central Brazil, *Trans. R. Soc. Trop. Med. Hyg.* **80**:462–470.

25. Olson, C. L., Nadeau, K. C., Sullivan, M. A., Winquist, A. G., Donelson, J. E., Walsh, C. T., and Engman, D. M., 1994, Molecular and biochemical comparison of the 70-kDa heat shock proteins of *Trypanosoma cruzi*, *J. Biol. Chem.* **269**:3868–3874.

26. McDaniel, J. P., and Dvorak, J. A., 1993, Identification, isolation and characterization of naturally occurring *Trypanosoma cruzi* variants, *Mol. Biochem. Parasitol.* **57**:213–222.

27. Gibson, W., and Miles, A. A., 1986, The karyotype and ploidy of *Trypanosoma cruzi*, *EMBO J.* **5**:1299–1305.

28. Sher, A., Hieny, S., and Joiner, K., 1986, Evasion of the alternative complement pathway by metacyclic trypomastigotes of *Trypanosoma cruzi*: Dependence on the developmentally regulated synthesis of surface protein and N-linked carbohydrate, *J. Immunol.* **137**:2961–2967.

29. Teixeira, M. L., and Dvorak, J. A., 1985, *Trypanosoma cruzi*: Histochemical characterization of parasitized skeletal muscle fibers, *J. Protozool.* **32**:339–343.
30. von Kreuter, B. F., Sadigurski, M., and Santos-Buch, C. A., 1988, Complementary surface epitopes, myotropic adhesion and active grip in *Trypanosoma cruzi* host cell recognition, *Mol. Biochem. Parasitol.* **30**:197–208.
31. von Kreuter, B. F., and Santos-Buch, C. A., 1989, Modulation of *Trypanosoma cruzi* adhesion so host muscle cell membranes by ligands of muscarinic cholinergic and beta adrenergic receptors, *Mol. Biochem. Parasitol.* **36**:41–50.
32. von Kreuter, B. F., Walton, B. I., and Santos-Buch, C. A., 1995, Attenuation of parasite cAMP levels in *T. cruzi*–host cell membrane interactions *in vitro*, *J. Eukaryotic Microbiol.* **42**:21–27.
33. Hall, B. F., 1993, *Trypanosoma cruzi*: Mechanisms for entry into host cells, *Semin. Cell Biol.* **4**:323–333.
34. Teixeira, A. R. L., Figueiredo, F., Rezende Filho, J., and Macedo, V., 1983, Chagas disease: A clinical, parasitological, immunological, and pathological study in rabbits, *Am. J. Trop. Med. Hyg.* **32**:258–272.
35. Teixeira, A. R. L., Cordoba, J. C., Souto-Maior, I., and Solorzano, E., 1990, Chagas disease: Lymphoma growth in rabbits treated with benznidazole, *Am. J. Trop. Med. Hyg.* **43**:146–158.
36. Trieu-Cuot, P., Carlier, C., and Courvalin, P., 1988, Conjugative plasmid transfer from *Enterococcus faecalis* to *Escherichia coli*, *J. Bacteriol.* **170**:4388–4391.
37. Heinemann, J. A., and Sprague, G. F., 1989, Bacterial conjugative plasmids mobilize DNA transfer between bacteria and yeast, *Nature* **340**:205–209.
38. Bakkeren, G., Koukolikova-Nicola, Z., Grimsley, N., and Hohn, B., 1989, Recovery of *Agrobacterium tumefaciens* T-DNA molecules from whole plants early after transfer, *Cell* **57**:847–857.
39. Yamazoe, M., Nakai, S., Ogasawara, N., and Yoshikawa, H., 1991, Integration of woodchuck hepatitis virus (WJV) DNA at two chromosomal sites (Vk and gag-like) in a hepatocellular carcinoma, *Gene* **100**:139–146.
40. Aufiero, B., and Schneider, R. J., 1990, The hepatitis B virus X-gene product transactivates both RNA polymerase II and III promoters, *EMBO J.* **9**:497–504.
41. Caselman, W. H., Meyer, M., Kekule, A. S., Lauer, U., Hofschneider, P. H., and Koshy, R., 1990, A transactivator function is generated by integration of hepatitis B virus pre s/s sequences in human hepatocellular carcinoma DNA, *Proc. Natl. Acad. Sci. U.S.A.* **87**:2970–2974.
42. Houck, M. A., Clark, S. B., Petersen, K. R., and Kidwel, M. G., 1991, Possible horizontal transfer of Drosophila genes by the mite *Proctolaelaps regalis*, *Science* **253**:1125–1128.
43. Teixeira, A. R. L., Lacava, Z., Santana, J. M., and Luna, H., 1991, Insercao de DNA de *Trypanosoma cruzi* no genoma de celula hospedeira de mamifero por meio de infeccao, *Rev. Soc. Bras. Med. Trop.* **24**:55–58.
44. Teixeira, A. R. L., Junqueira, L. F., Solorzano, E., and Zapala, M., 1983, Doenca de Chagas experimental em coelhos isogenicos III/J. Fisiopatologia das arritmias e da morte subita do chagasico, *Rev. Assoc. Med. Bras.* **29**:77–83.
45. Hontebeiyrie-Joskowicz, M., Said, G., Milon, G., Marshall, G., and Eisen, H., 1987, L3T4+ T-cells able to mediate parasite specific delayed-type hypersensitivity play a role in the pathology of experimental Chagas disease, *Eur. J. Immunol.* **17**:1027–1033.
46. Teixeira, A. R. L., Arganaraz, E. R., Freitas Jr., L. H., Lacava, Z. G. M., Santana, J. M., and Luna, H., 1994, Possible integration of *Trypanosoma cruzi* kDNA minicircles into the host cell genome by infection, *Mutat. Res.* **3305**:197–209.
47. Zwirner, N. W., Malciodi, E. L., Chiaramonte, M. G., and Fossati, C. A., 1994, a lytic

monoclonal antibody to *Trypanosoma cruzi* blood stream trypomastogotes which recognizes an epitope expressed in tissues affected in Chagas disease, *Infect. Immun.* **62**:2483–2489.

48. Searle, A. G., 1987, Mouse chromosome mapping, *Genomics* **1**:3–18.
49. Cerisola, J. A., Lugones, H., and Rabinovich, L. B., 1972, *Tratamiento de la Enfermedad de Chagas* (Fundacion Rizuto, ed.), Buenos Aires.
50. Lugones, H. S., 1979, Enfermedad de Chagas en la infancia, *An. Sanidad* **13**:1.
51. Ledesma Patino, O. S., Ribas Meneclier, C. A., Kalalo, E., Lugones, H., de Marteleur, A. E., and Barbieri, G., 1992, Epidemiologia, clinica y laboratorio de la enfermedad de Chagas aguda en Santiago del Estero, in: *Actualizaciones en la Enfermedad de Chagas* (R. Madoery, C. Madoery, and M. I. Camera, eds.), Libro del Organismo Oficial del Congreso Nacional de Medicina, Simposio Satelite, Cordoba.
52. Braverman, J., 1970, Acotaciones a proposito del complejo oftalmologico en la enfermedad de Chagas-Mazza en el nino, *Sem. Med.* **16**:440–445.
53. Jorg, M. E., 1974, *Trypanosoma cruzi* humana o enfermedad de Chagas-Mazza, Reedicion de *Actualizacion de Tratamientos*, No. 380, Ano XXIX, Publicacion Laboratorios Roche-Argentina, Buenos Aires.
54. Jorg, M., and Natula, O., 1980, Inoculacion de *Trypanosoma cruzi* por picadura de *Triatoma infestans*, *Rev. Argent. Parasitol.* **1**:28–33.
55. Basso, G., Basso, R., and Bailoni, A., 1978, *Investigaciones sobre la Enfermedad de Chagas-Mazza*, Universitaria de Buenos Aires, Buenos Aires.
56. Mazza, S., Basso, G., and Basso, R., 1940, *Comprobacion en adulto de citoesteatonecrosis subcutanea chagasica por siembra hematogenea (Chagomas hematogenos) de Schizotrypanum cruzi*, Publicacion No. 48 de la MEPRA, Buenos Aires.
57. Andrade, Z. A., Andrade, S. G., Correa, R., Sadigurski, M., and Ferrans, V. J., 1994, Myocardial changes in acute *Trypanosoma cruzi* infection. Ultrastructure evidence of immune damage and the role of microangiopathy, *Am. J. Pathol.* **144**:1403–1411.
58. Sadigurski, M., von Kreuter, B. F., Ling, P.-Y., and Santos-Buch, C. A., 1989, Association of elevated anti-sarcolemma, anti-idiotype antibody levels with the clinical and pathologic expression of chronic Chagas myocarditis, *Circulation* **80**:1269–1276.
59. Gazzinelli, R. T., Morato, M. J., Nunes, R. M. E., Cancado, J. R., Brener, Z., and Gazinelli, G., 1988, Idiotype stimulation of T lymphocytes from *Trypanosoma cruzi* infected patients, *J. Immunol.* **140**:3167–3172.
60. Felix, J. C., von Kreuter, B. F., and Santos-Buch, C. A., 1993, Mimicry of heart cell surface epitopes in primary anti-*Trypanosoma cruzi* Lyt2+ lymphocytes, *Clin. Immunol. Immunopathol.* **68**:141–146.
61. Rivetti, E., Moragas, A., and Ripoll, C., 1994, Chagas connatal, aspectos clinicos, *Rev. Soc. Bras. Med. Trop.* **27**(Supl. I):775.
62. Rivetti, E., Ripoll, C., Moragas, A., and Esteban, M., 1994, Tratamiento del recien nacido con Chagas connatal, *Rev. Soc. Bras. Med. Trop.* **27**(Supl. I):776.
63. Lauria-Pires, L., Bogliolo, A. R., Numes, M. H. C., Santana, J. M., Tinoco, D. L., and Teixeira, A. R. L., 1993, Efeito de superinfeccoes com estoques e clones de *Trypanosoma cruzi* bioquimica e molecularmente caracterizados na evolcao da doenca de Chagas experimental, *Rev. Soc. Bras. Med. Trop.* **26**(Supl. II):62–64.
64. Porto, C. C., 1962, O eletrocardiograma no prognostico da doenca de Chagas, *Arq. Bras. Cardiol.* **15**:349–350.
65. Minoprio, P., Coutinho, A., Spinella, S., and Hontebeyirie-Joskovics, M., 1991, XID immunodeficiency impairs increased parasite clearance and resistance to pathology in experimental Chagas disease, *Immunology* **1**:176–182.
66. Soong, L., and Tarleton, R. L., 1992, Selective suppressive effects of *Trypanosoma cruzi* infection on IL-2, c-mic and c-fos gene expression, *J. Immunol.* **20**:2095–2102.

67. Minoprio, P., 1991, Chagas disease: CD5 B cell dependent Th2 pathology? *Res. Immunol.* **142**:137–143.
68. Santos Lima, E. C., Mombaerts, P., Coutinho, A., Tonegawa, S., and Minoprio, P., 1993, *Trypanosoma cruzi* infection in mice with disrupted γ-Tcr genes, *Mem. Inst. Oswaldo Cruz* **88**:182.
69. Sun, I., and Tarleton, R. L., 1993, Predominance of CD8⁺ T lymphocytes in the inflammatory lesions of mice with acute *Trypanosoma cruzi* infection, *Am. J. Trop. Med. Hyg.* **48**:161–165.
70. Laguens, R. P., Meckelt, P. M. C., and Chambo, J. C., 1988, Origin and significance of anti-heart and anti-skeletal muscle autoantibodies in Chagas disease, *Res. Immunol.* **142**:160–164.
71. Borda, E. S., Sterin-Borda, L. J., Pascual, J. O., Gorelik, G., Felix, J. C., von Kreuter, B. F., and Santos-Buch, C. A., 1991, *Trypanosoma cruzi* attachment to lymphocyte muscarinic cholinergic and beta adrenergic receptors modulates intracellular signal transduction, *Mol. Biochem. Parasitol.* **47**:91–100.
72. Ferguson, T. H., and Iverson, G. M., 1986, Isolation and characterization of an antigen specific suppressor inducer molecule from serum of hyperimmune mice by using a monoclonal antibody, *J. Immunol.* **136**:2896–2903.
73. Sato, S., Qian, I. H., Koludo, S., Ikegami, K., Suda, T., Hamaoka, T., and Fujiwara, H., 1988, Studies of the induction of tolerance to alloantigen-specific delayed-type hyper-sensitivity T cells by a single injection of allogeneic lymphocytes via portal venous route, *J. Immunol.* **140**:717–722.
74. Cossio, P. M., Laguens, R. P., Diez, C., and Arana, R. M., 1974, Chagasic cardiopathy. Antibodies reacting with the plasma membrane of striated muscle and endothelial cells, *Circulation* **50**:11272–11274.
75. Cossio, P. M., Damilano, G., Vega, M. T., Laguens, R. P., Meckelt, P. C., Diez, C., and Arana, R. M., 1976, In vitro interaction between lymphocytes of chagasic individuals and heart tissue, *Medicina* **36**:287–289.
76. Acosta, A. M., Sadigursky, M., and Santos-Buch, C. A., 1985, Anti-striated muscle anti-body produced by *Trypanosoma cruzi*, *Proc. Soc. Exp. Biol. Med.* **172**:364–369.
77. Sadigurski, M., Acosta, A. M., and Santos-Buch, C. A., 1982, Muscle sarcoplasmic reticulum antigen shared by a *Trypanosoma cruzi* clone, *Am. J. Trop. Med. Hyg.* **31**:934–941.
78. Wood, J. N., Hudson, L., Jesell, T. M., and Iamamoto, M., 1980, A monoclonal antibody defining antigenic determinants on subpopulations of mammalian neurones and *Trypanosoma cruzi*, *Nature* **296**:34–36.
79. Teixeira, M. L., Rezende Filho, J., Figueiredo, F., and Teixeira, A. R. L., 1980, Chagas disease: Selective affinity and cytotoxicity of *Trypanosoma cruzi*–immune lymphocytes to parasympathetic ganglion cells, *Mem. Inst. Oswaldo Cruz* **75**:33–45.
80. Goin, J. C., Borda, E., Leiros, C. P., Storino, R., and Sterin-Borda, L., 1994, Identification of antibodies with muscarinic cholinergic activity in human Chagas disease: Patholog-ical implications, *J. Auton. Nerv. Syst.* **47**:45–52.
81. Iosa, D., Casadei Massari, D., and Dorsey, F., 1991, Chagas cardioneuropathy: Effect of ganglioside treatment in chronic dysautonomic patients. A randomized, double-blind, parallel, placebo-controlled study, *Am. Heart J.* **122**:775–780.
82. Tafuri, W. L., Maria, T. A., Lopes, E. R., and Chapadeiro, E., 1978, Microscopia ele-tronica do miocardio na tripanosomiase cruzi humana, *Rev. Inst. Med. Trop. São Paulo* **15**:347–370.
83. Acosta, A. M., and Santos-Buch, C. A., 1985, Autoimmune myocarditis induced by *Trypanosoma cruzi*, *Circulation* **71**:1255–1261.

84. Cunha Neto, E., Moliterno, R., Coelho, V., Guilherme, L., Bocchi, E., Higuchi, M. de L., Stolf, N., Pileggi, F., Steinman, L., and Kalil, J., 1994, Restricted heterogeneity of T cell receptor variable alpha chain transcripts in hearts of Chagas disease cardiomyopathy patients, *Parasite Immunol.* **16**:171–179.

85. Cunha Neto, E., Duranti, M., Gruber, A., Zingales, B., Sato, M., Messias, I., Higuchi, M., Stolf, N., Belotti, G., Pileggi, F., Jatene, A., and Kalil, J., 1993, Molecular identification of myosin cross-reactive Tc 140/116 *Trypanosoma cruzi* protein: Cross-reactive antibodies show 100% association with cardiomyopathy patients, *Mem. Inst. Oswaldo Cruz* **88** (Supl.):187.

86. Levin, M. J., Kaplan, D., Ferrari, I., Arteman, P., Vazquez, M., and Panebra, A., 1993, Humoral response in Chagas disease: *Trypanosoma cruzi* ribosomal antigens as immunizing agents, *FEMS Immunol. Med. Microbiol. (Netherlands)* **7**:205–210.

87. Sterin-Borda, L. J., Gorelik, G., Genaro, A. M., Goin, J. C., and Borda, E. S., 1990, Human chagasic IgG interacting with lymphocyte neurotransmitter receptors triggers intracellular signal transduction, *FASEB J.* **4**:1661–1667.

88. Cunha Neto, E., Renesto, P. G., Coelho, V., Guilherme, L., Bocchi, E., Higuchi, M. L., Stolf, N., Belotti, G., Pileggi, F., Jatene, A., and Kalil, J., 1993, Primary T cell growth from heart biopsies of Chagas cardiomyopathy patients, *Mem. Inst. Oswaldo Cruz* **88** (Supl.):186.

12

Infection and Endocrine Autoimmunity

ANTHONY P. WEETMAN

1. INTRODUCTION

Autoimmunity is among the most common causes of endocrine disease, and indeed the first demonstration that autoimmune responses could be induced in animals came with studies on the thyroid gland. Since then it has become apparent that the body normally avoids autoimmune disease by a hierarchy of defense mechanisms, ranging from clonal deletion through clonal anergy to active suppression, and the molecular mechanisms responsible are now being unraveled.[1] It is also clear that autoimmune endocrine disorders arise in genetically predisposed individuals, but additional, probably environmental factors also contribute to susceptibility.[2]

This is perhaps best illustrated by the fact that genetically identical twins have only around 30–50% concordance for type 1 diabetes mellitus and Graves' disease, and this falls to less than 10% for human lymphocyte antigen (HLA)-identical siblings, indicating that non-HLA genes are important and that these genes interact with one or more environmental factors. The balance of importance between genes *versus* environment depends at least in part on the age of the individual: genetically determined susceptibility is more obvious in juveniles with thyroid autoimmun-

ANTHONY P. WEETMAN • Department of Medicine, Clinical Sciences Centre, Northern General Hospital, Sheffield S5 7AU, England.

Microorganisms and Autoimmune Diseases, edited by Herman Friedman *et al.* Plenum Press, New York, 1996.

ity, while continued exposure to environmental agents leads to their increasing role in later life.[3]

However, the nongenetic factors involved in precipitating the disease process remain unclear. It is commonly assumed that endogenous events, such as changes in the balance of hormones, may be of key importance: the influence of sex steroids on thyroid autoimmunity is well described, and rapid fluctuations in these and other hormones during and after delivery may account for postpartum thyroiditis.[2] Moreover, monozygotic twins are not identical in their complement of rearranged T-cell receptor and immunoglobulin gene segments, due to the stochastic nature of rearrangement. Despite these caveats, exogenous factors responsible for autoimmune endocrine disease are being identified increasingly frequently and include dietary components, toxins, and infections. This chapter will consider briefly how microorganisms may cause autoimmune disease and then review the evidence that infection is important in thyroid disorders, pituitary autoimmunity, and autoimmune polyendocrinopathy. The role of viruses in type 1 diabetes mellitus is considered elsewhere in this book.

2. THE POTENTIAL ROLE OF INFECTION IN AUTOIMMUNITY

Viruses are the infectious agents most frequently suggested to play a role in autoimmune disease. Infection leads to activation of specific and nonspecific immune responses, resulting in clearance of the virus or, in some cases, persistence of infection. These immunological responses depend on the immunogenetic background of the host and on the virus itself, including its immunogenicity, and the route and size of the initial infection. It is important, although often difficult, to distinguish between direct, virus-induced immunopathology, for example, the lethal meningoencephalitis resulting from intracerebral inoculation of lymphocytic choriomeningitis virus (LCMV) in the mouse,[4] and the induction of autoimmune disease such as postmeasles encephalitis.[5]

There are a number of ways that viruses can initiate autoimmune disorders (Table I). Some of these are theoretical or have only been established in animal models of varying complexity. Release of sufficient, previously sequestered (e.g., intracellular) endogenous antigen by viral damage may activate untolerized T cells or be sufficient to tip the balance in a smoldering autoimmune response in favor of disease. Altered expression of cellular constituents has been demonstrated in murine pancreatic beta cells infected with Coxsackie B4 virus, which, depending on mouse strain, can increase expression of the autoantigenic enzyme glutamic acid

TABLE I
Mechanisms of Virus-Induced Autoimmunity

Direct effect on cells
  Cytopathic effect with release of autoantigens
  Altered expression of enzymes and receptors
  Release or expression of virus gene product in persistently infected target cells
  Direct infection or interaction with lymphocytes, leading to polyclonal T-cell or B-cell
    activation, or immunosuppression
Molecular mimicry
  Cross-reactive viral epitope triggers autoreactive T cells
Breaking tolerance to self antigens
  Viral epitope engenders T-cell help for a second viral epitope identical to host epitope

decarboxylase.[6] In mice that are persistent carriers of chronic LCMV infection (due to LCMV-specific CD8+ T-cell deletion), shedding of large amounts of soluble viral antigen can form immune complexes with LCMV antibodies, leading to glomerulonephritis. Epstein-Barr virus (EBV) infection may cause polyclonal B-cell activation, and although this usually results in the formation of low-affinity, natural autoantibodies, such activation may be critical in initiating disease in predisposed individuals.

Epstein-Barr virus infection has also been implicated in the pathogenesis of rheumatoid arthritis via molecular mimicry, in which there is close homology between epitopes in the viral and host proteins, in this case between EBV EBNA-1 and a synovial cell autoantigen. The concept of molecular mimicry arose from the hypothesis that parasites may evade elimination by expressing similar antigens to the host. It remains difficult to evaluate the significance of homologies between viral and host proteins obtained from database matches, as in many cases the viruses identified by such searches have not been implicated in the autoimmune disease, and the mere existence of such homologous sequences does not indicate that they are truly T-cell epitopes that can be generated from the intact protein *in vivo*. This latter point is not addressed in studies which have used peptides (often in large amounts) to immunize animals and shown that virus-derived sequences can induce disease. For molecular mimicry to work, the viral epitope must be similar enough to the host sequence for cross-reactivity and yet different enough to be recognized by nontolerized T cells. Once initiated, persistent infection is not then required to sustain the autoimmune response.[7]

As an alternative, viral infection may break tolerance, as shown in transgenic mice expressing the LCMV glycoprotein gene in pancreatic beta cells. These animals are healthy and do not react to the LCMV glycoprotein until infected with LCMV, which leads to insulitis and dia-betes.[8] It is probable that the glycoprotein-specific CD8+ T cells that cause these lesions are not deleted during ontogeny but do not cause disease as there is no help from specific CD4+ T cells, which require a costimulatory signal for activation; this cannot be supplied by the beta cell. During LCMV infection, presentation of glycoprotein by dendritic cells and macrophages provides the second signal to activate CD4+ T cells, which in turn leads to CD8+ T-cell stimulation and diabetes. In the nontransgenic situation, this mechanism may operate if one viral epitope initiates T-cell activation which leads to help for a response to a second viral epitope, identical to one in a host autoantigen. While this may seem an unlikely event, it is possible that the pathogenic epitopes on self antigens are not major T-cell determinants (which induce tolerance) and that responses to these minor determinants can be triggered by viral proteins.[9] Once initiated, these responses may be restimulated by self antigen, as the conditions to activate memory T cells are less stringent than for naive T cells.

Similar considerations apply to bacterial and parasitic infections which could also initiate autoimmune disease. Indeed, there has been elegant confirmation that infection may break T-cell tolerance with the demonstration that the nematode *Nippostrongylus brasiliensis* can cause reversal of anergy by activation of alternative reaction pathways.[10] However, two bacteria-specific mechanisms have been highlighted recently: molecular mimicry in heat-shock proteins and T-cell activation by super-antigens. Heat-shock or stress proteins are highly conserved between species and have a number of intracellular functions, particularly as molecular chaperones; in conditions of cell stress they facilitate degradation and prevent interaction of abnormal proteins.[11,12] The close sequence homology between bacterial and human heat-shock proteins has suggested that these may be involved in autoimmune disease, supported by observations particularly in adjuvant arthritis induced in rats by *Mycobacterium tuberculosis* and in the analogous rheumatoid arthritis. Secondly, exogenous bacterial superantigens, such as the staphylococcal entero-toxins, may initiate polyclonal T-cell and B-cell activation by cross-linking major histocompatibility complex (MHC) class II antigens (on the antigen-presenting cell or B cells) and the Vβ chain of the T-cell receptor.[13]

3. AUTOIMMUNE THYROID DISEASE

The role of infectious agents in thyroid autoimmunity can be reviewed by first examining the autoimmune sequelae of thyroid infection and then considering the evidence for a role of infection in two classical autoimmune thyroid disorders, Graves' disease and Hashimoto's thyroiditis.

3.1. Autoimmune Sequelae of Thyroid Infection

Acute thyroiditis, usually due to bacterial infection gaining entry via an intact pyriform sinus tract, does not lead to thyroid autoimmunity or permanent thyroid dysfunction.[14] However, subacute or de Quervain's thyroiditis has been associated with a number of autoimmune phenomena (Table II). The etiology of subacute thyroiditis is almost certainly viral, but direct proof is rarely obtained. There is usually a prodrome characterized by muscle aches, fever, and an upper respiratory tract infection. The disease is usually self-limiting and there are seasonal clusters of disease. Histopathology reveals an infiltrate by neutrophils, then lymphocytes with histiocytic giant cells and thyroid follicular changes; viral inclusion bodies have not been identified. Several agents have been implicated, including the viruses causing mumps (one of the few viruses to be cultured from thyroid tissue), measles, and influenza, as well as EBV, Coxsackievirus, and adenovirus.[15] This heterogeneity suggests that subsequent autoimmune reactions may be diverse, but there has been no correlation of these with the presumed initiating agent.

The reported prevalence and titer of thyroid antibodies in subacute thyroiditis varies but are generally regarded as low.[17] In one study from Denmark, however, thyroglobulin (TG) antibodies were found in four of

TABLE II
Autoimmune Phenomena in Subacute Thyroiditis

Phenomenon	Reference
Humoral	
Infrequent development of low titer antibodies against thyroglobulin and thyroid peroxidase	16–18
TSH receptor–stimulating antibodies documented in varying proportions of patients	16, 19, 20
Multiple thyroid antibodies identified by immunoblotting	21
Circulating immune complexes	16
Cellular	
Circulating T-cell sensitization to thyroid antigens	22, 23
Increased numbers of circulating activated T cells	24
Thyroid infiltration by activated T cells and class II expression on thyroid epithelium	24, 25
Elevated serum IL-6 levels	26

eight patients and persisted in three,[16] whereas 24% of 62 Italian patients had transient TG antibodies.[18] Microsomal/thyroid peroxidase (TPO) antibodies were found in 64% of the Italian patients but in none of the patients from Denmark. In the positive patients, these quickly disappeared. Subacute thyroiditis may result in permanent hypothyroidism in patients with preexisting thyroid autoimmunity, but it is unclear whether this is due to a direct, cytopathic effect of the virus or an exacerbation of the autoimmune process.[17]

Thyroid stimulating hormone (TSH)-receptor (TSH-R) antibodies, measured by the ^{125}I-TSH binding inhibition assay, have been found in a high proportion (33–100%) of patients, sometimes persisting after recovery.[16,19] The frequency of TSH-R stimulating antibodies in these studies was lower and did not coincide with the hyperthyroid state. Others have broadly confirmed these findings of transient TSH-R antibodies which generally do not cause receptor stimulation in conventional bioassays.[15,20] By immunoblotting, eight of nine sera from patients with subacute thyroiditis reacted with a range of uncharacterized thyroid proteins, persisting for up to 3 years after diagnosis.[21] This is consistent with the low thyroid reserve and intrathyroidal iodine content detectable even 5 years after the onset of disease.[27,28] although it is not clear as yet whether autoimmunity is responsible for this chronic thyroid injury.

Transient T-cell responses to thyroid antigens have been detected in patients with subacute thyroiditis using proliferation and migration inhibition assays,[22,23] and there is an increase in the number of activated (HLA-DR-expressing) CD8+ T cells.[24] These cells are also prominent in the thyroid and release interferon gamma. This presumably accounts for the induction of HLA-DR expression on thyroid follicular cells.[24,25] Such results clearly demonstrate that factors other than expression of MHC class II molecules on thyroid cells are required to initiate Graves' disease and Hashimoto's thyroiditis, in contrast to initial predictions.[29] High circulating levels of interleukin (IL)-6 have recently been described in subacute thyroiditis, presumably derived in part from the thyroid lymphocytic infiltrate.[26] This pleiotropic cytokine causes B-cell stimulation, and this may explain the appearance of the wide variety of thyroid antibodies in some of these patients.

More convincing evidence for a viral role in etiology comes from studies on congenital rubella infection. Both TG and TPO antibodies have been found in 23–34% of such patients, two to three times higher than expected.[30-32] Moreover, congenital rubella and Hashimoto's thyroiditis (including TG and TPO antibodies) has been reported in a 5-year-old patient, in whom the viral antigen was detected in germinal centers of lymphoid follicles within the thyroid,[33] and several case reports have linked both Graves' disease (with ophthalmopathy) and autoimmune hypothyroidism to this infection.[34,35] However, in these juvenile patients, there was a long interval between viral infection and the onset of the thyroid disorder, and the mechanisms involved in these associations are unclear. However, it is noteworthy that type 1 diabetes mellitus is also significantly increased in congenital rubella,[32] suggesting the possibility of a rather nonspecific effect of the virus on the immune system, leading to an increased general risk of developing autoimmune disease.

3.2. Infection in Thyroid Autoimmunity

3.2.1. Experimental Animal Models

As yet, there is no satisfactory animal model for Graves' disease. However, there are several types of experimental autoimmune thyroiditis (EAT) with varying degrees of similarity to the human counterpart, Hashimoto's thyroiditis. The two models most closely resembling this disorder are induced in rats and mice by thymectomy, with or without sublethal irradiation, or occur spontaneously, the best studied model being the obese strain (OS) chicken.[1]

PVG/c rats are considerably less susceptible to EAT after thymectomy and irradiation when raised under specific pathogen–free (SPF) conditions, compared to conventional rearing (Fig. 1), and the same applies to other strains in which the prevalence of thyroiditis is lower than in the PVG/c strain.[36] The incidence and titer of TG antibodies is also reduced in SPF rats. Interestingly, and so far unexplained, the same procedure of thymectomy and irradiation induces insulitis and diabetes in PVG/c rats under SPF conditions, but this is much less common in conventional animals.[37] Transfer of gut contents from normal animals to newly weaned SPF rats after administration of antibiotics augmented their susceptibility to EAT.[36] These experiments suggest that intestinal microflora are involved in the susceptibility to thyroid disease in this model, but the mechanisms involved are unclear. Possibilities include polyclonal activation of B cells or the passage of cross-reactive antigen across the irradiation-damaged gut. The depressed T-cell responses after thymectomy and irradiation may facilitate the breaking of tolerance.

The avian leukosis retrovirus Rous-associated virus type 7 causes obesity, stunting, and ataxia when inoculated into white leghorn SC chick embryos, but in addition, these birds have a lymphocytic thyroiditis with hypothyroidism, similar to the spontaneous thyroiditis in OS chickens.[38]

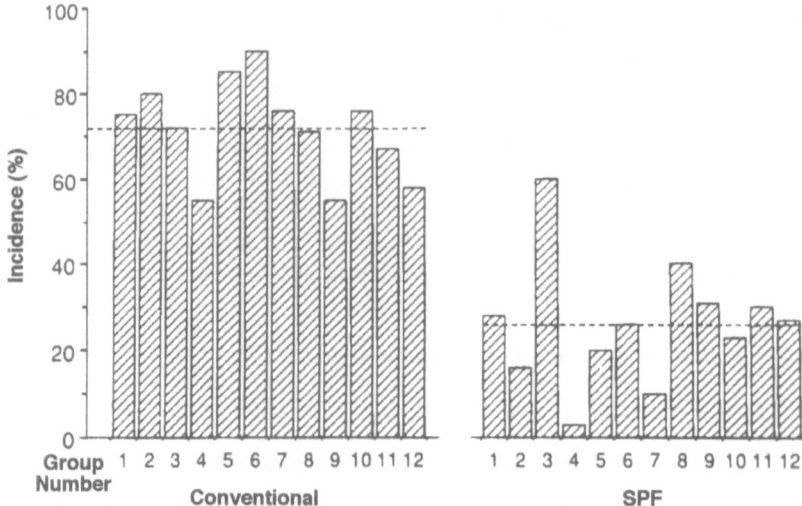

FIGURE 1. Incidence of thyroiditis in sequential groups of PVG/c rats following thymectomy and irradiation and reared under conventional and specific pathogen free (SPF) conditions. (Reproduced with permission from W. J. Penhale and P. R. Young, *Clin. Exp. Immunol.*, 1988:72:288.)

Thyroglobulin antibodies are not present. Analysis by Southern blotting of genomic DNA using a probe containing avian endogenous virus (*ev*) loci has revealed a new locus *ev*22 which is unique to the OS strain.[39] Although this might be of relevance in OS thyroiditis, genetic and endogenous hormonal influences seem more important, and backcross experiments have shown that the presence of *ev*22 does not correlate with the severity of thyroiditis or the presence of TG antibodies.[40] This retrovirus may have little, if any, direct role in the disease, although it could influence the altered immunoendocrine feedback regulation of glucocorticoids found in OS chickens. Using 2′,5′-oligoadenylate synthase, which is only activated by double-stranded RNA, as a probe for the presence of viruses, initial studies have shown increases of the enzyme and 2′,5′-oligomers in OS chicken thyroid cells but not in the spleen.[40] This suggests the presence of a thyrotropic virus, but conclusive proof is awaited, and, even then, its role in the autoimmune process is unknown.

Following earlier studies, in which reovirus infection was shown to induce a polyendocrine disease (see below), newborn mice were infected with reoviral type 2 and their thyroids were studied.[41] About 20% of the mice developed a mild, focal leukocytic infiltrate, with severe focal thyroiditis in 75% at 4 weeks: TG antibodies were present in a third of these animals but TSH levels were normal. Reovirus antigens were detected in only a few of the thyroid cells themselves, and the results seem due to the virus initiating thyroid autoimmunity. Subsequently, the S1 segment of the reovirus genome was shown to be responsible for the formation of TG antibodies.[42] The S1 segment encodes a polypeptide which influences the tissue selectivity of the virus. In contrast, persistent infection of thyroid cells with LCMV, following inoculation of neonatal mice, results in reduced levels of TG mRNA and circulating thyroid hormones but does not cause thyroid necrosis or infiltration by lymphocytes.[43] Only a single animal in these studies developed TG antibodies.

3.2.2. Graves' Disease and Hashimoto's Thyroiditis

Several exogenous factors have been implicated in the initiation of these diseases, including dietary iodide uptake,[44] stress,[45] and infection. In a 10-year, community-based survey of patients with Graves' disease, there was a tendency for cases to present during the summer months but there was no evidence of clustering of cases in space or time.[46] We have observed a similar seasonal effect (Fig. 2). These data suggest that infec-

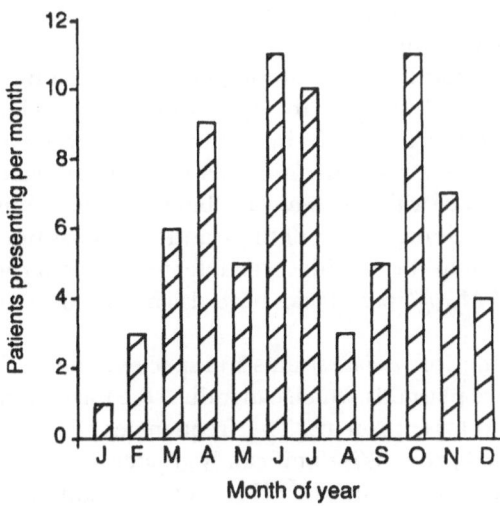

FIGURE 2. Month of presentation in sequential patients with Graves' disease from a single endocrine clinic. (N.B. the number of patients seen in clinic is reduced in August.)

tions which behave in an epidemic manner are unlikely to cause Graves' disease, and the seasonal variation in presentation could simply be due to worsening of heat intolerance during the summer months. However, endemic infections or those with long latency would not be detected by these types of analyses. There has been no formal study of the prevalence of thyroid autoimmunity in married couples, but anecdotal evidence suggests that conjugal Graves' disease is likely to be the result of chance, rather than shared infection.[47] One report found serological evidence for recent viral or bacterial infections in over a third of Graves' patients compared to 10% of controls.[48] However, the initiation of Graves' disease probably predates presentation by several months, if not years, and the relevance of such recent infections may lie in their role as stressors rather than in initiation.

More persuasive evidence for a role of infection has come from several studies linking infection with *Yersinia enterocolitica* to Graves' disease. Patients with *Yersinia* infections more frequently have ill-defined thyroid antibodies than controls,[49] and Graves' patients have a high prevalence of *Yersinia* antibodies, although it is important to note that the same is true of other thyroid disorders.[50,51] Stimulated by these observations, attempts have been made to identify cross-reactive antigens in *Yersinia* and the thyroid. Lysozyme-treated extracts of *Y. enterocolitica* were shown to contain a saturable, TSH-binding moiety with similar specificity and affinity to the TSH-R,[52] and antibodies from Graves' patients were also found to recognize these binding sites under physiological conditions.[53] This suggests that an epitope in *Y. enterocolitica* may initiate Graves' disease by molecular mimicry, although operating via B-cell rather than T-cell recognition of antigen. In support of this, polyclonal antibodies against the TSH-R, produced by immunization of rabbits, reacted with *Y. enterocolitica* envelope proteins, and TSH-R antibodies could be induced by immunizing mice with *Y. enterocolitica* but not other organisms.[54] Unfortunately, the functional activity of these antibodies was not assessed. This cross-reactivity may be a widespread feature of microorganisms, as *Leishmania* and *Mycoplasma* species have also been shown to contain TSH-binding sites.[55]

The virulence of *Yersinia* strains and species is determined by closely related virulence plasmids, which can induce the formation of at least five immunogenic release proteins. The frequencies of antibodies to these plasmid-encoded proteins is significantly greater in patients with Graves' disease than in healthy controls.[56] However, these antibodies appear late in the disease and are also found with high frequency in Hashimoto's thyroiditis patients, suggesting that they are the nonspecific result of polyclonal activation. Moreover, a recent detailed study has been unable to confirm this, although intrathyroidal T cells from Graves' patients were stimulated by these release proteins.[57]

Such conflicting results demonstrate that the association between *Yersinia* infection and Graves' disease is unproven at present: there are few cases where there is a close temporal relationship between *Y. enterocolitica* infection and Graves' disease,[58] and the majority of infected patients do not develop thyroid autoimmunity. Although genetic susceptibility could determine which individuals develop Graves' disease after *Yersinia* infection, it is also possible that the positive results in antibody binding assays reflect low-affinity, natural autoantibodies.[30]

Several viruses have been implicated in the initiation of thyroid autoimmunity. Three patients were described in whom there was an association between recent EBV infection (determined serologically) and Hashimoto's thyroiditis, transient autoimmune thyroiditis, or Graves' disease.[59] Transformation by EBV of normal human peripheral blood lymphocytes results in thyroid antibody production, most likely natural autoantibodies,[60] and such autoreactive B cells are more frequent in old age,[61] suggesting that polyclonal activation by viruses could explain the rise in thyroid antibodies with age. As for *Yersinia*, genetic susceptibility may determine whether autoimmune disease will result from viral infections. Residues 463–481 of TPO have homology with an EBV-encoded protein, but no heightened T-cell reactivity was seen with a peptide covering this region in patients with thyroid autoimmunity compared to controls.[62]

Endogenous and exogenous (infectious) retroviruses in mice have direct immunological effects by inducing the production of endogenous superantigens and immunosuppressive and other active proteins, and they may also cause autoimmunity by indirect mechanisms, such as molecular mimicry and encoding proteins which themselves act as autoantigens.[63,64] Their potential involvement in human disorders such as systemic lupus erythematosus and rheumatoid arthritis is the subject of ongoing investigation, but attempts have also been made recently to identify retroviral sequences in thyroid autoimmunity. DNA extracted from five Graves' thyroids hybridized with two probes for HIV-1 *gag* sequences, in contrast to negative signals from nonautoimmune thyroid tissue.[65] As these patients did not have HIV-1 or HTLV-1 antibodies, it was assumed that this represented infection with a novel, exogenous retrovirus. However, two other groups have been unable to confirm these findings, despite including in their methods attempts at DNA amplification using primers from the HIV *gag* and LTR genes,[66,67] and there has been no follow-up from the group that originally reported the presence of retrovirus-like sequences. Despite this, the HIV regulatory protein *nef* shares significant homology with a region of the TSH-R, and Graves' sera contain antibodies reacting to *nef-1*.[68] The relevance of these results is entirely speculative at present.

Spumaretroviruses (or foamy viruses) are cytopathic exogenous retroviruses which have been implicated in subacute thyroiditis.[69] Leukocyte DNA from 66% of 29 patients with Graves' disease contained sequences related to the human spumaretrovirus regions, *gag*, *bel*$_1$, *bel*$_2$, and LTR.[70] All 23 healthy blood donors tested were negative, suggesting that these results are not due to an endogenous retrovirus. No antibodies against spumaretroviruses could be detected in the Graves' patients, which remains difficult to explain. Recently, positive immunofluorescence for spumaretrovirus *gag* protein was detected in the thyroid cells of all seven Graves' patients tested, but was not seen in 14 controls and in four of five Hashimoto's thyroiditis patients.[71] Surprisingly, the lymphocytes in the Graves' thyroids were not stained, at odds with the high frequency of positive *gag* sequences in Graves' peripheral blood.[70] Therefore, this particular retrovirus requires further investigation as a thyrotropic virus in Graves' disease; if confirmed, it will then be worth exploring mechanisms by which this may induce autoimmunity.

Finally, a role for heat-shock proteins (hsp) has been suggested by observations of increased expression of hsp72 on thyroid cells in autoimmune thyroiditis as well as on fibroblasts from the orbit and pretibial skin in Graves' ophthalmopathy and dermopathy.[72,73] Moreover, hsp60 has sequence homology with TG.[74] Thus, it could be postulated that infection leading to an immune response against microbial hsp could induce cross-reactivity to hsp within the thyroid and in fibroblasts that are the target of the unexplained complications of Graves' disease. However, T-cell responses to hsp are not prominent in these patients,[75] suggesting that expression of hsp is probably the result rather than cause of the autoimmune injury.

4. PITUITARY AND POLYENDOCRINE AUTOIMMUNITY

Unlike thyroid autoimmunity and type 1 diabetes mellitus, the other major autoimmune endocrine disorders, Addison's disease, autoimmune oophoritis, and hypophysitis, are uncommon, and their immunological characteristics are less clear.[2] There is an established association between these various endocrinopathies, most frequently in the type 2 polyendocrine autoimmune syndrome (Table III). It remains unexplained why these disorders occur together so frequently, as the autoantigens recognized by the immune system appear to be unique to each endocrine organ (except for the cross-reactivity between ovarian and adrenal antibodies in some patients with Addison's disease).

SJL/J mice infected with reovirus type 1 5–7 days after birth develop a polyendocrine disease characterized by mild diabetes, insulitis, hypophysitis, retarded growth, and the presence of antibodies against insulin and growth hormone and cells in the islets, anterior pituitary, thymocytes, and gastric mucosa.[75] Treatment with antilymphocyte serum to induce immunosuppression prevented these changes but did not have an antiviral effect, indicating that the autoimmune phenomena are unlikely to be due simply to a cytopathic effect of the reovirus. Monoclonal antibodies derived from the spleens of infected mice react with multiple organs, particularly gastric mucosa, anterior pituitary, and pancreatic islets, with some also reacting with cells in the small intestine.[76] The monoclonals do not react with thyroid, adrenal, or other tissue, ruling out a common cytoskeletal antigen as a source of cross-reactivity. In an alternative approach, more than 600 monoclonal antiviral antibodies were screened for reactivity with various tissues by indirect immunofluorescence, and single or multiple reactivity was detected with 3.5% of the tissues, anterior pituitary cells being the tissue most frequently recognized.[77] There was no clear association with the viruses recognized by these antibodies, which included LCMV, measles, rabies, and vaccinia. Finally, multiple autoantibodies which react with thyroid, ovary, and

TABLE III
Major Features of the Polyendocrine Autoimmunity Syndromes[a]

Type 1 polyendocrine autoimmunity
  Addison's disease
  Hypoparathyroidism
  Chronic mucocutaneous candidiasis
  Less common: Ovarian failure, alopecia, vitiligo, pernicious anemia, diabetes mellitus
Type 2 polyendocrine autoimmunity
  Graves' disease or autoimmune hypothyroidism
  Type 1 diabetes mellitus
  Addison's disease
  Premature ovarian failure
  Vitiligo
  Less common: Pernicious anemia, myasthenia gravis, alopecia, autoimmune hypophysitis

[a]From Ref. 2.

adrenal appeared after infection with mouse cytomegalovirus; acute inflammatory changes also occurred in the adrenal.[78]

Such studies have been extended to man. Monoclonal antibodies, derived from peripheral blood lymphocytes taken from patients with type 1 diabetes mellitus, were shown to react with multiple tissues, including anterior pituitary, thyroid follicular cell cytoplasm, gastric mucosa, and pancreatic islets as well as ducts.[79] However, these were all IgM class, and it seems most likely that they represent low-affinity, cross-reactive natural autoantibodies, in turn casting doubt on the findings in the reovirus-infected mice. None really resembles the reactivity of antibodies in the human endocrinopathies.

More direct evidence for viral involvement in endocrine autoimmunity has come from experiments in which golden Syrian hamsters were injected with recombinant E1 and E2, which are glycosylated, membrane-associated rubella virus proteins.[80] Transient pituitary antibodies were detected in almost all of the animals but none developed thyroid, adrenal, or islet cell antibodies. All of the hamsters also developed pituitary lymphocytic infiltration. These phenomena were not seen in control animals but occurred in hamsters infected with rubella virus. Furthermore, neonatal thymectomy prevented disease, indicating a crucial role for T cells. As both intact virus and recombinant protein induced disease in this model, the most likely explanation for these results is molecular mimicry, but the pituitary autoantigen is unknown. This study highlights the propensity for rubella to induce endocrine autoimmunity, and it is intriguing that growth hormone deficiency has been reported in the congenital rubella syndrome.[31,32]

5. CONCLUSION

The best evidence for a role of infection in endocrine autoimmunity comes from type 1 diabetes mellitus, considered elsewhere in this book. However, there is accumulating data suggesting that viruses and bacteria may play a role in Graves' disease and, to a lesser extent, in Hashimoto's thyroiditis. Animal models have shown that infection may have multiple effects that allow autoimmunity to develop. The recent observations on retroviruses in the thyroid are of great interest, but further work is clearly required to identify the viruses involved and the nature of this involvement in the autoimmune process. Less is known about the role of viruses in other endocrinopathies, although rubella may contain proteins which can induce pituitary autoimmunity by molecular mimicry.

REFERENCES

1. Sinha, A. A., Lopez, M. T., and McDevitt, H. O., 1990, Autoimmune diseases: The failure of self tolerance, *Science* **248**:1380–1388.
2. Weetman, A. P., 1991, *Autoimmune Endocrine Disease*, Cambridge University Press, Cambridge.
3. Burek, C. L., Hoffman, W. H., and Rose, H. R., 1982, The presence of thyroid autoantibodies in children and adolescents with autoimmune thyroid disease and in their siblings and parents, *Clin. Immunol. Immunopathol.* **25**:395–404.
4. Buchmeier, M. J., Welsh, R. M., Dutko, R. J., and Oldstone, M. B. A., 1980, The virology and immunobiology of lymphocytic choriomeningitis virus infection, *Adv. Immunol.* **30**:275–322.
5. Johnson, R. T., and Griffin, D. E., 1986, Virus induced autoimmune demyelinating disease of the central nervous system, in: *Concepts in Viral Pathogenesis II* (A. L. Notkins and M. B. A. Oldstone, eds.), Springer-Verlag, New York, pp. 203–209.
6. Hou, J., Sheikh, S., Martin, D. L., and Chatterjee, N. K., 1993, Coxsackievirus B4 alters pancreatic glutamate decarboxylase expression in mice soon after infection, *J. Autoimmunity* **6**:529–542.
7. Dyrberg, T., 1991, Molecular mimicry in autoimmunity, in: *Molecular Autoimmunity* (N. Talal, ed.), Academic Press, London, pp. 197–203.
8. Oldstone, M. B. A., Nerenberg, M., Southern, P., Price, J., and Lewicki, H., 1991, Virus infection triggers insulin-dependent diabetes mellitus in a transgenic model—Role of anti-self (virus) response, *Cell* **65**:319–331.
9. Gammon, G., and Sercarz, E., 1989, How some T cells escape tolerance induction, *Nature* **342**:183–184.
10. Röcken, M., Urban, J. F., and Shevach, E. M., 1992, Infection breaks T-cell tolerance, *Nature* **359**:79–82.
11. Latchman, D. S., 1991, Heat shock proteins and human disease, *J. R. Coll. Phys. London* **25**:295–299.
12. Kaufmann, S. H. E., 1990, Heat shock proteins and the immune response, *Immunol. Today* **11**:129–135.
13. Cole, B. C., and Atkin, C. L., 1992, The mycoplasma arthritides T cell mitogen, MAM: A model superantigen, *Immunol. Today* **12**:271–275.
14. Weetman, A. P., 1992, Thyroiditis, *Curr. Pract. Surg.* **4**:118–122.
15. Volpé, R., 1979, Subacute (de Quervain's) thyroiditis, *Clin. Endocrinol. Metab.* **8**:81–95.
16. Bliddal, H., Bech, K., Feldt-Rasmussen, U., Hoier-Madsen, M., Thomsen, B., and Nielsen, H., 1985, Humoral autoimmune manifestation in subacute thyroiditis, *Allergy* **40**:599–604.
17. Tikkanen, M. J., and Lamberg, B. A., 1982, Hypothyroidism following subacute thyroiditis, *Acta Endocrinol.* **101**:348–353.
18. Lio, S., Pontecorvi, A., Caruso, M., Monaco, F., and D'Armiento, M., 1984, Transitory subclinical and permanent hypothyroidism in the course of subacute thyroiditis (de Quervain), *Acta Endocrinol.* **106**:67–70.
19. Wall, J. R., Strakosch, C. R., Bandy, P., and Bayly, R., 1982, Nature of thyrotropin displacement activity in subacute thyroiditis, *J. Clin. Endocrinol. Metab.* **54**:349–353.
20. Strakosch, C. R., Joyner, D., and Wall, J. R., 1978, Thyroid stimulating antibodies in patients with subacute thyroiditis, *J. Clin. Endocrinol. Metab.* **46**:345–349.
21. Weetman, A. P., Smallridge, R. C., Nutman, T. B., and Burman, K. D., 1987, Persistent thyroid autoimmunity after subacute thyroiditis, *J. Clin. Lab. Immunol.* **23**:1–6.

22. Wall, J. R., Fang, S.-L., Ingbar, S. H., and Braverman, L. E., 1976, Lymphocyte transformation in response to human thyroid extract in patients with subacute thyroiditis, *J. Clin. Endocrinol. Metab.* **43**:587–589.
23. Galluzzo, A., Giordano, C., Adronico, F., Filardo, C., Andronico, G., and Bompiani, G., 1980, Leukocyte migration test in subacute thyroiditis: Hypothetical role of cell-mediated immunity, *J. Clin. Endocrinol. Metab.* **50**:1038–1041.
24. Karlsson, F. A., Tötterman, T. H., and Jansson, R., 1986, Subacute thyroiditis: Activated HLA-DR and interferon-γ expressing T cytotoxic/suppressor cells in thyroid tissue and peripheral blood, *Clin. Endocrinol.* **24**:487–493.
25. Leclere, J., Faure, G., Bene, M.-C., Thomas, J.-L., Paul, J.-L., and Hartemann, P., 1986, In situ immunologic disorders in de Quervain's thyroiditis, in: *Frontiers in Thyroidology* (G. Medeiros-Neto and E. Gaitan, eds.), Plenum Publishing Corporation, New York, pp. 1365–1368.
26. Bartalena, L., Brogioni, S., Grasso, L., and Martino, E., 1993, Increased serum interleukin-6 concentration in patients with subacute thyroiditis: Relationship with concomitant changes in serum T4-binding globulin concentration, *J. Endocrinol. Invest.* **16**: 213–218.
27. Teixeira, V. L., Romaldini, J. H., Rodrigues, H. F., Tanaka, L. M., and Farah, C. S., 1985, Thyroid function during the spontaneous course of subacute thyroiditis, *J. Nucl. Med.* **26**:457–460.
28. Smallridge, R. C., de Keyser, F. M., Van Herle, A. J., Butkus, N. E., and Wartofsky, L., 1986, Thyroid iodine content and serum thyroglobulin: Clues to the natural history of destruction-induced thyroiditis, *J. Clin. Endocrinol. Metab.* **62**:1213–1219.
29. Bottazzo, G. F., Pujol-Borrell, R., Hanafusa, T., and Feldmann, M., 1983, Role of aberrant HLA-DR expression and antigen presentation in induction of endocrine autoimmunity, *Lancet* **ii**:1115–1119.
30. Tomer, Y., and Davies, T. F., 1993, Infection, thyroid disease, and autoimmunity, *Endocrine Rev.* **14**:107–115.
31. Clarke, W. L., Shaver, K. A., Bright, G. M., Rogol, A. D., and Nance, W. E., 1986, Autoimmunity in congenital rubella syndrome, *J. Pediatr.* **104**:370–373.
32. Fellner-Ginsberg, F., Witt, M. E., Yagihashi, S., Dobersen, M. J., Taub, F., Fedun, B., Roman, S. H., Davies, T. F., Cooper, L. Z., Rubinstein, P., and Notkins, A. L., 1984, Congenital rubella syndrome as a model for type 1 (insulin-dependent) diabetes mellitus: Increased prevalence of islet cell surface antibodies, *Diabetologia* **27**:87–89.
33. Ziring, P. R., Gallo, G., Finegold, M., Buimovici-Klein, E., and Ogra, P., 1977, Chronic lymphocytic thyroiditis: Identification of rubella virus antigen in the thyroid of a child with congenital rubella, *J. Pediatr.* **90**:419–420.
34. Ziring, P. R., Fedun, B. A., and Cooper, L. Z., 1975, Thyrotoxicosis in congenital rubella, *J. Pediatr.* **87**:1002.
35. Avruskin, T. W., Brakin, M., and Juan, C., 1982, Congenital rubella and myxedema, *Pediatrics* **69**:495–496.
36. Penhale, W. J., and Young, P. R., 1988, The influence of the normal microbial flora on the susceptibility of rats to experimental autoimmune thyroiditis, *Clin. Exp. Immunol.* **72**:288–292.
37. Stumbles, P. A., and Penhale, W. J., 1993, IDDM in rats induced by thymectomy and irradiation, *Diabetes* **42**:571–578.
38. Carter, J. K., Owl, C. L., and Smith, R. E., 1983, Rous-associated virus type 7 induces a syndrome in chickens characterised by stunting and obesity, *Infect. Immun.* **39**:410–421.
39. Ziemiecki, A., Kroemer, G., Mueller, R. G., Hala, K., and Wick, G., 1988, *Ev22*, a new

endogenous avian leukosis virus locus found in chickens with spontaneous autoimmune thyroiditis, *Arch. Virol.* 100:267–271.

40. Wick, G., Hu, Y., Gruber, J., Kühr, T., Wozak, E., and Hála, K., 1992, The role of modulatory factors in the multifaceted pathogenesis of autoimmune thyroiditis, *Rev. Immunol.* 9:77–89.

41. Srinivasappa, J., Garzelli, C., Onodera, T., Ray, S., and Notkins, A. L., 1988, Virus-induced thyroiditis, *Endocrinology* 122:563–566.

42. Onodera, T., and Awaya, A., 1990, Anti-thyroglobulin antibodies induced with recombinant reovirus infection in BALB/c mice, *Immunology* 71:581–585.

43. Klavinskis, L. S., Notkins, A. L., and Oldstone, M. B. A., 1988, Persistent viral infection of the thyroid gland: Alteration of thyroid function in the absence of tissue injury, *Endocrinology* 122:567–575.

44. McGregor, A. M., Weetman, A. P., Ratanachaiyavong, S., Owen, G. M., Ibbertson, K., and Hall, R., 1985, Iodine: An influence on the development of autoimmune thyroid disease, in: *Thyroid Disorders Associated with Iodine Deficiency and Excess* (R. Hall and J. Köbberling, eds.), Raven Press, New York, pp. 209–216.

45. Winsa, B., Adami, H.-O., Bergiström, R., Gamstedt, A., Dahlberg, P. A., Adamson, U., Jansson, R., and Karlsson, A., 1991, Stressful life events and Graves' disease, *Lancet* 338:1475–1479.

46. Cox, S. P., Phillips, D. I. W., and Osmond, C., 1989, Does infection initiate Graves' disease?: A population based 10 year study, *Autoimmunity* 4:43–49.

47. Ebner, S. A., Badonnel, M.-C., Altman, L. K., and Braverman, L. E., 1992, Conjugal Graves' disease, *Ann. Intern. Med.* 116:479–481.

48. Valtonen, V. V., Ruutu, P., Varis, K., Ranki, M., Malkamaki, M., and Makela, P. H., 1986, Serological evidence for the role of bacterial infections in the pathogenesis of thyroid diseases, *Acta Med. Scand.* 219:105–111.

49. Lidman, K., Eriksson, U., Norberg, R., and Fagraeus, A., 1976, Indirect immunofluorescence staining of human thyroid by antibodies occurring in *Yersinia enterocolitica* infections, *Clin. Exp. Immunol.* 23:429–435.

50. Bech, K., Larsen, J. H., Hansen, J. M., and Nerup, J., 1974, *Yersinia enterocolitica* infection and thyroid disorders, *Lancet* 2:951–952.

51. Shenkman, L., and Bottone, E. J., 1976, Antibodies to *Yersinia enterocolitica* in thyroid disease, *Ann. Intern. Med.* 85:735–739.

52. Weiss, M., Ingbar, S. H., Winblad, S., and Kasper, D. L., 1983, Demonstration of a saturable binding site for thyrotropin in *Yersinia enterocolitica*, *Science* 219:1331–1333.

53. Heyma, P., Harrison, L. C., and Robins-Browne, R., 1986, Thyrotropin (TSH) binding sites on *Yersinia enterocolitica* recognized by immunoglobulins from humans with Graves' disease, *Clin. Exp. Immunol.* 64:249–254.

54. Luo, G., Fan, J.-L., Seetharamaiah, G. H., Desai, J. K., Dallas, J. S., Wagle, N., Doan, R., Niesel, D. W., Klimpel, G. R., and Prabhakar, B. S., 1993, Immunization of mice with *Yersinia enterocolitica* leads to the induction of antithyrotropin receptor antibodies, *J. Immunol.* 151:922–928.

55. Sack, J., Zilberstein, D., Barile, M. F., Lukes, Y. G., Baker, J. R., Wartofsky, L., and Burman, K. D., 1989, Binding of thyrotropin to selected *mycoplasma* species: Detection of serum antibodies against a specific mycoplasma membrane antigen in patients with autoimmune thyroid disease, *J. Endocrinol. Invest.* 12:77–86.

56. Wenzel, B. F., Heesemann, J., Wenzel, K. W., and Scriba, P. C., 1988, Patients with autoimmune thyroid diseases have antibodies to plasmid encoded proteins of enteropathogenic Yersinia, *J. Endocrinol. Invest.* 11:139–140.

57. Arscott, P., Rosen, E. D., Koenig, R. J., Kaplan, M. M., Ellis, T., Thompson, N., and Baker, J. R., 1992, Immunoreactivity to *Yersinia enterocolitica* antigens in patients with autoimmune thyroid disease, *J. Clin. Endocrinol. Metab.* **75**:295–300.
58. Legoux, J. L., Spielmann, D., Portier, H., Legoux, A., and Cortet, P., 1980, Hyperthyroidism one month after acute enterocolitis due to *Yersinia enterocolitica*, *Lancet* **1**:482–483.
59. Coyle, P. V., Wyatt, D., Connolly, J. H., and O'Brien, C., 1989, Epstein-Barr virus infection and thyroid dysfunction, *Lancet* **i**:899.
60. Garzelli, C., Taub, F. E., Jenkins, M. C., Drell, D. W., Ginsberg-Fellner, F., and Notkins, A. L., 1986, Human monoclonal autoantibodies that react with both pancreatic islets and thyroid, *J. Clin. Invest.* **77**:1627–1631.
61. Fong, S., Tsoukas, C. D., Frincke, L. A., Lawrance, S. K., Holbrook, T. L., Vaughan, J. H., and Carson, D. A., 1981, Age-associated changes in Epstein-Barr virus–induced human lymphocyte autoantibody responses, *J. Immunol.* **126**:910–914.
62. Tandon, N., Freeman, M., and Weetman, A. P., 1991, T cell responses to synthetic thyroid peroxidase peptides in autoimmune thyroid disease, *Clin. Exp. Immunol.* **86**:56–60.
63. Krieg, A. M., and Steinberg, A. D., 1990, Retroviruses and autoimmunity, *J. Autoimmunity* **3**:137–166.
64. Krieg, A. M., Gourley, M. F., and Perl, A., 1992, Endogenous retroviruses: Potential etiologic agents in autoimmunity, *FASEB J.* **6**:2537–2544.
65. Ciampolillo, A., Marini, V., Mirakian, R., Buscema, M., Schulz, T., Pujol-Borrell, R., and Bottazzo, G. F., 1989, Retrovirus-like sequences in Graves' disease: Implications for human autoimmunity, *Lancet* **1**:1096–1099.
66. Tominaga, T., Katamine, S., Namba, H., Yokoyama, N., Nakamura, S., Morita, S., Yamashita, S., Izumi, M., Miyamoto, T., and Nagataki, S., 1991, Lack of evidence for the presence of human immunodeficiency virus type 1–related sequences in patients with Graves' disease, *Thyroid* **1**:307–314.
67. Humphrey, M., Baker, J. R., Carr, F. E., Wartofsky, L., Mosca, J., Drabick, J. J., Burke, D., Djuh, Y. Y., and Burman, K. D., 1991, Absence of retroviral sequences in Graves' disease, *Lancet* **337**:17–18.
68. Burch, H. B., Nagy, E. V., Lukes, Y. G., Cai, W. Y., Wartofsky, L., and Burman, K. D., 1991, Nucleotide and amino acid homology between the human thyrotropin receptor and the HIV-1 nef protein: Identification and functional analysis, *Biochem. Biophys. Res. Commun.* **181**:498–505.
69. Werner, J., and Gelderblom, H., 1979, Isolation of foamy virus from patients with de Quervain thyroiditis, *Lancet* **ii**:258–259.
70. Lagyae, S., Vexiau, P., Morozov, V., Guénebaut-Claudet, V., Tobaly-Tapiero, J., Canivet, M., Cathelineau, G., Périès, J., and Emanoil-Ravier, R., 1992, Human spumaretrovirus-related sequences in the DNA of leukocytes from patients with Graves' disease, *Proc. Natl. Acad. Sci. U.S.A.* **89**:10070–10074.
71. Wick, G., Grubeck-Loebenstein, B., Trieb, K., Kalischnig, G., and Aguzzi, A., 1992, Human foamy virus antigens in thyroid tissue of Graves' disease patients, *Int. Arch. Allergy Immunol.* **99**:153–156.
72. Heufelder, A. E., Wenzel, B. E., Gorman, C. A., and Bahn, R. S., 1991, Detection, cellular localisation and modulation of heat shock proteins in cultured fibroblasts from patients with extrathyroidal manifestations of Graves' disease, *J. Clin. Endocrinol. Metab.* **73**:2483–2489.
73. Heufelder, A. E., Goellner, J. R., Wenzel, B. J., and Bahn, R. S., 1992, Immunohistochemical detection and localization of a 72-kilodalton heat shock protein in autoimmune thyroid disease, *J. Clin. Endocrinol. Metab.* **74**:724–731.

74. Jones, D. B., Coulson, A. F. W., and Duff, G. W., 1993, Sequence homologies between hsp60 and autoantigens, *Immunol. Today* 14:115–119.
75. Onodera, T., Ray, U. R., Melez, K. A., Suzuki, H., Toniolo, A., and Notkins, A. L., 1982, Virus-induced diabetes mellitus: Autoimmunity and polyendocrine disease prevented by immunosuppression, *Nature* **297**:66–68.
76. Haspel, M. V., Onodera, T., Prabhakar, B. S., McClintock, P. R., Essani, K., Ray, U. R., Yagihashi, S., and Notkins, A. L., 1983, Multiple organ-reactive monoclonal autoantibodies, *Nature* **304**:73–76.
77. Srinivasappa, J., Saegusa, J., Prabhakar, B. S., Gentry, M. K., Buchmeier, M. J., Wiktor, T. J., Koprowski, H., Oldstone, M. B. A., and Notkins, A. L., 1986, Molecular mimicry: Frequency of reactivity of monoclonal antiviral antibodies with normal tissues, *J. Virol.* **57**:397–401.
78. Bartholomaeus, W. N., O'Donoghue, H., Foti, D., Lawson, C. M., Shellam, G. R., and Reed, W. D., 1988, Multiple autoantibodies following cytomegalovirus infection: Virus distribution and specificity of autoantibodies, *Immunology* **64**:397–405.
79. Satoh, J., Prabhakar, B. S., Haspel, M. V., Ginsberg-Fellner, F., and Notkins, A. L., 1983, Human monoclonal autoantibodies that react with multiple endocrine organs, *N. Engl. J. Med.* 1:217–220.
80. Yoon, Y.-W., Choi, D.-S., Liang, H.-C., Baek, H.-S., Ko, I.-Y., Jun, H. S., and Gillam, S., 1992, Induction of an organ-specific autoimmune disease, lymphocytic hypophysitis, in hamsters by recombinant rubella virus glycoprotein and prevention of disease by neonatal thymectomy, *J. Virol.* **66**:1210–1214.

13

Infection as a Precursor to Autoimmunity

NOEL R. ROSE

1. INTRODUCTION

The ability to distinguish self from nonself is a cardinal feature of the normal immune system. Nevertheless, the development of autoimmune responses is a common occurrence. Many mechanisms have been proposed to explain the failure of self/nonself discrimination. Of the explanations offered, the intervention of infection is among the most conspicuous. Indeed, the first example of an autoimmune disease, paroxysmal cold hemoglobinuria, was originally attributed to syphilitic infection.[1] In the recent literature, insulin-dependent diabetes mellitus has been associated with Coxsackievirus Group B infection.[2] Despite intensive research in many laboratories, however, scanty information is available on the mechanisms by which infection may function to precipitate autoimmune disease. Some investigators have cited examples of molecular mimicry, based on homology between an amino acid sequence of some endogenous antigen and an antigen of a pathogen.[3] Other investigations pointed to an effect of infection on the regulation of the immune response.[4] Still others have suggested that an infectious agent may produce changes in self-antigens.[5] None of these mechanisms, however, has yet been proven to cause an actual autoimmune disease.

NOEL R. ROSE • Departments of Pathology and of Molecular Microbiology and Immunology, The Johns Hopkins Medical Institutions, Baltimore, Maryland 21205.

Microorganisms and Autoimmune Diseases, edited by Herman Friedman *et al.* Plenum Press, New York, 1996.

My colleagues and I undertook a few years ago a program to explore in depth one example of infection-induced autoimmunity. We felt that such knowledge might serve as the impetus for other studies in this important area. As a model for our studies, we selected Coxsackievirus B3-induced myocarditis in the mouse.

2. MYOCARDITIS

Myocarditis in the human is characterized by cardiac myocyte necrosis and an inflammatory cell infiltrate of mononuclear and polymorphonuclear cells.[6] In North America, myocarditis is most frequently associated with Coxsackievirus Group B infection and viral RNA can be demonstrated in cardiac tissues of many myocarditis patients.[7] Many patients demonstrate circulating autoantibodies to heart antigens, such as the myocyte sarcolemma, the adenine nucleotide translocator, the β-adrenergic receptor, and the myosin heavy chain. These observations have spawned the suggestion that the initial viral assault may give rise to a secondary immunopathogenetic response, producing chronic myocarditis. Moreover, circumstantial evidence associates myocarditis with ensuing dilated cardiomyopathy, which accounts for a significant number of cases of heart failure each year.[8]

3. AN ANIMAL MODEL OF MYOCARDITIS

In order to investigate the immunopathogenesis of Coxsackie-induced myocarditis, we decided to develop a model of the disease in mice. Presuming that the response of animals to the viral infection is in part genetically determined, we examined a number of different mouse strains.[9] Marked differences in the susceptibility of different strains, both to the initial virus infection and to the subsequent course of disease, were demonstrated. Most strains of mice developed an acute myocarditis, beginning 2 or 3 days after infection with a cardiotropic strain of Coxsackievirus B3 (CB3), reaching a peak on day 7, and beginning a process of resolution on day 9. By day 21, hearts of most mice were completely healed. In a few strains, however, the myocarditic process failed to resolve. This continuing form of myocarditis also took on a different histological appearance. Whereas the early disease, as represented on days 7 and 9, was characterized by myocardiocyte necrosis and mixed mononuclear and polymorphonuclear infiltration, often accompanied by calcification, the continuing disease observed on days 15 and 21 in a few strains of mice showed relatively little myocyte necrosis and a diffuse mononuclear infiltration. This later stage of disease was, moreover, characterized by the production of heart-specific autoantibodies.[10] Detailed investigation showed that these autoantibodies were directed to cardiac myosin.[11]

Following the clue provided by the autoantibodies, we prepared purified cardiac myosin and immunized mice of a number of different strains.[12] Myocarditis, resembling that seen in the continuing form of postviral disease, appeared 9 to 15 days after two injections of cardiac myosin with complete Freund's adjuvant. This disease was present only in those strains of mice genetically susceptible to the progressive form of postviral myocarditis; other mouse strains showed no such changes in their hearts. Moreover, this form of myocarditis was seen only following immunization with cardiac myosin; injection with skeletal myosin produced no cardiac inflammation.

Based on these results, we concluded that most mice develop an acute, self-limited myocarditis following infection with CB3, but that a few strains go on to produce a subsequent autoimmune response to cardiac myosin, resulting in autoimmune myocarditis.

4. GENETICS OF AUTOIMMUNE MYOCARDITIS

Much of our earlier research focused on the role of genes of the major histocompatibility complex (MHC) in susceptibility to autoimmune disease.[13,14] Helper T cells recognize peptide fragments in the context of MHC class II molecules. Since each class I or class II molecule will accommodate peptides with particular amino acid motifs, it is logical to assume that the susceptibility of mice to myosin immunization depends on the particular MHC haplotype. Analysis of susceptibility to virus-induced autoimmune myocarditis in MHC-congenic mice showed that MHC plays a role in determining the severity of disease. For example, A/J and A/CA mice developed severe myocarditis, A/SW had moderate disease, whereas A/BY showed only mild lesions.[11] On the other hand, the major degree of susceptibility of mice to autoimmune myocarditis resides not at the MHC locus, but in the "A" background. Mice that are identical at H-2 but different in background genes show dramatic variations in their susceptibility to autoimmune myocarditis. For example, A/J mice (H-2^a) develop severe disease, whereas B10.A (H-2^a) are resistant to both the virus-induced and the myosin-induced disease. The location and nature of these non-MHC susceptibility genes are yet to be determined.

Indirect evidence suggests, however, that local production of cytokines may be influential in the pathogenic autoimmune process.

5. THE ROLE OF CYTOKINES

Resistant B10.A mice were made susceptible to Coxsackievirus-induced myocarditis by giving lipopolysaccharide at the time of virus infection.[15] We undertook experiments to examine the response of susceptible and resistant B10.A mice in terms of cytokine production.[16] Marked differences were found in serum levels of the proinflammatory cytokines, IL-1-β and TNF-α. The levels of both cytokines rose following CB3 infection. In B.10A mice, they reached a peak on day 7 and then declined, coinciding with a diminishing inflammatory response. In contrast, serum levels of IL-1 and TNF continued to rise in the lipopolysaccharide-treated B10.A mice. Immunocytochemical studies suggested that much of this additional cytokine product arose from the mononuclear cells infiltrating the heart.

To determine more directly the role of proinflammatory cytokines in the induction of autoimmune myocarditis, we carried out experiments administering recombinant IL-1 and TNF to genetically nonsusceptible B10.A mice.[17] Fifteen days after infection with CB3 or immunization with cardiac myosin, these animals developed a severe characteristic myocarditis. Autoantibodies to myosin were also present.

The question then arose whether cytokine production is involved in the induction of disease in genetically susceptible A/J mice. In order to carry out these experiments, we administered an antagonist of IL-1 receptor.[18] We found that this peptide inhibited or delayed the onset of autoimmune myocarditis in A/J mice, but had no discernible effect on the early, viral phase of disease. Based on these results, therefore, we concluded that IL-1 and TNF are critical factors in the immunopathogenesis of autoimmune myocarditis.[19]

6. MYOSIN AS AUTOANTIGEN

As previously pointed out, autoimmune myocarditis is induced by cardiac myosin and not by skeletal myosin.[11] Since these two isoforms differ only slightly in amino acid sequence, this finding has provided an important clue in identifying the distinct epitopes of myosin responsible for the autoimmune response. Three groups of investigators have now identified sequences in the mouse that are capable of inducing myocarditis.[20–22] Another team has identified a sequence in the rat.[23] These findings suggest that multiple epitopes on the myosin heavy chain are capable of inducing myocarditis, depending in part on the animal species and strain. It is not unlikely, moreover, that more than one epitope on myosin is responsible for the autoimmune response.

Since myosin is generally considered an intracellular antigen, the question arises of how it may induce autoimmune disease. The first conjecture is that an epitope is shared between CB3 and cardiac myosin. Indeed, some cross-reaction between cardiac myosin, Group A streptococcal M-protein, and CB3 virion peptide has been shown, using an IgM monoclonal antibody.[24] Whether such a shared conformational epitope can mediate a T-cell-dependent autoimmune disease, however, is doubtful. Thus far, none of the sequences of myosin capable of inducing myocarditis have any homology with CB3 virion peptides. At the moment, therefore, it does not appear that molecular mimicry accounts for the initiation of the autoimmune response.

A second possibility is that the virus infection permits release of the sequestered intracellular myosin. This protein then may be taken up either by resident dendritic cells in the heart or by infiltrating mononuclear cells, and then presented, in conjunction with class II MHC, to antigen-reactive CD4[+] helper T cells. In the experimental model using myosin immunization, the antigen is taken up by antigen-presenting cells that migrate to the draining lymph nodes where encounters with myosin-reactive T cells may occur. If CD8[+] cytotoxic T cells are generated, they may circulate to the heart. Myosin peptides, like other intracellular antigens, may be presented at the myocyte surface, even in otherwise normal hearts. The administration of complete Freund's adjuvant might actually enhance this process by intensifying expression of MHC class I as well as adhesion molecules on vascular endothelia. The enhancement of these MHC molecules points to a continuing inflammatory process.[25] Our finding that myosin-specific antibody can be eluted from the heart tissue of A/J mice with severe myocarditis provides additional evidence that epitopes of cardiac myosin are indeed available on the myocyte cell surface.[26]

7. DISCUSSION

CB3-induced myocarditis in mice has proved to be a fruitful model for understanding the mechanisms by which infection can initiate a pathogenic autoimmune process. It is the first instance where the antigen responsible for the autoimmune response has been defined, isolated, and shown to induce the disease when injected in pure form.[12] Recent reports defining the myosin sequences capable of inducing the disease substantiate and extend our previous conclusion that myosin itself is the requisite antigen. It must be added that in the virus-induced model additional autoantibodies can be demonstrated. In A/J mice infected with CB3, we have found antibodies to the adenine nucleotide translocator, to ketoacid dehydrogenase and β-adrenergic receptor, all resembling antibodies found in the human disease.[27] The presence of these antibodies suggests that the total pathological process induced by the virus is more complicated than that resulting from immunization with myosin alone. It opens the question of whether these antibodies may be the cause of cardiac failure, since the loss of cardiac function is often out of proportion to the degree of myocardial inflammation.

The mechanism by which virus acts to initiate disease is still uncertain. The preponderance of evidence, in our view, is against the role of molecular mimicry. Despite repeated efforts, we have been unable to induce myocarditis by immunization of genetically susceptible mice with inactivated, concentrated pellets of CB3 incorporated in complete Freund's adjuvant. Moreover, as pointed out previously, none of the pathogenic amino acid sequences of myosin described thus far are analogous to the sequences in viral proteins. We are, therefore, more inclined to believe that the role of virus is to liberate the sequestered intracellular myosin and make it available in an immunogenic form. In addition, the virus infection provides a rich source of potential antigen-producing cells.

The relationship of our mouse model to human myocarditis must also be considered. The antemortem diagnosis and treatment of myocarditis remain clinical problems. In our view, the problem of therapy is complicated by the likelihood that in humans the disease often represents a mixture of viral and autoimmune factors. It is possible that a predominantly autoimmune myocarditis will respond favorably to immunosuppressive treatment, whereas disease at the viral end of the spectrum would not benefit from such treatments.[28] We believe, therefore, that increased effort should be devoted to seeking diagnostic signs in human patients of a predominantly autoimmune form of myocarditis. Such signs might employ the HLA haplotypes of the subject, the expression of MHC-gene products in the heart and, most important, upon the production of heart-specific antibodies in the serum. We have already demonstrated the presence of myosin-specific antibodies in about half of the patients with myocarditis. Other autoantibodies to cardiac antigens have been referred to above. Whether any of these, or any combination of them, will be useful for identifying subpopulations of myocarditis patients who are likely to respond favorably to immunosuppressive treatment should be determined.

ACKNOWLEDGMENTS. The author's research was supported by NIH grant HL33878. He is grateful to Mrs. Hermine Bongers for expert editorial assistance.

REFERENCES

1. Donath, J., and Landsteiner, K., 1904, Ueber paroxysmale Hämoglobinurie, *Muench. Med. Wochenschr.* **51**:1590–1593.
2. Solimena, M., and De Camilli, P., 1995, Coxsackieviruses and diabetes, *Nature Med.* **1**:25–26.
3. Fujinami, R. S., 1988, Viruses and molecular mimicry, in: *Molecular Mimicry in Health and Disease. Interactions of Biological Substances with Neural. Endocrine and Immune Cells* (Å. Lernmark, T. Dyrberg, L. Terenius, and B. Hökfelt, eds.), Elsevier, Amsterdam, pp. 237–244.
4. Mitchison, N. A., 1990, Chronic infection, immunopathology and immune suppression, in: *The Role of Micro-Organisms in Non-Infectious Disease* (R. R. P. de Vries, I. R. Cohen, and J. J. van Rood, eds.), Springer-Verlag, Berlin, pp. 101-109.
5. Ascher, M. S., and Sheppard, H. W., 1988, AIDS as immune system activation: A model for pathogenesis, *Clin. Exp. Immunol.* **73**:165–167.
6. Rose, N. R., Neumann, D. A., and Herskowitz, A., 1992, Coxsackievirus myocarditis, *Adv. Intern. Med.* **37**:411–429.
7. Kandolf, R., Klingel, K., Zell, R., Canu, A., Fortmüller, U., Hohenadl, A., Albrecht, M., Reimann, B.-Y., Franz, W. M., Heim, A., Raab, U., and McPhee, F., 1993, Molecular mechanisms in the pathogenesis of enteroviral heart disease: Acute and persistent infections, *Clin. Immunol. Immunopathol.* **68**:153–158.
8. Cetta, F., and Michels, V. V., 1995, The autoimmune basis of dilated cardiomyopathy, *Ann. Med.* **27**:169–173.
9. Wolfgram, L. J., Beisel, K. W., Herskowitz, A., and Rose, N. R., 1986, Variations in the susceptibility to Coxsackievirus B$_3$-induced myocarditis among different strains of mice, *J. Immunol.* **136**:1846–1852.
10. Wolfgram, L. J., Beisel, K. W., and Rose, N. R., 1985, Heart-specific autoantibodies following murine Coxsackievirus B$_3$ myocarditis, *J. Exp. Med.* **161**:1112–1121.
11. Neu, N., Beisel, K. W., Traystman, M. D., Rose, N. R., and Craig, S. W., 1987, Autoantibodies specific for the cardiac myosin isoform are found in mice susceptible to Coxsackievirus B$_3$-induced myocarditis, *J. Immunol.* **138**:2488–2492.
12. Neu, N., Rose, N. R., Beisel, K. W., Herskowitz, A., Gurri-Glass, G., and Craig, S. W., 1987, Cardiac myosin induces myocarditis in genetically predisposed mice, *J. Immunol.* **139**:3630–3636.
13. Vladutiu, A. O., and Rose, N. R., 1971, Autoimmune murine thyroiditis. Relation to histocompatibility (H-2) type, *Science* **174**:1137–1139.

14. Bacon, L. D., Kite, J. H., Jr., and Rose, N. R., 1974, Relation between the major histocompatibility (B) locus and autoimmune thyroiditis in obese chickens, *Science* **186**:274–275.

15. Lane, J. R., Neumann, D. A., Lafond-Walker, A., Herskowitz, A., and Rose, N. R., 1991, LPS promotes CB3-induced myocarditis in resistant B10.A mice, *Cell. Immunol.* **136**: 219–233.

16. Lane, J. R., Neumann, D. A., Lafond-Walker, A., Herskowitz, A., and Rose, N. R., 1992, Interleukin 1 or tumor necrosis factor can promote Coxsackie B3-induced myocarditis in resistant B10.A mice, *J. Exp. Med.* **175**:1123–1129.

17. Lane, J. R., Neumann, D. A., Lafond-Walker, A., Herskowitz, A., and Rose, N. R., 1993, Role of IL-1 and tumor necrosis factor in Coxsackie virus-induced autoimmune myocarditis, *J. Immunol.* **151**:1682–1690.

18. Neumann, D. A., Lane, J. R., Allen, G. S., Herskowitz, A., and Rose, N. R., 1993, Viral myocarditis leading to cardiomyopathy: Do cytokines contribute to pathogenesis? *Clin. Immunol. Immunopathol.* **68**:181–190.

19. Rose, N. R., and Hill, S. L., in press, The pathogenesis of post-infectious autoimmune myocarditis, in: *New Frontiers in the Immunoregulation of Allergic and Autoimmune Disease* (M. Ballow, ed.), *Clin. Immunol. Immunopathol.*

20. Donermeyer, D. L., Beisel, K. W., Allen, P. M., and Smith, S. C., 1995, Myocarditis-inducing epitope of myosin binds constitutively and stably to I-A^k on antigen-presenting cells in the heart, *J. Exp. Med.* **182**:1291–1300.

21. Liao, L., Sindhwani, R., Leinwand, L., Diamond, B., and Factor, S., 1993, Cardiac α-myosin heavy chains differ in their induction of myocarditis. Identification of pathogenic epitopes, *J. Clin. Invest.* **92**:2877–2882.

22. Neu, N., 1995, Personal communication.

23. Wegmann, K. W., Zhao, W., Griffin, A. C., and Hickey, W. F., 1994, Identification of myocarditogenic peptides derived from cardiac myosin capable of inducing experimental allergic myocarditis in the Lewis rat, *J. Immunol.* **153**:892–900.

24. Quinn, A., Adderson, E. E., Shackelford, P. G., Carroll, W. L., and Cunningham, M. W., 1995, Autoantibody germline gene segment encodes V$_H$ and V$_L$ regions of a human anti-streptococcal monoclonal antibody recognizing streptococcal M protein and human cardiac myosin epitopes, *J. Immunol.* **154**:4203–4212.

25. Herskowitz, A., Ahmed-Ansari, A., Neumann, D. A., Beschorner, W. E., Rose, N. R., Soule, L. M., Burek, C. L., Sell, K. W., and Baughman, K. L., 1990, Induction of major histocompatibility complex antigens within the myocardium of patients with active myocarditis: A nonhistologic marker of myocarditis, *J. Am. Coll. Cardiol.* **15**:624–632.

26. Neumann, D. A., Lane, J. R., LaFond-Walker, A., Allen, G. S., Wulff, S. M., Herskowitz, A., and Rose, N. R., 1991, Heart-specific autoantibodies can be eluted from the hearts of Coxsackievirus B3-infected mice, *Clin. Exp. Immunol.* **86**:405–412.

27. Neumann, D. A., Rose, N. R., Ansari, A. A., and Herskowitz, A., 1994, Induction of multiple heart autoantibodies in mice with Coxsackievirus B3- and cardiac myosin-induced autoimmune myocarditis, *J. Immunol.* **152**:343–350.

28. Mason, J. W., O'Connell, J. B., Herskowitz, A., Rose, N. R., McManus, B. M., Billingham, M. E., Moon, T. E., and the Myocarditis Treatment Trial Investigators, 1995, A clinical trial of immunosuppressive therapy for myocarditis, *N. Engl. J. Med.* **333**:269–275.

Index